Be a Successful Building Contractor

Be a Successful Building Contractor

R. Dodge Woodson

McGraw-Hill

New York San Francisco Washington, D.C. Auckland Bogotá
Caracas Lisbon London Madrid Mexico City Milan
Montreal New Delhi San Juan Singapore
Sydney Tokyo Toronto

Library of Congress Cataloging-in-Publication Data

Woodson, R. Dodge (Roger Dodge), 1955–
    Be a successful building contractor / R. Dodge Woodson.
      p.   cm.
    Includes index.
    ISBN 0-07-071828-8 (hc). — ISBN 0-07-0718296 (softcover)
    1. Contractors.   2. Building trades—Management.   I. Title.
TH438.W6597   1997
690' .68—dc21
                                       96-49568
                                          CIP

McGraw-Hill

A Division of The McGraw·Hill Companies

1 2 3 4 5 6 7 8 9 0 DOC/DOC 9 0 2 1 0 9 8 7

ISBN hc 0-07-071828-8
ISBN pb 0-07-071829-6

The sponsoring editor for this book was Zoe G. Foundotos, the editing supervisor was Scott Amerman, and the production supervisor was Pamela Pelton. It was set in Garamond by McGraw-Hill's Professional Book Group Composition Unit, in Hightstown, N.J.

Printed and bound by R. R. Donnelley & Sons Company.

McGraw-Hill books are available at special quantity discounts to use as premiums and sales promotions, or for use in corporate training programs. For more information, please write to the Director of Special Sales, McGraw-Hill, 11 West 19th Street, New York, NY 10011. Or contact your local bookstore.

This book is dedicated to Kimberley,
Afton, and Adam, the best family
anyone could ever ask for.

Contents

Introduction   ix

Acknowledgements   xi

1   Becoming an after-hours builder   1

2   Getting your feet wet   7

3   Deciding what you want   13

4   Taking the plunge   21

5   Choosing a business structure   35

6   Avoiding 15 common builder mistakes   41

7   Building model homes: Pros and cons   49

8   Courting bankers   55

9   Selling from the hood of your truck   69

10   Opening your own office   75

11   Finding the best building lots   81

12   Controlling desirable lots   89

13   Developing your own building lots   95

14   Building on speculation   101

15 Working with real estate brokers   107

16 Making the most of your time and money   129

17 Getting computerized   137

18 Keeping track of your cash   153

19 Buying trucks, tools, equipment, and inventory 183

20 Dealing with subcontractors, suppliers, and
    code officers   191

21 Bidding: Methods that really work   223

22 Running smooth and profitable jobs   247

23 Keeping your customers happy   261

24 Creating and promoting an attractive business
    image   271

25 Hiring employees   287

26 Planning for your future   305

    Appendix: Federal tax forms   319

    Glossary   349

    Index   355

    About the author   367

Introduction

Becoming a successful building contractor is a goal that can lead to a wonderful life. You can make more money than you might imagine, and the quality of your life can improve. Being a builder is a not only potentially lucrative, it can be a lot of fun. Driving around town and seeing houses that you built should give you a good feeling and a lot of self-esteem.

Is it difficult to become a home builder? Not really. Opening your own building business is probably a lot easier than you think. Can you really make a lot of money building houses? The answer to this question depends on several factors, but it's not at all uncommon for builders to pocket a profit from each house that is equivalent to 20 percent of the home's value. In other words, if you build a modest $100,000 home, you could walk away with $20,000 for a few months' work. A $200,000 home could pay you $40,000 for the same amount of time. Builders who take a hands-on role in the construction of home, for example, by acting as the carpenter, can make much more.

It's not fair to say that all builders make a 20-percent profit from building houses. Some make more and some make less. But, when you consider how few hours you might invest to build a house as a general contractor, even $10,000 per house is a high hourly rate. You can definitely make some very good money as a builder.

You don't have to be a carpenter to become a home builder. I started out as a plumber and became a builder who built as many as 60 single-family homes in a year. Lawyers have become builders, and so have people from many other walks of life. You need some knowledge of construction trades, but you can gain much of that from books like this one.

All in all, getting into business as a builder is easy. Making money and staying in business is not so simple, but it's something that people with average intelligence can do. You don't need a college degree or 10 years of field experience to become a general contractor.

I've worked in the trades for about 22 years. My career started with plumbing and grew into remodeling. Then I moved into building houses on speculation. My next move was to become a custom builder, where the money is better and the risk is less. From there, I got into real estate sales as a part of my building business. In recent years, I've worked as a consultant to contractors who needed help setting up businesses or straightening out businesses that are going badly. My experience with all aspects of home building runs deep, and I'm going to share it with you.

By reading this book, you can learn from my experience and mistakes. Just one piece of advice from these pages could save your financial life. I'm not going to candy-coat it for you; building can be a tough, competitive, and risky business.

Until you learn the ropes, you could hang yourself. But you won't, because you are smart enough to be reading this book. The builders who get in trouble are usually people who think they know it all. After 22 years, I'm still learning, and I don't think anyone ever knows it all.

Take a few minutes to scan the table of contents. You'll notice that just about every topic you'll need to help you to start your own building business is covered. If you thumb through the pages, you'll see sample forms and illustrations that make the learning curve shorter for you. I don't stand on a soap box and preach to you. Instead, I talk with you as if I were sitting across the desk from you. My writing style is easy to understand, and I'm talking from experience gained over more than two decades.

I've been through recessions and economic slow-downs. Those times weren't pleasant, but I survived. You can turn adversity to your advantage if you have the right attitude and knowledge. If you are interested in becoming a successful building contractor, I believe you absolutely owe it to yourself to read this book. I don't say this just because I wrote the book, but because I know how much a book like this would have helped me in my early years as a builder. I don't think you will be disappointed with the advice you find in this guide to becoming a successful building contractor.

Acknowledgments

I would like to acknowledge my parents, Maralou and Woody, for encouraging me to enter the wonderful world of construction and for all of their support over the years.

I would also like to thank the following government agencies for their permission to use their illustrative materials:

- U.S. Department of Agriculture
- U.S. Department of the Army
- U.S. Department of the Treasury—Internal Revenue Service

Be a Successful Building Contractor

1
Becoming an after-hours builder

When most people think of moonlighting for extra money, they don't consider becoming an after-hours builder. Lots of people know that plumbers, electricians, carpenters, painters, and other tradespeople work evenings and weekends to make extra cash. Almost every tradesperson I've known, myself included, has moonlighted at one time or another. Many of them do it as a way of getting into business for themselves on a full-time basis. But can a home builder get started by working nights and weekends? Yes, and this chapter shows you how.

Home builders come from all walks of life. Many of them start out working as carpenters and grow into being contractors. They sometimes start with repairs and remodeling as they work their way up to being full-scale home builders. I started out as a plumber and grew into remodeling. From there, I went on to build as many as 60 homes a year. Who else might become a builder?

I've met builders who started in some professions that you might not think would lead to a career as a builder. Lawyers, real estate brokers, electricians, farmers, and police officers are just some of the backgrounds of builders I have met. So, nearly anyone can make the transition from whatever they are currently doing to home building, although it is easier for some than it is for others.

If your background is in construction, you have an obvious advantage over someone who has never set foot on a construction site. While you might have never built a house, working in new houses, around other trades, gives a person a good idea of what goes on during the construction process. Field experience alone isn't enough to make someone a good builder, but it sure helps.

So, how can you get away from what you are doing for a living now and enjoy being a home builder?

To tackle the job right, you need money and experience, not to mention a lot of contacts with subcontractors, suppliers, lenders, and assorted other people. I'm sure that some people have the financial resources to start a building business in a first-class manner. I never enjoyed this luxury. I had to start at the bottom and crawl up the mountain. At times, it seemed as if the mountain was made of gravel, because every time I would near the top, I would slide back down the hill. But I persevered and made it. I think you can too.

After running my plumbing and remodeling business for a while, I wanted more. Home building seemed like a logical path to take. Since I didn't own a home, it made sense to try my hand at building by creating a new home for myself, and that is what I did. By building the home myself, I was able to use my sweat equity as a down payment, so I essentially built the house without using any of my own money. I was even able to pocket a little cash as profit from the deal. That first house was the stepping stone that led to the top of my mountain.

When I say that I built my first house, I should qualify my terminology so that you understand that I did not drive every nail personally. I acted as a general contractor. My wife and I did the plumbing, drywall work, painting, tile work, and some other odds and ends, but we used subcontractors for the rest of the work. The house wound up costing more than I had projected it would, and the drywall work wasn't all that great. Since that house, I've always hired others to do drywall work. After building my own home in my spare time, I decided to sell it and start another new house.

For three years my wife and I built a new house each year. After building our second new home, we started building houses on speculation. By the time we built our third house, we were averaging about 12 new homes a year, in addition to our plumbing and remodeling business. When we really got rolling, we were doing 60 homes a year, running the plumbing business, doing a little selective remodeling, and operating a real estate sales business and a property management business. All of this came about due to the income made from building. The money from plumbing and remodeling paid our bills, and the moonlighting we did with spec houses gave us the cash to jump into the big leagues. I suspect that you could do something similar if you put your mind to it.

IDENTIFYING YOUR BASIC NEEDS

What are the basic needs for becoming a part-time builder? They are less than you might imagine. If you operate as a general contractor, you don't need much in the way of trucks, tools, and equipment. The subcontractors you engage provide for their own needs. All you have to do is schedule and supervise their work. This, however, is not as simple as it seems.

Since you will probably be working your day job when you get started, you need an answering service or machine to take your phone messages. A license to build might be required in your region, and a business license is normally required. You can work from home and meet your prospective customers in their homes. You should invest in liability insurance, and you need to advertise. On the whole, the financial requirements for becoming a part-time builder are minimal.

It is best for builders to have reserve capital to get past financial hardships, but if you're good and if you're lucky, you can make do with very little cash. If you bid jobs accurately and profitably, you have no need for a lot of cash. After all, you have your regular employment to pay your routine bills, your overhead is low as a builder, and construction loans can provide your cash flow for building.

CLEARING YOUR FIRST HURDLE

The first hurdle you are likely to have to clear is your lack of a track record. Your customers will probably want the names of references, and they might even want to see examples of your work. When you are starting out, you can't provide references or work samples. This obstacle can be difficult to clear, but there are some ways to work around the problem.

If you are building your first house on speculation, you won't have to worry about references or work samples. Potential buyers of the home can see what they are getting. This makes your job a little easier. If you can't afford to take the risk of building on spec, you have to be more creative in coming up with a way to get customers to accept you.

I've already told you that the first houses that I built and sold were homes that I built without having buyers already lined up for them. Doing this created my references for me. But when I moved to Maine from Virginia, I didn't have any references. Getting my first couple of contract homes to build was a bit more difficult. People would come into my office and talk about building, but they wanted to see tangible proof of my abilities, which I, of course, could not show them. I had to come up with a plan.

To overcome the problem of not having a model or sample home to show people, I changed the types of advertisements I was running. The ads offered people a chance to have a new house built at a reduced cost by a builder who would pay them a monthly fee for the first year they owned the home in exchange for allowing the home to be shown periodically as a reference. I discounted my prices by 5 percent and offered buyers $150 a month for 12 months as a model-home fee. The strategy worked. Buyers came in, houses were built, and I was back up and running.

The discounted price and the $1200 per house for model fees cut my profit down on the first two houses I built, but after those two homes were established as references, I didn't need to continue offering my special deal. By giving up a little in the beginning, I was able to get off to a fast start. You could try something similar, or maybe you can come up with a better idea. The point is, you can find ways around the problem of not having references and sample homes.

GAINING KNOWLEDGE

How much knowledge of construction do you need to succeed as a home builder? The more you have, the better off you are, but you can get started with a basic level of knowledge and learn while you earn. As a general contractor, you don't have to do any of the physical work involved with construction. Your primary function once a job is underway is to schedule and supervise workers. It is obviously much easier to supervise people when you understand what they are doing and how the work should be done. But you don't have to be a drywall finisher to supervise drywall work. If a job looks good, you know it. When one looks bad, you can see it.

Code enforcement officers check the work being done for code violations. In theory, you don't have to know much about construction to be a builder who subs all of the work out to independent contractors. In reality, however, builders with the most knowledge of construction are generally much more successful than people who don't know or don't attempt to learn all they can about the home-building process.

If you have good organizational skills and can manage people, budgets, and schedules, you should be able to become a viable home builder. What you need to know about various trades can be learned from reading and watching videos. Numerous books and videos available for do-it-yourselfers give step-by-step instructions for everything from plumbing to tile work. You can educate yourself by reading and watching, without ever having to work on a construction site. It is best, however, to have some practical field experience.

AVOIDING DANGERS

One of the biggest dangers for rookie builders is lack of experience in pricing homes. Even experienced carpenters often don't know how to work up prices for complete homes. They are not accustomed to figuring in the cost of septic systems or sewer taps or the cost of floor coverings and finish grading. How can you get the best estimates possible when you've never done one before?

Take a set of blueprints to your supplier of building materials. Ask the manager to have someone do a take-off of your material needs and price it. Many suppliers provide this service free of charge, but some don't. Circulate copies of the blueprints to every subcontractor that you need. Have the subs give you prices for all the work they are expected to do. While your subs and suppliers are working up their prices, you can start doing some homework of your own.

Take a set of blueprints to a reputable real estate appraiser. Ask the appraiser to work up either an opinion of value or a full-blown appraisal. This costs you some money, but it is well worth it.

Once the appraiser has given you a figure, you know about what the market value of the home is. You can consult some pricing guides to determine what various phases of work should cost. These guides are available in most bookstores, and they have multiplication factors that allow you to adjust the prices to coincide with prices in your particular region.

When you get your prices back from suppliers and subcontractors, you can compare them with the numbers you came up with from the pricing guides. You can also look at the difference between the bid prices of your subs and suppliers as they are compared to the finished appraisal figure. The spread between the bids and the market value represents your potential profit. It should normally relate to about a 10- to 20-percent gross profit. A 15-percent profit might be an average, but the amount varies with economic conditions and the prices given by subs and suppliers. I'll talk more about pricing and estimating later in the book, but the procedures I have just discussed are the basics.

FIGURING CAPACITY

How many houses can you build in a year as a moonlighting builder? The answer to this question varies with experience level, sales opportunities, lot availability, individual time commitments, and so forth. You should certainly be able to build at least two houses, and probably four. Since you are using subcontractors, you can have multiple houses under construction at the same time.

I used to do about 12 houses a year as a part-time venture, but I had help from my wife. A goal of six houses a year seems like it would be realistic for someone who had a construction background and a good stable of subcontractors. Once you get into the building game, you can assess your time needs and adjust your goal as needed.

DETERMINING HOW MUCH MONEY YOU CAN MAKE

How much money can you make as a part-time builder? It depends on the types and sizes of houses you are building and how well you manage your production schedule and financial budget. If, for example, you are building $100,000 homes and your gross profit is 20 percent, that means that you are making $20,000 for each house you build. Four houses a year at this rate would be $80,000. Can you really do this? It can be done, but you will probably make less money per house until you get your act refined. But even at $15,000 a house, you're still pulling in $60,000, and that's not bad for a part-time job.

If you want to look at the worst case, assume that you make a 10-percent profit. That's $10,000 a house, or $40,000 a year. There really is a lot of money to be made as a builder once you work the bugs out of your system. If you can survive the first few houses you build, you stand a very good chance of enjoying a long and profitable career.

2
Getting your feet wet

Getting your feet wet in the building business without getting in over your head takes planning. A person who decides one day to be a home builder should not just run ads in the local newspaper and wait for the phone to ring. It can take months of preliminary planning to prepare for opening a home-building business. The money to be made as a builder is rich, but so are the risks. To avoid failure, you must have some solid plans.

The first thing you need to do is make sure you are ready to assume the role of a builder. Do you have enough general knowledge to perform the functions of a general contractor? If you don't, start reading, attending classes, or working on some construction sites. Gain as much experience and knowledge as you can before you offer your services as a builder to the public.

There are many ways to prepare yourself for becoming a general contractor. One low-impact way is to read every book you can find on building and related trades. Read books that have been written for homeowners and do-it-yourselfers. Seek out titles that have been written for professionals, such as the *Builder's Guide* series published by McGraw-Hill. Absorb the wealth of knowledge provided by seasoned professionals that can be found in these books.

In addition to reading, you can attend classes that pertain to various trades and home building. Some places offer workshops for people hoping to build their own homes. Look into the possibilities of attending workshops or vocational classes if you feel you need more training than you can get from a book.

Videos have become extremely popular, and videos are available that show viewers how to perform certain tasks, such as hanging cabinets or installing plumbing. Your local library or video store might have some of these learning tools on hand. If not, I'm sure they can help you order videos that can boost your knowledge level.

Even if you are an experienced construction worker, you might still need to attend some classes. There is more to being a builder than just constructing a home. The business side of building gives some tradespeople trouble. If you are not comfortable with your office skills and management ability, take some courses that cover these subjects. Administrative skills are sorely needed in the building business.

Now, assuming that you feel ready to become a builder, you must find a way to tap into this lucrative market. If you have some money and good credit, you could start by building a spec house. This can be extremely risky, however, and I wouldn't recommend it as a starting point, even though it is how I got started. The safest way to break into building is with houses that are presold.

DECIDING WHAT TYPE OF HOUSE TO BUILD FIRST

What type of house should you build first? Any type someone is willing to pay for, of course. But really, you'll find advantages to building some styles of houses over others. A ranch-style house is the easiest to build. Straight-up two-story homes give buyers the most square footage for their dollar. Cape Cods are very popular with people who are looking to build their first home on a budget. Elaborate designs make a builder's job tougher. So where and how do you begin?

As long as your subcontractors are all experienced workers, you should be able to build any type of house. But logic dictates that simple designs are faster and easier to build. It is reasonable that you should try to start with a house that can be built quickly, so that you can generate a cash flow and a profit as soon as possible. Keeping your starting designs simple makes it easier to figure material and labor needs, so you run less risk of cost overruns. With all this in mind, I would come up with three sets of house plans to start off with. The houses would be a ranch, a Cape Cod, and a two-story.

You're probably wondering why you should come up with specific house plans if you are going to be doing custom building. This is a fair question. You will be building to explicit specifications supplied by your customers, but you need something to advertise. If you simply run an ad that says you are a home builder and that you are open for business, you're not going to get a lot of buyer activity. But if you run a picture of a particular house, list features and benefits of the home, and include a price, your phone should start to ring. Obviously, to do this, you need to have a plan to work with.

Part of the decision on what styles of homes to choose for your ads should be based on local conditions. Economic factors come into play, as do public preferences for particular house styles. You're going to be competing with established builders, so you need an edge or an angle to work with. Gaining this advantage is often easier than you might imagine.

PICKING THE RIGHT HOUSE PLAN

Picking the right house plan is not something that should be done based on personal preference. Your decision on what size and style of home to offer the public should be the result of research. Your investigation won't require a lot of high-tech surveillance or weeks of your time. All you have to do is ride around and see what types of houses are popular in your area. Are a lot of split-foyers being built? How many newer homes have only one level of living space? Do most new homes have attached garages? Are front porches fashionable? Looking for these types of design features can bring you up to speed quickly on what the public wants. If you see that only one out of eight houses is a ranch, you can shuffle your ranch plans to the bottom of your pile.

It is not wise to try reinventing the wheel. The odds of success when you seek radical change are low. Your chances of success are much better if you follow the lead of your competitors. If colonial two-story homes are abundant in your area, look for a nice set of plans for a colonial. Go with the flow, but with a twist.

CREATING YOUR EDGE

What is your edge as the new builder on the block? It's something you have to create. It might be low prices. Your claim to fame could be outstanding designs or superior workmanship at affordable prices. Something as obvious as marketing and advertising could be what sets you above your competition. The edge can be almost anything, but you need it to survive and prosper. If you are just a carbon copy of all the other builders, you are at a disadvantage. Finding what works best for you is a personal thing, but I can give you some ideas.

Price

Price is a factor that many businesses use as a lever. Trying to beat the prices of your competitors would not be my first recommendation. If you become known as the cheap builder or the discount builder, you might have trouble moving up to higher-priced homes. But getting a reputation as a value-conscious builder is a different story. Everyone likes getting a good value, but some people look down on cheap or discount deals. The result is the same, but the image you present is different.

To create the aura of a value-based builder, you have to make your homes a little different. You can put your two-story colonial up against your competitors, but you have to create some subtle differences in the plan. Your goal is to make customers compare apples to oranges, rather than apples to apples. This way, your homes don't appear to be a cheap version of what the competition is offering. If your competitor has a laundry room built onto the side of a home, as many colonials do, you might consider putting the laundry hook-up in a basement or closet to eliminate the cost of the attached room. This changes the outside appearance of the home, but it also lowers the cost in a way that can't be construed as a discount.

Customer base

Before you jump into blueprint selections, you must identify your customer base. Do you want to deal with first-time buyers and starter homes or do you want to work with more affluent buyers? You might make less money on a per-deal basis with starter homes, but these entry-level houses are a good place to begin your building business, for several reasons. Number one, first-time buyers are often ignored by larger, more established contractors. This opens the market up for you. Second, first-time buyers are not usually too choosy when making a commitment to purchase a house. Any house they can get is better than what they have, so they might jump at whatever you have to offer them. Another big advantage to first-time buyers is that they are not encumbered with a house to sell. This makes it possible for you to make quick sales that are not contingent on the sale of other properties.

Second-time home buyers often have to liquidate their existing homes before they can make a building commitment. Having owned a home previously, these buyers are frequently more selective than first-time buyers. The price of houses purchased by move-up buyers is more than that of the homes for most first-time buyers, but the difference can be made up in volume and quick turnover with first-time buyers.

I can't tell you absolutely what might be best for you. But I can tell you that I have catered to first-time buyers as both a builder and a broker, and I've done well in doing so. If I were you, I'd give serious consideration to focusing on first-time buyers.

Financing

Financing programs can create an edge for you. You don't have to provide the financing personally. If you establish relationships with various lenders and can advertise certain financing plans that are available for the homes you build, customers should respond favorably. Financing is usually the key to whether or not a building deal flies. Your competitors might have access to the same loans, but if you're the one making people aware of what programs are available, you're likely to win the jobs.

BRINGING IT ALL TOGETHER

Bringing it all together to offer the public a fast, easy package is a sure way to success. Home buyers are usually excited and often naive. Most buyers respond to advertising, and almost anyone should listen to a sincere presentation from a caring builder. You don't have to be the biggest builder in town to capture your share of the market. But you do have to be professional and persistent.

So many options exist for potential home builders that it would take several chapters of a book to list most of them. Since I can't afford to use all the space on possible options, let me give you an example of how you might bring your building business into reality with minimal risk. You can use the example as a template for other types of approaches, if you don't want to pursue the specific example given.

Assume that you have done your homework and you've decided to go after first-time buyers. Cape Cods with unfinished upstairs and small ranch-style homes are your intended offerings. You will be competing against modular homes and a few established builders. With the money you have available, you plan to do a little advertising on cable television, some print advertising in the local paper, and a small direct-mail campaign. You've got your house styles selected and all of your construction costs calculated. It's now time to launch your attack.

You've decided to start by running ads in the local newspaper and on cable television. Once you get some name recognition, you're going to use your direct-mail strategy. The mailing list you rent must be segregated to include only people who live in rental property and who have adequate income levels to afford the houses you plan to build.

As a format for your new company, you decided to play up certain strong points. You are focused on first-time buyers. Easy financing is available. House designs have been chosen with young families in mind. Quality construction and energy efficiency are cornerstones of your business. Flexibility and freedom of design is in the hands of your customers. You can guarantee a quick turnaround on building that puts customers in their new homes in less than 90 days from the date that trees are cleared and ground is broken. You're offering a 10-year home warranty, endorsed by a major building association. All of your features and benefits are listed in your advertisements. From start to finish, you make home ownership easy and enjoyable—that's your motto.

You start running your ads, and you're amazed at the response. Why are so many people calling you, the new builder in town, when they could be calling established professionals? They're doing it because you identified a need, filled it, and made the public aware of what you were doing. I've done this type of thing time and time again.

Established builders usually become complacent. They make enough money and get enough work from word-of-mouth referrals that they don't have to go out looking for work. When first-time buyers approach some builders and real estate brokers, they are shunned. I've heard this from buyers on countless occasions. The buyers feel as if the builders and brokers don't want their business. This group of buyers is a prime target for your approach. Since you are willing to coddle them, the buyers will flock to you and spread the word of how great you are to their friends.

TESTING THE WATERS

When you are ready to get into building, you have to test the waters. If you have plenty of money and don't mind losing it, you can do this with advertising. For people like myself, who don't have enough money to throw around on feasibility studies and advertising, research is the key. Talking with competitive builders, real estate brokers, and appraisers is a fast way to get some inside information. Looking through comparable sales books distributed by multiple listing services to brokers is also a valid way to see what's selling and how much it's selling for. Comp books also let a person know how long a property was on the market before it sold.

You have to carve your own niche in the world of building. Mine has been first-time buyers. Some builders specialize in expansive, expensive homes. What the builders of big homes make on one job might take me four jobs to make, but I have the volume. It would not be fair for me to say that my way is the right way, but it's worked for me, and I think it would be a good place for you to get started.

3
Deciding what you want

Whether you are thinking of opening your own business or trying to determine how to make your existing business better, you must know what it is that you want. One way to accomplish this is to list your goals and objectives. Business goals provide many advantages. They can pave the way to higher income and a more enjoyable life. Without goals, you have no direction. A business without direction is almost certain to go down.

Many people start a business without goals, and then, after spending considerable sums of start-up money, realize the business is not what they want. These people have two basic options. They can shut down the business, or they can go on with a business that does not make them happy. If they close the business, they lose money. If they continue with the business, they are no happier than they were before they started it. Neither option is desirable.

It is not uncommon for people to put themselves into business situations that they regret. For some, the stress of owning and operating a business is too much. For others, the financial ups and downs are more than they care to deal with. Being in business for yourself is not all leisure time and big bank accounts. Being self-employed requires discipline, long hours, dedication, and persistence. Owning a business is not the glorified cakewalk some people fantasize it to be.

While self-employment is rarely easy, it can be rewarding, both financially and mentally. This chapter is going to show you how to use goals to reach the results you want. If you are thinking that you don't need to set goals, you are wrong. It is easy to say that your goals are to be rich, successful, and happy, but you are not likely to make it to any of these points without realistic, step-by-step goals.

To say that you want a business with 20 employees is a goal, but it is not enough. If you want to build a business with 20 employees, you have to obtain the goal with the use of smaller, progressional goals. Unless you are extremely wealthy to begin with, you cannot afford to open your shop with 20 employees standing by to take care of business. You must develop the business to accommodate the employees.

Have you ever noticed how some businesses flourish while others falter? Maybe you have worked for companies that always seemed busy but never seemed satisfied with the profits or production. Why do you suppose that is? Have you wondered why some successful business people are satisfied to keep their business small, even when it appears they could expand? The answers to these questions lie in goals or the lack of them.

When your job becomes your business, your life changes. Everyone hopes the change is for the better, but that is not always the case. Owning and running your own business is not the same as going to your old job. You don't have a company supervisor to answer to, but you still have a boss; in fact, you now have many supervisors. Your new supervisors are your customers. If you don't do your job to the satisfaction of the customers, you won't have your new, self-employed job for long.

The truth is, being in business for yourself can be much more demanding than holding down a job. For example, let's say you are a carpenter. When you work for someone else, all you have to worry about is the quality of your carpentry work and the basic responsibilities of an employee: showing up for work on time, giving a fair day's production, and so on. You go to work, do your carpentry, and go home. Once you're home, the rest of the day, evening, and night is yours. This is not usually the case when you are in business for yourself.

As a self-employed carpenter, you have to perform all the normal carpentry duties, but your job does not stop there. Paperwork must be done. Phone calls must be returned. Estimates must be made. Complaints must be answered and solved. Marketing strategies must be planned. Accounts receivable and payable are a routine chore. The list for additional duties goes on and on.

As you can see, running your own business is not the same as working a regular job. When your job becomes your business, you have many more job-related responsibilities. Time with your family is at a premium. Weekend outings might have to be forfeited so you can catch up on business matters that were not completed during the week.

Opening your own business is no small undertaking. The time and financial requirements of starting a business can be overpowering. Before you jump into the deep, and sometimes turbid, water of becoming self-employed, you should give careful consideration to your goals and desires.

WHAT DO YOU WANT FROM YOUR BUSINESS?

What do you want from your business? This seems to be a simple question, but many people can't answer it. As a business consultant, I talk with a wide variety of people and businesses. When I go in to troubleshoot a business, the first question I ask is, What does the owner want from the business? More often than not, the owners don't know what they want. Generally, the owners who give an answer give a broad and unfocused reply. To be successful and ensure the survival of your business, you must have a business plan.

When I opened my first business, a plumbing business, I wanted to be my own boss. I wanted to work my own hours and not be worried about putting in 18 years only to be let go before retirement. My dream called for building a powerful business that would take care of me in my old age. Well, I started the business and I was relatively successful. However, looking back, I can see countless mistakes that I made.

Since my first business, I have gone on to open many new businesses. Each time I start a new venture, I seem to find new faults with my procedures. It is not that my methods don't work; I just always seem to find ways to improve upon them. I wouldn't begin to tell you that I have all the answers or can tell you exactly what you need to know to make your business work. But I can give you hundreds of examples of what not to do, and I can tell you what has worked in my business endeavors and those of my clients.

I don't believe you ever finish refining your business techniques. Even if the business climate was stable, you could always find ways to enhance your business. The business world changes frequently, forcing changes in business procedures and policies. What worked 10 years ago might not be effective today. If you are going to start and maintain a healthy business, you must be willing to change.

Now, back to the question: What do you want from your business? This question is applicable to people contemplating starting a business as well as present business owners. Take some time to think about the question. Then write down your desires and goals on a sheet of paper. You must *write down* your goals and desires. For years I refused to believe that writing my goals and desires on paper would make a difference, but it does.

With your list compiled, check it over. Break broad categories into more manageable sizes. For example, if you wrote down that you want to make a lot of money, define how much is a lot. Is it $30,000, $50,000, or $100,000 a year? If you jotted down a desire to work your own hours, create your potential work schedule. Will you work 8-hour days or 10-hour days? Will you work weekends? Are your scheduled hours going to comply with the needs of customers? This type of detailed planning improves your chances of success and happiness.

WHERE DO YOU WANT YOUR BUSINESS TO BE IN FIVE YEARS?

Where do you want your business to be in five years? Most new businesses don't last five years. Many businesses fail before the end of their first year. By the third year, a high percentage of new businesses are defunct. A key step in securing a good future for your business is the development of goals and plans.

How big do you want your business to become? How many employees to you want? Are you willing to diversify your business? These questions are typical of the type of questions you should be asking yourself. Let's take a moment to look closer at some thoughts for your business future.

HOW BIG DO YOU WANT YOUR BUSINESS TO BECOME?

How big do you want your business to become? Do you want a fleet of trucks and an army of employees? If you answer yes to this question, you must ask yourself more questions. Are you willing to pay the high overhead expenses that go hand in hand with a large group of employees? Do you need to take classes on human resources to manage your employees? Do you have the knowledge to oversee accounting procedures, safety requirements, and insurance needs?

You see, in business, almost every answer raises new questions. As a business owner, you must be prepared to answer all the questions. It is all but impossible for an individual to have the experience and knowledge to answer so many diversified questions correctly. For example, let's say you are about to hire your first employee. Do you know what questions you can ask without violating the potential employee's rights? Are you aware of the laws pertaining to discrimination and labor relations? The chances are good that you don't, so what do you do? You could play it by ear and hope for the best, but that type of action could result in a lawsuit and a loss of your assets. You should consult with professionals in the field of expertise pertaining to your questions. Hmm, maybe you should keep the business small and work it only by yourself. This employee stuff could get complicated.

ARE YOU WILLING TO DIVERSIFY?

Are you willing to diversify? Should you consider hiring your own in-house trades and using them on the houses you build to save money? Could you sub the trades out for general services to the public in addition to using them on your own jobs? It's possible that hiring your own plumber or electrician could make sense, but it is a big commitment.

I think it is healthy for businesses to diversify, but I believe the conditions must be right. In my opinion, you should not hire people to do a job that you have no knowledge of. For example, you could hire a master electrician and expand your business base. It would be appealing to have your own in-house electrician, and the money made from service and repair calls could be quite good. So why not do it? There are many good reasons not to put a specialized tradesperson on your payroll. Overhead cost is certainly one good reason to keep your payroll roster lean. Also, if you don't know much about electrical work, you could be setting yourself up for long-range trouble by jumping into the unknown.

HOW BIG IS BIG ENOUGH?

How big is big enough? When you are planning the destiny of your business, you must know what measuring stick to use in your planning. When you think of the size of a business, what scale of measurement do you use? Do you think in terms of gross sales, number of employees, net profits, tangible assets, or some other means of comparison? Gross income is one of the most common measurement factors of a business. However, gross income can be very deceiving.

Theoretically, a higher gross income should translate into a higher net income, but it doesn't always work that way. Having a fleet of new trucks might impress people and create a successful public image, but it might also cause your business to fail. The best measurement of your business is the net profit. Having 10 carpenters in the field might allow you to deposit large checks in the bank, but after your expenses, how much is left? It has been my experience and the experience of my clients that you must base your growth plans on net income, not gross income. Determine how much money you want to make, and then create a business plan that allows you to reach your goal.

WHAT TYPE OF CUSTOMER DO YOU WANT TO SERVE?

What type of customer do you want to serve? In the early stage of your business, any paying customer is welcome. However, it is important for you to determine the type of clientele you wish to work with. The steps you make in the early months of your business influence the character of the business for a long time to come.

During the initial start-up of a business it is easy to justify taking any job that comes along. The same is true when you are trying to survive poor economic times. While this type of approach might be necessary, you should never lose sight of your business goal. For example, if you have chosen to specialize in new construction, make every effort to concentrate in this field of work. When you are forced into re-modeling or repair work, do it to pay the bills but continue to pursue new con-struction. If you bounce back and forth between different types of work, it is more difficult to build a strong customer base and streamline your business. Let me give you an example from my past.

When I opened my first business, it was the only source of income I had. I wanted to be known as a remodeling plumber. After research, I had determined I could make more money doing high-scale remodeling than I could in any other field of plumbing that I had an interest in. I reasoned that remodeling was more stable than new con-struction and required less running around and lost time than service work. So, I had a plan: I would become known as the best remodeling plumber in town.

During the development of my business, finding enough remodeling work to make ends meet was tough. I took on some new-construction plumbing, cleaned drains, and repaired existing plumbing. It was tempting to get greedy and try to do it all, but I knew that wouldn't work, at least not with me being the only plumber in the company. Why wouldn't it work? It wouldn't work because of the nature of the different types of work.

With new construction, bid prices were very competitive. To win the job and make money, I had to work fast and eliminate lost time. If I was plumbing a new house and my beeper went off, I would have to pick up my tools and stock and leave the job to call the answering service. Then I would have to call the customer. Then I would have to either respond to the call or try to put the customer off until I left the new-construction work. Anyway I looked at it, I was losing money on the house I was plumbing.

The time I spent picking up and responding to service calls was eating into my narrow profit margin on the house. Service customers were annoyed if I didn't re-spond within an hour. I was losing money and running the risk of making customers angry. The same was true if I left a remodeling job to answer a service call. The re-modeling customer would become distressed because I left to take a service call. It didn't take long for me to see the potential for problems.

I set my sights on remodeling and put all my effort into getting remodeling jobs. In a matter of months, I was busy, and my customers were happy. My net income rose because I had eliminated wasted time. In time, I added more plumbers and built a solid service and repair division. Then I added more plumbers and took on more new-construction work. But you see, you must be careful in structuring your business plan, or it can get away from you and cause you to work harder, only to make less money.

When you consider your desired customer base and work type, do it judiciously. If you live in a small town, you might not be able to specialize in building only cus-tom homes. You might have to do framing, roofing, or remodeling. You might have to do a little of everything to stay busy. But if you want to specialize in a certain field, never lose your perspective in pursuing your desired type of work and customers.

WHAT POSITION WILL YOU PLAY IN THE BUSINESS?

What position will you play in the business? Most entrepreneurs starting out must play all positions. However, just like setting a goal for the type of work your business will do, you should establish a goal for the type of work *you* want to do. Do you want to work in the field or in the office? Will you trust important elements of your business to employees or will you want to do it all yourself?

The delegation of duties and proper management are difficult for many first-time business owners. Most people fall into one of two traps. The first group believes that they must do everything themselves. This group hesitates to delegate duties, and even after assigning the task to competent employees, they must keep their fingers in the work. The second group believes everything can be delegated, and they pass on responsibilities to people who are not in a position to make the call. Somewhere between these two extremes is where most successful business owners are.

It is counterproductive to hire employees if you are not going to allow them to do their jobs. You should supervise and inspect the employees' work, but you should not be looking over their shoulders every five minutes. If you did a good job in screening and hiring your employees, they should be capable of working with limited supervision. When your time is spent hovering over employees, you are neglecting many of your management and ownership duties.

If you decide to do everything yourself, you must recognize the fact that a time will come when you have to turn away business; one person can only do so much. It is better to politely refuse work than it is to take on too much work and not get it done.

Will you be content to stay in an office? If your nature tends to keep you outside doing physical work, being office-based can be a struggle. It has its advantages, but office work can be a real drag to the person accustomed to being out and about.

If you don't want to work in the office, you must make arrangements to allow your freedom of movement. Answering services and machines afford some relief. A receptionist is another way to keep the office staffed while you are out, but this is an expensive option. If you will be in the field, a pager and cellular telephone might be your best choice for keeping in touch with your customers. It is up to you to devise the most efficient way to stay on the job and out of the office, but don't overlook this question when planning your business.

Now for the reverse situation: Suppose you want to be in the office. Who will be in the field? The solution to this problem is not easy for the new business owner to solve. If you hire employees to do the field work, you need enough work to keep them busy. Getting steady work is rarely simple, and for a new business, it can be nearly impossible. I have always run into the same problem with this situation. I either have too many employees and not enough work or too much work and not enough workers.

Most new businesses are run by the owners. Tradespeople commonly do field work during the day and office work at night. While this is usually mandatory, it doesn't have to stay this way forever. Decide where you want to be, in the field or in the office, and work up a plan to meet your goal.

HAVE YOU EVALUATED YOUR CASH RESERVES?

Have you evaluated your cash reserves? Any business needs cash reserves to get over the humps in the road to prosperity. A large number of businesses fail each year due to limited cash reserves. Without backup money, what will you do when a scheduled draw payment is held up and your bills are due? If you begin paying your bills late, your credit will be damaged.

How much money do you need in reserve? The answer to this question relates to the nature of your business and your ability to project and maintain budgets and schedules. The degree to which you wish to launch a marketing campaign is another factor. Advertising takes money, and new businesses need to advertise.

Some people say you shouldn't start your own business until you have at least one year's salary in savings. I must admit, this would be a comfortable way to get started, but for most people, saving up a year's salary isn't feasible. When I opened my first business, many years ago, I borrowed $500 for tools and advertising. I had less than $200 in my savings account. Looking back, I was probably stupid to try such a venture, but I tried it, and it worked. I'm not, however, suggesting that you follow in those footsteps.

I am a risk-taker. For me, trying a new venture is an adventure. I have learned never to gamble more money than I can afford to lose, and I think this line of thinking is good. As I mentioned in the previous paragraph, I wasn't always so cautious, but having a wife and children can change your perspective on worthwhile gambles.

In evaluating how much money is enough, I suggest you play the worst-case game. In this game, you draw out scenarios of how your business venture might go. You are looking for what could be the worst possible outcome of your decision to start or stay in a business. Once you have played the "what-if" game enough, you start to develop some insight into what you stand to lose. The next step is to determine how much you are willing or can afford to lose.

I don't think there is any clear-cut answer to how much reserve capital is enough. Each individual has different needs. If I were forced to give an opinion, I would suggest having enough money to last at least four months without income. I also suggest that when you feel you have established your monthly money needs, add 20 percent to it. You always run into unforeseen expenses. Even with this reserve, you must monitor your success on a frequent basis. If the business is not going well or your budget spending is running high, look for alternative sources of income.

The value of setting goals is known by most successful businesspeople. While it might seem silly to set goals, goals make it easier to achieve desired results. It is important to break your desires and goals down into realistic segments. Setting standards that are too hard to obtain only leads to disappointment and disillusionment. Take your goals one step at a time, and you can probably reach most of them.

4

Taking the plunge

Taking the plunge into your own business is a big financial responsibility. There's really no way around this fact. You simply can't become a professional home builder without running the risk of losing money. While some of the risks are basically unavoidable, you can curtail many common expenses. By trimming the fat where you can, it's possible to lessen the odds of failure for your business.

Many home builders get caught up in the large amounts of money that they handle. It's not uncommon for a builder to make bank deposits in excess of $50,000. Most of the money is not theirs to spend on personal effects. Generally, the large sums of money are disbursements from construction loans. This money is often deposited into a builder's operating account and then paid out to suppliers and subcontractors. But some builders see this type of large money and begin to feel rich. They become extravagant in their expenses. This is a sure-fire way to destroy a building business.

To reap the highest profits, you have to work with a streamlined overhead. But it's possible to be so consumed with keeping expenses low that you damage your chances for big success. You have to know what expenses to cut out and which ones to keep.

CUTTING EXPENSES

Cutting the right expenses and projecting the future are integral parts of refining a business that can survive into later years. No business can stand stagnant and survive. Times change, and businesses must change with them. Your time is valuable, but what you do with the money you make means the difference between success and failure.

Cutting the right expenses is one way to hone the edge on your profits. However, cutting the wrong expenses can cost you much more than you save. How do you know which expenses to cut? Can you predict the future? You can if you look back at history and ahead to cyclic changes. While you might not be able to tell which horse will win the next race, you can make projections for your business that are more fact than fiction.

BEATING HEAVY OVERHEAD EXPENSES

Beating heavy overhead expenses is one of the best ways to ensure the success of your business. Overhead expenses can be enough to drive you out of business. If you fail to investigate and rate your operating costs, you might find yourself looking for a job. Since this is such an important aspect of your business management, let's take a closer look at how you can get a handle on your overhead expenses.

What are overhead expenses? Overhead expenses are expenses that are not directly related to a particular job. Examples of these expenses could include rent, utilities, phone bills, advertising, insurance, office help, and so on.

Rent

Office rent might not be much of a factor in your business. If you work out of your home, you might not notice any new financial strain by converting one of your rooms into a designated office. However, if you rent a commercial office space, you will be aware of increased money demands.

How much you pay in rent depends on how much space you lease and where the space is located. Prime locations are expensive for an office of any size. While you might rent a small upstairs office in an average location for less than $400 a month, the same space in a fashionable part of town could cost upwards of $800 a month. The size of your office also affects its cost.

Should you pay any rent or just work out of your home? If you can operate your business from your home, you can save the cost of rent, but you make sacrifices. A home office doesn't get you away from your family and household. This can be a disadvantage, especially for people lacking self-discipline. It can be easy to leave your home office to take care of that household chore you put off last Saturday. It is equally easy to leave your desk to watch your favorite show on television. If your spouse or children come into the office, you are likely to lose production time. Background noise can be distracting when you are talking business either in the office or over the phone.

If clients will be coming to your office, a home office can have a negative effect on the client's perception of your business success. However, many customers might envy you for being able to work from home. In general, a home office is a good idea for a new business, if the circumstances won't have an adverse effect on your ability to get and conduct business.

If a home office won't cut it, you need to turn to commercial space. With rent being an overhead expense, how much can you afford to pay? This question creates two questions: How much money do you have to spend on rent? How much can the company pay in rent and still maintain a desirable profit level?

If you have built your business budget and done your business forecasting, you can make some assumptions on how much rent is too much. When you know you must rent commercial space, you could decide on how much to spend by seeing what's available. You should know how much space you need and what locations are acceptable to you. With this knowledge, shopping for an office is a simple matter. You simply respond to various for-rent ads and check prices and amenities.

Once you know what office space is renting for, you are in a position to make a decision. Some of the deciding factors might include the term of the lease, the amount of the security deposit, if you must pay for utilities or if they are included in the rent, and other expenses related to the office.

For most contracting firms, a plush office is not necessary. As long as the office space is clean, well-organized, and accessible, it should be fine. In deciding on rental overhead, keep your costs as low as you can while still getting what you need (but not necessarily what you want).

Utilities

Utilities are expenses you can hardly do without. These expenses include heat, hot water, electricity, air conditioning, public water fees, and sewer fees. You might be able to make minor cuts in these expenses, but you will have difficulty in slashing the costs of these necessities.

Phone bills

Phone bills for a busy business can amount to hundreds of dollars each month. While it is often imperative to make business calls, you might be able to trim your phone bill down into a more manageable range.

One way to reduce your expenses is to take advantage of all the discount programs offered by the many phone services. Another possibility is to make your long-distance calls after normal business hours. Reducing idle chit-chat can have a favorable impact on your phone bill. If you have directory advertising that is billed on the invoice for your telephone service, you might consider altering the size of your advertisement. However, don't be too quick to cut back on your directory advertising; you might lose more business than the saving is worth.

Advertising

Advertising is a must-do expense for most businesses. However, you can make your advertising dollar stretch further by making your advertising more effective. Keep records on the pulling power of your ads. Track responses to your ads and the number of responses that turn into paying work. Target your advertising to bring in the type of work that is the most profitable for your firm. By refining your advertising, you can make it pay for itself.

Insurance

You cannot reasonably avoid the burden of insurance. Most business owners resent this expense, but they can't afford to be without it. There are two keys to controlling your insurance expenses. The first key is to avoid overinsuring the company. Don't pay premiums for more insurance than you need. The second key is to shop for rates and services. Insurance is a volatile market; the rates change often and quickly. The insurance you had last month might need to be reassessed this month. By doing periodic evaluations and shopping, you can maintain maximum control over your insurance expenses.

Professional fees

Professional fees might not be incurred on a monthly basis, but they can amount to hundreds, possibly thousands, of dollars a year. Certainly you should engage professional help when you need it, but you can lower the cost of these services by doing some of the work yourself.

If you have a CPA do your taxes, and you probably should, you can save money by doing some preparation work. If you have all your documents organized and properly labeled, you save the CPA time. By submitting a complete tax package to the accountant, you reduce the number of phone calls, visits, and time you spend with the professional. All the time you save for the accountant results in a lower fee.

You should consult attorneys on legal matters, and they should draft and review legal documents. However, if you know what you want, you can reduce lost time. Better yet, if you draft an outline for the attorney to go by, you can save even more time for the lawyer. When you meet with your attorney, have your questions prepared, preferably in writing, and organized. The quicker you get in and out, the less you have to pay.

Office help

Your business might or might not require office help. If you have personnel in your office, they are generally an overhead expense. Employees can be one of your most expensive overhead items. Don't generate this type of overhead expense until your business can't function properly without it.

Office supplies

Office supplies might not seem like a large expense, but they can add up. One way to reduce the cost of office supplies is to buy them in bulk. Instead of purchasing one legal pad, buy a case. Buying in bulk from wholesale distributors can reduce the money you spend on disposable office supplies.

Office equipment and furniture

Every office needs some equipment and furniture. You need a desk, a chair, and a telephone. You should have a filing cabinet, and you might want other pieces of furniture and equipment.

Be selective in what you purchase. Before you buy anything, make sure you need it and that the cost is justified. Office equipment and furniture are expenses in which many business owners go overboard.

Do you need a copier? Every business has a need from time to time to make copies of documents. Some businesses do enough volume to justify buying or leasing a copier, but most small businesses don't. When you consider that you can go to the local print shop and make copies for about a dime each, you see that it takes a lot of copies to pay for owning or leasing a copier. Sure, going out to make copies is inconvenient, but it can save you a considerable amount of money.

Do you need a fax machine? Unless you deal with a large number of commercial clients, you probably don't. If you use a fax infrequently, you can go to a print shop and pay a few bucks per page to have your documents faxed. Most of these pay-as-you-go fax places allow you to use their fax number to receive incoming documents. When you consider that a fax machine costs $300 or more, you can send a lot of documents from the print shop before your investment would be returned on a purchase.

Don't get caught up in the gadget trap. Many business owners like to buy gadgets for their offices, and there are plenty of gadgets available. How often would you use a globe of the world? Do you really need a binding machine to bind your reports and proposals? Can you live without an electric stapler? Before you spend precious money on items that do little more than get in your way, consider what you are buying.

Vehicles

Company vehicles can be considered overhead expenses. While it's true you need transportation, you don't have to have the ultimate in automotive engineering. If you can do your job in a $9500 mini pickup truck, don't buy a full-size truck at twice the price. Cutting your overhead is a matter of common sense and logic. Buy what you need and don't buy what you don't need.

LEARNING WHICH EXPENSES TO CUT

You must learn which expenses to cut. Cutting the wrong expenses might be worse than not cutting any expenses. I have just finished telling you about normal overhead expenses. What other expenses should you cut back on?

Field supervisors

Field supervisors can be a significant expense. If you have a field supervisor, ask yourself if you could do the field supervision and use that individual as an income producer. Some business owners are unwilling to delegate duties to anyone, and others push too much responsibility off on employees. Field supervision is mandatory for many contractors, but unless you have numerous jobs going simultaneously, you should be able to supervise the field work.

You gain many advantages by being your own field supervisor. You see, first-hand, how your jobs are progressing. Customers see you on the job and are more comfortable that they are getting a good job and special attention. The cost of having an employee as a supervisor is eliminated or reduced. Before you pay high wages to a field supervisor, consider doing the work yourself.

Leftover materials

Leftover materials are common in the contracting business. Since the quantity of these leftovers is usually minimal, many contractors put the material in storage. If the items will be used within a month, putting them in storage is not a bad idea. However, if you don't know when you will have an occasion to use the materials, return them to the supplier for credit.

By returning leftover materials, you don't have your money tied up in unneeded inventory and you don't need a large storage area. Another consideration in storing materials is the wages you pay or the time you lose in handling materials. When the materials are moved from the job to your storage facility, time is spent. When the materials are taken out of storage and transported to a job, more time is spent.

Most suppliers that you deal with regularly are happy to pick up your leftovers and credit your account. By taking this route, you don't pay to have the materials removed from the job site. When you need the materials again, the supplier delivers them for you. You save money both ways.

Travel expenses

You can cut your travel expenses by keeping mileage logs for all your vehicles. When tax time comes, you can deduct the cost of mileage from your taxes. The amount you may deduct is set by the government, but it is enough to make keeping a mileage log worthwhile.

Cash purchases

Most contractors have occasions when they purchase small items with cash. The items might be nails, photocopies, stamps, or any number of other business-related items. If you keep the receipts for these items, they can become tax deductions. While a receipt for less than a dollar might not seem worth the trouble of recording, if you collect enough of them, you will appreciate the savings.

TESTING YOUR EXPENSES

You can decide which costs to cut by testing your expenses. The two examples I gave earlier on copiers and fax machines indicate one method of testing. Other ways to test your expenses are as varied as the expenses themselves.

Before you incur or cut an expense, be sure you are doing the right thing. Taking action too quickly can result in costly mistakes. For example, taking your advertisement out of the phone directory will probably be a regrettable mistake. Dropping your health insurance might come back to haunt you. You must scrutinize all of your expenses and cut only the ones that do not hurt you or your business.

Knowing which expenses are justified

You should know which expenses are justified and should not be cut. Just like your ad in the phone book or your health insurance, some expenses are absolutely necessary to keep your business going.

How do you know which expenses are justified? If you test the expenses, you can tell if they are necessary. For example, a plumber would be lost without a right-angle drill. This tool expense is more than justified, it is a necessity. A painter's ladders would fall under the same classification. These expenses are obviously necessary.

The difficulty comes when looking at costs that are not so obvious. Does a plumber need a backhoe? Many times plumbers work with backhoes, but the occasions usually are not frequent enough to justify purchasing the piece of equipment. However, if the plumbing company does extensive work with a backhoe, it might be wise to purchase or lease one. You might consider yourself to be in a similar situation. As a builder, you no doubt have many occasions when a backhoe or excavator is needed. Does this mean you should rush out and buy one? I don't think so. You are normally much better off if you hire a subcontractor to do this type of work for you.

If you computerize your office, you need a printer. The printer is justifiable, but a laser printer might not be. A dot-matrix printer is capable of fulfilling most office needs, and its cost is a fraction of the price of a laser printer. So which printer should you buy? If you have defined needs for a laser printer, buy a laser, but if you don't, go with a dot-matrix.

Most business owners agree that advertising is a necessary expense. But spending money on ads that aren't producing is a waste. There is something to be said for quality being worth more than quantity. Some ads pull large responses but net few jobs. Other ads pull fewer responses but result in more work.

If you are placing ads in every medium available, you are probably wasting money. Advertisements should be keyed to let you know which ads are working and which ones are not. The odds are good that you can trim some of your advertising expenses without losing business.

You must walk a thin line when cutting expenses. If you don't cut out unnecessary spending, you are losing potential profits. If you cut the wrong expenses, you could be losing business. Knowing which expenses to cut must be learned with testing and experience.

Cutting the wrong expenses

Cutting the wrong expenses can be expensive. If you cut the wrong expenses, you might lose much more money than you save. Some of your actions might be difficult to reverse. For these reasons, you must use sound judgment in making cuts in your business expenses. Let's look at some of the expenses that you might not want to cut.

Directory advertising

Advertising in the phone book is one of the first expenses many business owners contemplate cutting. Before making cuts in your directory advertising, remember that you have to live with the changes for a full year. If you reduce the size of your ad or eliminate it, you can't reverse your actions until the next issue is printed.

Directory advertising does work. It is a proven fact that people let their fingers do the walking. Home builders can often do well with a much smaller ad than a service-and-repair company. People looking to have a house built are not dealing with an emergency situation, as they might be with an electrical or plumbing problem.

Being listed in the local phone book lends credibility to your company; it proves you have been in business for a while. The size of your ad in the directory can be influenced by your competition. If other builders are running big ads, it might be necessary for you to follow suit. However, many people believe that a box ad in the column listings is sufficient. You should remain listed in the phone directory in some capacity.

If you track your phone inquiries to determine which advertisements customers are responding to, you can determine what percentage of your business is coming from the phone book. I am sure you will get calls from directory advertising, but the cost of the ad could be more than the calls are worth. Track your calls for a year and decide if you are getting your money's worth from the phone book. You might be wise to have a modest listing in the directory and spend the money you save on a more targeted form of advertising. The main thing is to not make radical changes in your directory advertising before you are sure they are justified. A year is a long time to live with a mistake that hurts your business.

Answering services

Human answering services are another frequent target of business owners looking to cut expenses. While some businesses do all right with answering machines, most businesses do better when a person answers the phones.

Most answering services can be terminated and picked up the following month. If you are unsure of the value of your answering service, terminate it for a month and compare the number of leads you get for new work. If you don't notice a drop in business, you made a wise decision. If, however, you are losing business, you can reinstate the answering service.

Health insurance

Health insurance is very expensive, and many business owners consider eliminating their coverage at one time or another. You are taking a big risk to drop your health insurance. As expensive as the insurance is, if you have a major medical problem, the insurance is a bargain. The accumulation of big medical bills could drive you into bankruptcy.

If you feel you have to alter your insurance payments, look for a policy with a higher deductible. These policies have lower monthly premiums and still provide protection against catastrophic illness or injury.

Dental insurance

For most people, dental insurance is not as important as health insurance, but it is still comforting to have. If you have bad teeth, and crowns and root canals are likely to be needed, dental insurance can pay for itself. There are, of course, other types of dental services that can make your insurance premiums seem small. Try to avoid cutting out any of your insurance coverage.

Disability insurance

Disability insurance is not carried by all contractors, but it probably should be. How would you support yourself if you were disabled? Could you work with a broken leg? Disability insurance provides a buffer between you and financial disaster if you become disabled. You might get by without this type of coverage, but the gamble might not be worth the savings.

Inventory assets

Inventory assets often come under fire when money is tight. While you shouldn't carry a huge inventory that you don't need, you should stock your trucks with adequate supplies. If you cut back too far on inventory, your crews waste time running to the supply house, and your customers become frustrated by your lack of preparation for the job.

Retirement funding

Retirement plans are frequently one of the first expenses cut by contractors in money crunches. A short moratorium on retirement funding is okay, but don't neglect to reinstate your investment plans before it is too late.

Bid-sheet subscriptions

Bid-sheet subscriptions are sometimes put under the financial microscope. While these expenses are not monumental, they appear to be an easy cut to make. If all you do with your bid sheets is glance at them and trash them, by all means cancel your subscription. If, however, you bid work on the sheets and win some jobs, eliminating your subscription could have the same effect as turning work away.

Credit bureaus

The fees charged by credit bureaus can become a target for company cuts. Before making the decision to do without credit reports on your potential customers, weigh the risks you are taking. Doing work for one customer that doesn't pay would more than offset the savings you made by eliminating the credit bureau fees.

Advertising

Advertising is a tricky topic. When business is off, you need to advertise to get more business. But when business is down, so is your bank balance. Justifying the cost of advertising when your money machine is running on fumes is not easy. It is not wise to eliminate your advertising, but it is smart to target it. Do a marketing study to determine where and what to advertise, and then advertise. If your marketing research is accurate, the cost of your advertising is returned in new business.

Sales force

When times are tough, business owners look to the sales force. Even when the sales-people are paid only by commission, some business owners consider eliminating them. This makes no sense to me. If you have a sales force that only gets paid for sales made, why would you want to get rid of them? Unless you are going to replace the existing sales force with new, more dynamic salespeople, the move to eliminate sales is senseless.

LOOKING INTO THE FUTURE

You don't need a crystal ball to look into the future. What you need is determination, time, and skill. Time can be made, and skills can be learned, but you must already possess determination. If you aren't motivated to predict the future of your business, you won't be able to. On the other hand, if you are committed to making your business successful, you can do a fair job of projecting your future in business.

Your business will face many challenges over the coming years. If you aren't prepared for these obstacles, you might not get past them. But if you are prepared, your business has a good chance of lasting a lifetime. To be prepared, you must start making the preparations now.

Learning from the past

How can you judge what your business might encounter three years from now? Our economy runs in cycles. The businesses of most contractors are affected in some way by the real estate market. If the construction of new homes is down, most contracting fields suffer. When housing starts are up, contractors thrive. During interval periods, many companies make moves to improve business, but they often fail to survive the down periods of the economy.

Since the economy is cyclic, you can look back into history to project the future. The clues you find might not be right on the money, but they are likely to render a clear picture of what's in store for your business.

In the early 1970s, the real estate market was booming. People were making and spending money. Then, in the late '70s and early '80s, the business environment took a nose dive. Interest rates soared, and business production in most fields dropped. Contractors scurried to collect money due and to find new work. Times were tough, to be sure, but many contractors made it; I was one of them.

By the mid 1980s, business was good again. I was building as many as 60 homes per year, and most contractors were expanding their businesses, myself included. As time passed, the economy started shifting to a downward turn. By the late '80s, the business world was reeling again. Once again, the economy was sagging and businesses were closing their doors.

Now, we are in the '90s, and trends are pointing upward, slowly, but upward nevertheless. If you are just starting a business, what can you learn from this abbreviated lesson? You might see a trend for major slowdowns in the economy at least once every decade. You might also assume that financial failures over the last 20 years have occurred more often in the latter part of each 10-year period. Already, without much information, you can start to see that you might have to face a recession within the first 10 years of your business.

If you looked deeper into the history of the last 20 years, you would find some interesting facts. In the late '70s and early '80s, banks were quick to rise to the problems at hand. Creative financing blossomed, and the business world turned itself around.

In the more recent recession of the late '80s and early '90s, the banks did not rally to help; instead, many of them closed. Business owners in this recession didn't have the high interest rates to combat, but they also didn't have lenders willing to help them with their financial battles.

With interest rates low and efforts being made to get the economy back on track for the mid- and late-1990s, why aren't people spending money? I believe people are afraid to spend what money they have. Many people are without jobs, and the ones who have jobs don't know how long they will have them. Public confidence appears to be at an all-time low. Until confidence is restored, the rebirth of the economy will be painfully slow.

What does this tell you about your company? You can see immediately that there are different types of economic wars to win. If you have recently gone into business, you know what you are faced with when consumers are afraid to spend their money to improve their homes.

Why did so many lenders turn their backs on the recent recession? Could it be that the creative financing put in place by the banks in the earlier recession caused them to back away from the more recent recession? Why have some banks refused to make construction loans? How can an economy get back on its feet when lenders won't make construction financing available? When answered, all of these questions tell you something about the future.

Reading old newspapers at the library is one way to delve into the past. Talking with people who have lived through the tough times is a good way to gain insight into what happened and why it happened. Tracking past political performances can produce clues to the future. The key is spending the time and the effort to look back so that you can look ahead.

Once you are familiar with the past, you can see trends as they are forming. You can spot danger signals. These early-warning signs can be enough to save your business from financial ruin. If you like, you can compare your market study with the work of people tracking storms to warn of tornadoes and hurricanes. Certainly more people have been spared the pain of these vicious storms since scientists have studied and tracked past storms. Unless you have an astute business advisor, you have to learn to pick up on your own early-warning signals.

Making long-range plans

Long-range planning pays off in the end. By preparing for the worst, you can handle most situations that come your way. Your plans must focus on financial matters, but that's not all you have to plan for. As you grow older, how will your business be run? If you do your own field work, how will it get done when you are no longer physically able to do it? How quickly will you allow your business to grow? These are only some of the questions you want to answer when planning for the future.

If you bury your head in the sand, you won't be aware of the many changes going on around you. Can you imagine how business owners of only a few years ago felt about computers? Do you think they ever dreamed that computers would have such dominance in the marketplace? Change is inevitable and beyond your control. How you plan for the changes is within your grasp. To have a long-term business, you must have long-term plans.

ADJUSTING YOUR COMPANY

Adjusting your company to meet changing economic conditions and demands requires your attention. Businesses don't change themselves. People change them, people just like you. How will you change your business? You probably don't know yet, but you had better start making plans for the changes soon.

As you grow older, you might get tired of working in the field. What used to be fun might not be so enjoyable at an older age. The physical work that has kept you in good shape might become a bit much for you in 10 or 20 years. How will you adjust for the desire to get away from the physical work?

Some contractors plan to stay in the field until the day they retire. Many contractors anticipate hiring employees or subcontractors to pick up the slack in the physical work. Of the two options, I recommend planning on hiring help. You might well get tired of working in the field before you can afford to retire.

If you know that some day you plan to bring employees or subcontractors into your business, you can start planning for the change now. In your spare time, if you have any, you can read up on human resources and management skills. The knowledge you gain now will be valuable when you enlist the help of others in your business.

Preparing for company growth

Many business owners don't prepare properly for company growth. These owners go about their business and add to it as volume dictates. This is a dangerous way to expand your business. Allowing your company to grow too large too fast can put you out of business. I know it seems strange that having a bigger business could be worse than maintaining your present size, but it can.

When owners allow their companies to balloon with numerous employees, subcontractors, and jobs, management can become a serious problem. This is especially true for business owners with little management experience. The sudden wealth of quick cash flow and more jobs than you can keep up with is a company killer.

The operating capital that kept your small business floating over rough waters will not be adequate to keep your new, larger business afloat. Overhead expenses increase along with your business. These expenses might not be recovered with the pricing structure you are accustomed to using. All in all, growing too fast can be much worse than not growing at all. If you want to expand your company, plan for the expansion. Make financial arrangements in advance, and learn the additional skills you will need to guide the business along its growth path.

Continuing your education

Continuing education is a requirement for some businesses. Many licensed professionals are required by their licensing agencies to participate in continuing education. In New Hampshire, plumbers must attend an annual seminar before they can renew their licenses. In Maine, real estate brokers must complete continuing education courses to keep their licenses active. Whether your business forces you to pursue continuing education or not, you should. If you don't keep yourself aware of the changes in your industry, you will become outdated and obsolete.

You owe it to yourself and your customers to stay in touch with changes in your field. Read, attend seminars, go to classes, do whatever it takes to stay current on the changes affecting your business.

Taking bigger jobs

Bigger jobs can mean bigger risks. Before you venture into big jobs, make sure you can handle them. You need to consider several factors in your planning. Will you have enough money or credit to keep the big jobs and your regular work running smoothly? Do you have enough help to complete your jobs in a timely fashion? If you are required to put up a performance bond, can you? Will you be able to survive financially if the money you're anticipating from the big job is slow in coming? Do you have experience running large jobs? This line of questioning could go on for pages, but all the questions are viable ones to ask yourself. You shouldn't tackle big jobs until you are sure you can handle them.

Diversifying your company

Should you diversify your company? At some point you will probably ask yourself this question. To answer it, you have to spend some time thinking, evaluating, and researching.

There is no question that diversifying your company can bring you more income. But it can also cause your already successful business to get into trouble. Companies that diversify can fail for many reasons. When you split your interest into multiple fields, you are less likely to do your best at any one job. For this reason, many companies do better when they don't diversify.

In rural areas, it is sometimes necessary for a small business to fulfill many functions to survive. For example, a home builder might have to take on remodeling jobs. In areas with large populations, this type of spreading out is not needed. Builders can build, and remodelers can remodel; there is enough business to go around to all the specialists.

Your geographic location could be a factor in your decision to expand your business operations. The desire to make more money, however, is the most common reason for business alterations. Greed can be a very powerful destroyer. If you try to get your fingers into too many pies, they might all fall off the window sill. In other words, don't allow yourself to be driven by greed.

If you want to diversify, do so intelligently. Don't just decide one day you are going to hire a master electrician and expand your building business to include electrical services. What will you do if your master electrician quits? Without a master's license, you are out of business, at least in that phase of your business.

You must consider many factors before splitting your time and money into separate business interests. Most people struggle to keep one business healthy. If you get aggressive and open several business ventures, you might lose them all.

Adjusting your company for change is not a task you can complete and be done with. To maintain your business, you must occasionally change your plans. Routine adjustments are normal and should be expected. This demand on your time can cost you much more money than you make. The concepts needed to make the most of your time and money are simple. If you follow the advice in this chapter, you should be able to do very well. Getting too big too quickly is a common cause of business failure, so pace yourself carefully.

5
Choosing a business structure

This chapter helps you explore the different forms of business and decide which one best suits your needs. You might find that incorporating your business provides additional security from personal liability. You might discover that if you incorporate, your tax consequences are worsened. The lure of partnerships might not be so attractive after reading about what can happen to you in a partnership. In any event, by the time you finish this chapter, you should have a clear insight into what type of business structure is best for you.

Choosing your style of business structure is not a task that should be taken lightly. You have a multitude of considerations to evaluate before setting a business structure for your venture. Do you know the difference between a standard corporation and an S corporation? Are you aware of the tax advantages to running a sole proprietorship? Were you aware that having a partner can ruin your credit rating? If you answered no to these questions, you need to read this chapter.

First let me emphasize that I am not an expert on tax and legal matters. The information in this chapter is based on experience and research. As with all the information in this book, you should verify the validity of the information before using it. Laws change and different jurisdictions have different rules. Now, with that out of the way, let's dig into the various types of business structures.

WHAT ARE THE POSSIBLE BUSINESS STRUCTURES?

Corporations, sole proprietorships, and partnerships are all common forms of business structure. Each form of business structure has its own set of advantages and disadvantages.

Corporations

A corporation is a legal entity. To be a legal corporation, the corporation must be registered with and approved by the secretary of state of the state in which it operates. A corporation can live on in perpetuity. Corporations may operate under general management, may provide limited liability, and may transfer shares of stock.

Subchapter S corporations

A subchapter S corporation is not like a standard corporation. Commonly called an S corporation, these corporate structures must meet certain requirements. For example, an S corporation may not have more than 35 stockholders. An S corporation is one that chooses not to be taxed in the same way that a regular corporation is.

Additional requirements include having stockholders provide personal tax returns. The stockholders must show their share of capital gains and ordinary income. Tax preference must also be provided personally by the stockholders. The corporation files a tax return but does not pay income tax. This type of corporation allows the stockholders to avoid double taxation. However, subchapter S corporations may not receive more than 20 percent of their income from passive income. Passive income could be rent from an apartment building, book royalties, earned interest, and so on.

Partnerships

A partnership is an agreement between two or more people to do business together. In a general partnership, each partner is responsible for all of the partnership's debt. This is an important factor. If you have a bad partner, you could be held liable for partnership debts the partner incurs. Most partnerships don't pay income taxes, but they must file a tax return. The income from the partnership is taxed through the personal tax returns of the partners.

Partnerships can be comprised of all general partners or one general partner and any number of limited partners. General partners do not have limited liability. The general partner can be held accountable for all actions of the partnership. Limited partners can have limited risk. For example, a limited partner might invest $10,000 in the partnership and limit his or her liability to that $10,000. Then, if the partnership loses money, goes bankrupt, or has other problems, the most that the limited partner stands to lose is the initial investment of $10,000. The general partner, however, is held personally responsible for all actions of the partnership.

Sole proprietorship

A sole proprietorship is a business owned by an individual. The business must file a tax return, but any income tax is assessed against the individual's tax return. Sole proprietors are exposed to full liability for their business actions.

Now that you know what each type of business structure is, let's see what type fits your needs.

WHAT TYPE OF BUSINESS STRUCTURE IS BEST FOR YOU?

Now that you know what the different types of business structure are, what type would you choose, if you had to decide now? Well, the truth is, I haven't given you enough information to make a wise decision. But I will. Before you make a firm decision, you owe it to yourself to read and digest the remainder of this chapter. It might be interesting to see if your choice after reading the entire chapter is the same as it is now.

Should you incorporate your business? The primary advantage for most business owners from a corporation is the limited liability possible with a corporation, but the protection might not be as good as you think. Many business owners don't know how corporations work. These same owners pay the fees to incorporate without realizing they are not getting what they are paying for. Oh yes, they are getting a corporation, but they are not buying the protection they think they are. What is this big liability misconception? There are several misconceptions about corporations. Let's explore some of them.

Protecting yourself from liability

Will a corporation protect you from liability? Yes and no—it depends on what type of liability you are seeking protection from and how your business operates. Let me get specific on these issues.

Protection from lawsuits

Protection from lawsuits is one interest of people forming corporations. It is almost impossible to protect yourself entirely from lawsuits, but a corporation can help. If your business is a corporation, your liability might be limited to the corporate assets, but don't count on it.

Assume that your business is a corporation. You are a home builder and your own lead carpenter. When you have a house to build, you take an active part in the construction as a carpenter. This means that you are nailing nails and doing much of the physical work involved with the construction of each home. On one particular job, you cut a hole in the subfloor to allow your mechanical tradespeople easy access to the basement. Stairs to the lower level have not yet been installed. Workers are moving up and down with the use of ladders. A piece of plywood is nailed in place over the opening at the end of each day. However, one day you forget to nail the cover in place. Your customer comes out to inspect the day's work and falls through the hole in the subfloor.

Your customer decides to sue. Can the customer sue the corporation and go after its assets? Yes, the customer can sue the corporation. Can the customer sue you, as the primary stockholder of the corporation? Not really. I suppose anyone can sue anyone else for any reason, but it would be unlikely that the customer would get far in a lawsuit against you as a stockholder. But, don't relax; you're not safe. Since you did the work yourself, the customer can sue you individually for the mishap. This is a fact that most business owners don't realize.

If you had sent an employee or a subcontractor out to do the work, your risk of being sued for your personal assets, other than your corporation, would be minimal. So, if you don't do your own field work, a corporation can help protect your personal assets.

Personal financial protection

Personal financial protection is usually another goal of people wanting to incorporate their businesses. Some of these people believe they can limit their financial liability with a corporation, and, to some extent, this is true. If a corporation gets in financial trouble, it can be bankrupted without the loss of personal assets belonging to stockholders. However, most lenders and businesses that extend credit to corporations require someone to sign personally for the debt. If you endorse a corporate loan personally, you are responsible for the loan, even if the corporation goes belly up. Before you spend needed money on false protection, talk to experts for details on how incorporating affects you and your business.

Dealing with partners and partnerships

Dealing with partners and partnerships can get tricky. My worst business experiences have involved partners. It would be easy to think that my bad track record with partners has been my fault. However, the clients that I consult with have shared similar bad experiences with partners. I can think of very few successful partnerships. If you are considering going into business with a partner, go into the arrangement with your eyes wide open.

Partnerships should be treated seriously. If you are a general partner, you can be held accountable for the business actions of your partner or partners. This fact alone is enough to make partnerships a questionable option. There are very few advantages to partnerships, but the potential problems are numerous.

If you decide you want to set up a partnership, consult an attorney. The money you spend in legal fees forming the partnership is well spent. It is much better to invest in a lawyer to form the partnership than it is to pay legal fees later to resolve partnership problems.

WHAT ARE THE PROS AND CONS OF EACH FORM OF BUSINESS?

Before you can make a final decision on what type of business structure to assume, you should learn the pros and cons of each form of business. Let's take each type of business structure and examine the advantages and disadvantages.

Advantages of a corporation

Corporate advantages are abundant for large businesses, but a standard corporation might not be a wise decision for a small business. What advantages does a standard corporation offer? Depending on the size and operational aspects of a business, a corporation can provide protection from personal lawsuits and financial problems.

Since corporations can issue stock, there is the possible advantage of generating cash with stockholders. When a corporation goes public, large sums of money can be generated, but this is not usually a feasible option for a small business.

While small businesses can suffer from double taxation with a standard corporation, they can benefit from some tax advantages. By incorporating your business, some expenses that are not deductible as a noncorporate entity become deductible. Insurance benefits could be an example of this type of deductible advantage.

Disadvantages of a corporation

In many ways the corporate disadvantages are more significant than the advantages for a small business. The first disadvantage is the cost of incorporating. Filing fees must be paid, and most people use lawyers to set up corporations. The cost of establishing a corporate entity can range from less than $200 to upwards of $1000, which can be a lot of money for a fledgling business.

There is more to having a corporation than setting it up and forgetting about it. To keep the advantages of a corporation effective, you must maintain certain criteria. You have to have a registered agent. Many people use their attorney as a registered agent. If you use an attorney in this position, you are spending extra money. Board meetings must be held, and the corporate book must be maintained. A board of directors must be established. Written minutes must be recorded at meetings. Annual reports are another responsibility of corporations. Again, many people have an attorney tend to much of this work.

Corporate officers must be appointed, and the fees for maintaining an effective corporation can add up. If the corporate rules are broken, the corporation loses much of its protection potential. If an aggressor can pierce the veil of your corporation, you have personal liability.

Small business owners who incorporate their business as a standard corporation face double taxation. These owners pay personal income tax and also pay corporate taxes. This extra taxation is a serious burden for most small businesses.

The advantages of a subchapter S corporation

The advantages of a subchapter S corporation are a little better than those of a standard corporation for a small business. One of the biggest advantages of a subchapter S corporation over a standard corporation is the elimination of the double taxation. Stockholders of subchapter S corporations pay personal taxes on the money they and the corporation earn, but they only pay taxes once. S corporations offer the other advantages you would receive with a standard corporation.

Disadvantages of S corporations

The disadvantages of S corporations are not too bad. The cost of setting up and maintaining the corporation are still there. You cannot use an S corporation once you have more than 35 stockholders. If more than 20 percent of your corporate income is passive income, you cannot use a subchapter S corporation.

The advantages of a partnership

The advantages of a partnership are not numerous. The cost of establishing a partnership is less than incorporating, but I don't see the need for partnerships in small business operations. Partnerships have their place in business ventures, but I don't like them for most types of business. If you are a real estate investor, partnerships can work, but why gamble with a partnership for a service business? Set up a corporation or investment agreements, but avoid partnerships.

Disadvantages of partnerships

In my opinion, the disadvantages of partnerships are abundant. When you are in a partnership, as a general partner, you share a portion of the profits and all of the risk. Hooking up with a bad partner can not only ruin your business, it can ruin your future. I know some people like business partnerships, but I'm not one of them. If you decide to set up a partnership, get legal counsel to understand what you might be getting into.

Advantages of a sole proprietorship

The advantages of a sole proprietorship are many. A sole proprietorship is simple and inexpensive to establish. You are your business, so there are no complicated corporate records to keep. You can get tax advantages as a sole proprietor as you can with other forms of business structure. You are the boss; there is no partner to argue with over business decisions, and there are no stockholders to report to. Tax filing is relatively simple, and you don't have to share your profits with others.

Disadvantages of a sole proprietorship

There are some disadvantages to a sole proprietorship. As a sole proprietor, you have to sign for all your business credit personally. If your business gets sued, you get sued. There might be some tax angles that you will miss out on as a sole proprietor.

There is no generic answer for which type of business structure is best. Circumstances surrounding individual business owners dictate the ideal type of organization to form. Talking with your attorney is a good idea when you are trying to decide which path to take in establishing your business identity. It's likely that you will find a corporate structure to be best, but you should get professional legal advice before making a final decision.

6

Avoiding
15 common
builder mistakes

Over the years, I've done a lot of building and I've gotten to know more builders than I can count. My work as a consultant to other contractors has exposed me to a host of problems. I've suffered with a multitude of difficulties personally, and my work as a consultant has shown me mistakes that I haven't made. The combination of my experience as both a builder and a consultant has given me a great deal of respect for the building business. It takes only one slip-up to put a builder out of business.

Builders who are just getting into business are probably the most likely to make mistakes that are financially fatal. But experienced builders often fall into traps of their own. As a matter of fact, I'm not sure that the rookies are at the highest risk. When people start doing something new, they are often careful of every step they take. Once they feel like they are in control, they tend to lighten up on their caution. This is usually when trouble strikes. So, even if you've been building houses for 15 years, you could mess up.

Everyone makes mistakes. My mother used to tell me that the first time I did something wrong was experience and that the second time I did the same thing wrong was a mistake. This makes sense. Unfortunately, business owners sometimes only get one chance to do their jobs right. This can be especially true for home builders.

The amount of money at stake when houses are being built can be more than an average person could ever reasonably pay back if something disastrous happened. I remember a time when I was leveraged out for more than $4 million in construction loans. If something had happened to me at that time of my life, there would have been no way that I could have ever paid the money back by working a normal job.

Contractors can make all sorts of mistakes. Most of the problems can be avoided. How do you avoid problems if you don't know what to look out for? Experience is often the only protection we have against mistakes. But how do you survive in a tough business long enough to get the experience you need? It is a difficult situation.

You have taken a major step in the right direction by reading this book. By learning from my experience and mistakes, you will be better prepared to avoid your own. Seeing the danger signs in time to react before you are in too deep is paramount to your success as a builder. Hopefully, you can gain enough knowledge from this book and other sources to make your way to the top of your profession.

I've thought over all the mistakes that I can remember either making or seeing made. My list is a long one. Many of the problems, however, are not so deadly as to sink your business. So, what I've decided to do is share 15 of the all-time mistakes that I'm aware of. If you study these topics, you can see how you might avoid mistakes that builders frequently make.

1. IT TAKES MORE THAN YOU THINK.

It takes more money to become a full-time builder than you might think. This mistake is one of the first builders often make. If you sit down and run numbers on what your start-up costs will be, the numbers probably look manageable. The hardmoney expense of opening a building business isn't very high. It's the hidden expenses that can put you out of business before you really get started.

A lot of people either don't know about or don't think about many of the expenses they might incur when going into business for themselves. For example, you still have all of your personal bills to pay after you quit your job while you are waiting to draw your first income from your business. That steady paycheck won't be there for you each week. If you don't have enough money saved to support yourself for several months, your time as a professional builder could be very short. Income is not the only issue. If you presently have health insurance benefits supplied by your employer, you might not think to factor the cost of insurance into your new business overhead projections. Forgetting this fact could put you at the unhappy end of a surprise when you realize that hundreds of extra dollars are needed each month for insurance premiums.

Failing to prepare properly for going into business is one of the major reasons why so many businesses never see their one-year anniversary. Take the time to plan out all aspects of your business before you cut your ties with an employer.

2. AVOID HEAVY OVERHEAD EXPENSES.

Avoiding heavy overhead expenses is something that every business owner should strive to do. Most builders know that high overhead expenses are not financially healthy, but many of them dive right in anyway. If you rush out and rent a lavish office, buy a new truck, lease a heavy-duty copy machine, and spend your money and credit on things that you don't truly need, you might find yourself in a cash bind. Getting out of long leases and installment debt can be very difficult. Not being able to dump your overhead can put you out of business.

It's easy to get caught up in the thrill of owning your own business. Part of the fun is getting to buy things that you want. But you must be reasonable about your purchases and expense commitments. Hiring a full-time assistant is something that you probably shouldn't do right away. There are dozens, if not hundreds, of ways to waste your money before you even earn it. Move cautiously, and don't put yourself in a financial box that you can't get out of.

3. DON'T BE TOO CAUTIOUS.

There is such a thing as being too cautious. If you tend to be a money miser, you might find that you're not cut out to be an entrepreneur. To make a new business work, you have to take risks and spend money. I know I just finished telling you to be cautious, but don't let fear grasp and immobilize you.

I've seen contractors start a business and refuse to pay one penny for advertising. How they expect to get business if no one knows that they are in business is beyond me. My personal experience and mistakes have made it abundantly clear to me that a person can hurt himself or herself by trying to save money. I once terminated a big, expensive ad I had in the phone book to save money. Well, I didn't have to pay the ad expense any longer, but my business dropped off dramatically. I lost much more money in reduced business than I saved. You have to balance your spending to achieve the best results.

4. SELECT YOUR SUBCONTRACTORS CAREFULLY.

Select your subcontractors carefully. Even though your subs are independent contractors, they have a lot to do with your business image. Sloppy workmanship by your subcontractors reflects on you. If your subs are rude to your customers, you'll take the heat. Many builders look for subcontractors who offer the lowest price. This can be a major mistake. Don't assume that one electrician is as good as another. Screen your subcontractors thoroughly before you put them on a job. It only takes a few bad jobs for word to get around that you are a builder to avoid.

5. SET UP A LINE OF CREDIT.

Set up a line of credit with your banker before you need it. People often wait until they need money to apply for a loan. This is usually the worst time to ask for money from a bank. My experience with bankers has shown that they are much more likely to approve me for loans when I don't need them than they are when I do need money.

Most builders run into cash-flow crunches from time to time. Having a line of credit established to float you over the rough spots until your next draw or payment comes in can make a lot of difference to both your building operation and your credit report. If you have to put off paying your suppliers for a couple of months until money that's due you comes in, the suppliers might not go out of their way to do business with you in the future. Sometimes houses don't close on schedule. A builder's money can be held up for several weeks or even months. Unless you have plenty of money in a reserve account, set up a line of credit in anticipation of problems that you will likely encounter at some point in your building career.

6. GET IT IN WRITING.

When you are doing business, it's always best to get all the details in writing. For a builder, it's especially important to have clear contracts with both customers and subcontractors. You should always insist on a letter of commitment that shows a loan has been approved for your customer before you start construction. Some builders start work on custom homes as soon as a customer signs a contract. You can do this, but it's very risky. If your customer is denied a loan, you might never get paid for any of the work or material you have supplied. I hate to say it, but don't take your customer's word for the fact that a loan has been approved. I've never known a lender who didn't issue a letter of commitment once a loan was approved. You should have a copy of the commitment letter in your job file. Even if you are working only as a construction manager, you should require your customer to sign a letter of engagement (FIG. 6-1).

7. STAY AWAY FROM TIME-AND-MATERIAL PRICES.

Stay away from time-and-material (T&M) prices when dealing with your subcontractors. It's possible to save money with T&M deals, but you can also lose your shirt. The lure of saving some money is a strong one, but be aware that your attempt to save could cost you plenty.

I built a house for a customer a couple of years ago and used a site contractor on a T&M basis. The contractor was recommended to me by one of my carpenters. After meeting with the contractor, I found that he was just going into business for himself and wasn't sure how to give me a firm price. As experienced as I was, I messed up and agreed to a T&M basis. I did this to save money and to help the guy out. The deal backfired on me. I'd gotten quotes from other contractors, and the T&M price wound up being much higher than any of the quotes. If you agree to pay as you go, you might be paying a lot more than you plan to. Flat-rate pricing is sometimes more expensive, but it's always a safer bet.

8. CHECK ZONING REGULATIONS.

Before you disturb the first tree or dig your first hole, check zoning regulations. I know of many occasions when seasoned builders have gotten into big trouble with zoning problems. A commercial building erected in Virginia was over a setback line and had to be moved. A builder in Maine just recently built a house where part of the foundation was on someone else's land. A carpenter who used to work for me went into business for himself and placed a well on a building lot that, due to the location he chose, made it impossible to install a septic system without getting an easement for the adjoining land owner. I could go on and on with these types of stories.

Zoning regulations are easy to check into. If you go to your local code enforcement office or zoning office, you can find out exactly what you need to know in the way of setbacks and such. If you build a house that winds up being in a restricted or prohibited space, you're likely to have a major lawsuit filed against you. You can avoid this problem with some simple investigative work.

9. CHECK ON COVENANTS AND RESTRICTIONS.

Many subdivisions have covenants and restrictions that protect the integrity of the development. Customers who buy a piece of land and come to you to have a custom home built might not be aware that these restrictions exist. Some builders buy lots without ever asking what restrictions apply. Some examples of common restrictions include a minimum amount of square footage, a prohibition on the use of some types of siding and roofing, and so forth. Don't build a house and then find out that the siding you used has to be torn off and replaced. Even more importantly, don't build a ranch-style house and then discover that the subdivision allows only two-story homes. This is scary stuff, but it can happen.

Letter of Engagement

Client _____

Street _____

City/State/Zip _____

Work phone _____ Home phone _____

Services requested_____

_____

_____

_____

_____

_____

Fee for services described above $_____

Payment to be made as follows:

_____

_____

_____

By signing this letter of engagement, you indicate your understanding that this engagement letter constitutes a contractual agreement between us for the services set forth. This engagement does not include any services not specifically stated in this letter. Additional services, which you may request, will be subject to separate arrangements, to be set forth in writing.

A representative of _____ has advised us that we should seek legal counsel prior to using information or material received from _____.

We the undersigned hereby release _____, its employees, officers, shareholders, and representatives from any liability. We understand that we shall have no rights, claims, or recourse and waive any claims or rights we may have against _____, its employees, officers, shareholders, and representatives. We further understand that we will pay all costs of collection of any amount due hereunder including reasonable attorney's fees.

_____   _____
Client   Date   Client   Date

_____
Company Representative   Date

6-1 Letter of engagement.

You can check for covenants and restrictions yourself by reading a copy of the deed to a piece of property. If you're buying a lot, have your attorney check the prohibitions prior to making a final purchase commitment. When a customer comes to you with a lot he or she already owns, ask to see a copy of the deed. Protect yourself. Nobody is going to do it for you.

10. BUY INSURANCE.

Some states require builders to carry a minimum amount of insurance; others don't. Whether your local laws require insurance or not, you should get it. At the very least, you need liability insurance. You might also need many other types of coverage, such as worker's compensation insurance. Talk to your insurance agent for advice on exactly what types of coverage you should have.

If you build without insurance, like a few builders I know, you are sitting on a time bomb. One lawsuit is all it takes to ruin your life. You could lose your business, your home, and your future. Insurance premiums are a pain to pay, but if you ever need the coverage, you'll be thankful that you have it.

11. MAKE SURE YOUR QUOTES ARE ACCURATE.

Inaccurate quotes can have you working for nothing. Most builders commit to building a house for a flat fee. If the price you give your customers is too low, it's just too bad for you. While you will probably never bid a job so low that you have to take money out of your savings account to pay for the cost overruns, it's very possible that you will give away some percentage of your profit due to mistakes in your pricing.

Home building is a business that often runs in cycles. You might go for months with no work in sight and suddenly get flooded with requests for quotes. This is a dangerous situation. You've been sitting around with nothing to do and wondering how you will ever pay the bills. When you finally get a chance to bid a job, you might tend to bid low to make sure you get it. This is bad enough, but it gets worse. When bid requests are piling up on your desk, you might rush through them, hoping to make up for lost time. You can make mental errors under these conditions.

If you make mistakes in your take-offs or estimates, you might not know it until a house is nearing completion. By then, it's usually too late to do anything other than suck up the losses. Many builders, myself included, have lost enough money with mistakes made when quoting jobs to lose sleep over the experience. Take your time and make sure your estimates are right. Double-check everything you do. If you have someone you can go to for a second opinion, such as one of your field supervisors, do it. Fresh eyes often catch omissions. Once you commit to a price in a contract, you'll have a hard time raising it.

12. INSPECT YOUR JOBS FREQUENTLY.

Inspect your jobs frequently. Some builders are reluctant to go out into the field. They're either too busy, don't want to run into customers, or are just too lazy. You need to visit your job sites often. I try to get to every job at least once a day. When I was doing volume building, I had two full-time supervisors checking every job twice a day.

A lot can happen in just one day. If you skip an inspection, your customer might know more about the condition of a house than you do. It's very embarrassing to have a customer come to your office making a complaint about a condition that you are not aware of. As the general contractor, you should stay on top of your jobs as best you can. Failure to inspect jobs results in poor quality control. When this happens, a builder's reputation can be tarnished. If you get a bad reputation in the building business, you might as well look for some other type of work.

13. WORK ON CUSTOMER RELATIONS.

Customer relations are a crucial part of a building business. People get emotional when they are having a house built. They sometimes become irrational. You have to be willing to talk with your customers, and you might have to settle them down from time to time. If you don't tend to your customers well, you won't get the coveted referral business that so many builders live off of. It's essential that you keep your customers happy.

14. USE WRITTEN CHANGE ORDERS.

If your customers ask for added services or changes in contract obligations, use change orders (FIG. 6-2) to document what is going on. Otherwise you could be left holding the bag on some expensive items. A customer who upgrades from a regular bathtub to a whirlpool tub could be increasing the cost of a home by thousands of dollars. If you don't have authorization, in writing, to make this change, you might never get paid for the increased cost of your plumbing contract.

I can't remember a single house that I've ever built where a customer didn't request some type of change or addition to the original contract. In my early years, I would accept a verbal authorization for changes, but not anymore. I've been stiffed for money too many times to take the chance on getting paid. Collecting your money for extras can be difficult enough when you have written documentation of authorization, and it can be darn near impossible without some type of proof. A lot of builders lose a lot of money by not putting all of their agreements in writing. Don't allow yourself to become another statistic in the book of builders who lost money.

15. NEVER GET TOO COMFORTABLE.

Never get too comfortable with your business. Builders who think they know it all often find out quickly that they don't. If you slack up, the competition will take advantage of your reduced effort. To enjoy continued success, year after year, you can never stop improving your business. Some businesses, like real estate brokerages, are required to maintain continuing education requirements. You might not fall under such regulations, but you owe it to yourself to always improve your knowledge and understanding of what it is that you do for a living. If you stop learning, you will likely see a decline in your profits. At the very least, you probably won't see any increase in your business. Regardless of how long you work as a builder, there are always new things to learn.

Change Order

This change order is an integral part of the contract dated_____, between the customer, _____ , and the contractor,_____, for the work to be performed. The job location is _____.
The following changes are the only changes to be made. These changes shall now become a part of the original contract and may not be altered again without written authorization from all parties.

Changes to be as follows:

_____

_____

_____

_____

_____

_____

These changes will increase/decrease the original contract amount. Payment for theses changes will be made as follows:_____. The amount of change in the contract price will be _____ ($). The new total contract price shall be _____ ($).

The undersigned parties hereby agree that these are the only changes to be made to the original contract. No verbal agreements will be valid. No further alterations will be allowed without additional written authorization, signed by all parties. This change order constitutes the entire agreement between the parties to alter the original contract.

_____   _____
Customer   Contractor

_____   _____
Date   Date

_____
Customer

_____
Date

6-2 Change order.

7

Building model homes: Pros and cons

What are the pros and cons of building model homes? There are many, and I discuss them in this chapter. Some of the advantages are obvious. The disadvantages are not quite as easy to spot. Different builders have different feelings with respect to model homes. But most builders and brokers agree that it is easier to sell new to-be-built homes when prospective customers can walk through a model home.

I've done business with and without model homes. During my many years as a builder, I've sold more homes without models to show than I have when working with show homes. This is not to say that a builder will normally sell more houses without a model than with one. The reason my numbers indicate this is that I have had to work without model homes for much of my career. Given a choice, I would normally want a model available. But at times, a model gets in the way of making a sale.

ADVANTAGES OF MODEL HOMES

Ease of selling

It's easier to sell to-be-built homes when a model home is available for inspection. People like to see and touch items that they plan to buy. This is never more true than when people are making one of the largest purchases of their lives. When you have a model home available, people are likely to walk in and look around out of curiosity. Some of these lookers can be converted into buyers, giving you sales that you would not have made without a model home.

Builders who sell from model homes don't normally have to be concerned with references. Customers see the model and use it as a comparative reference. This is an obvious advantage. Not only does a model home serve as a reference, it is a showcase of options that potential purchasers can see and fall in love with. Let me give you an example.

Let's say that you specialize in building log homes. If you have customers come into an office and make buying decisions from catalogs, you will probably miss out on some profit that you might have made if you had a model home. For example, if

I ask you if you want a fireplace, you might want to say yes but feel that you can't afford the expensive option. But if you walk into a model home and see a gorgeous stone fireplace with a hand-hewn mantle, it's much more difficult to say no. Seeing the fireplace in person makes you feel as though the house just wouldn't be complete without it.

Some salespeople can paint such vivid pictures with their words that they can sell people on nearly any option just by talking about it. Most salespeople are not so talented, and few builders possess the descriptive ability to make buyers act on impulse. Builders who use model homes often load them up with options. Instead of a drop-in lavatory in the bathroom, which is what is included in the advertised base price, builders install attractive, cultured-marble vanity tops with integral lavatory bowls. The price of the house goes up when a customer wants to upgrade from a drop-in lavatory to a marble top, but most customers have trouble accepting something less than what they see. This allows a builder to advertise low base prices and work buyers up on a per-item basis.

If you were to advertise your base house without a deck but built a dramatic deck on your model home, you would probably sell a lot more decks. You could just list decks as an option, but having one in place on your model should increase your sales. The same concept applies to all sorts of extras.

The sell-up factor that is possible when options are loaded into a home can add a considerable amount of profit to a job. The mark-up on extras is usually high, and once customers get to the point at which they are selecting options, they're pretty well sold. This is how many builders can advertise rock-bottom prices and still make good money.

Credibility

Builders who have model homes have more credibility than contractors who don't have sample homes available for showing. Customers assume that a contractor who has model homes is stable and in business for the long haul. This isn't always the case, but the perception is there and that helps builders with model homes make more sales. You don't have to have model homes to appear credible, but they do add to your public image.

Extra attention

Model homes can bring a lot of extra attention to a builder. As you probably know, most model homes are dressed up to capture the attention of people passing by. It's not unusual for an array of multicolored plastic flags to be strung from the roof of a model home. Large banners and signs often identify models. The promotional fanfare that commonly surrounds a model home attracts attention. If the attendant of the model is a good salesperson, a percentage of the people who stop in for a quick look at the house will become paying customers.

Builders who are building in new subdivisions often sell from model homes. A contractor who is trying to compete in a subdivision where model homes are being used is at a great disadvantage without a model. People driving into the subdivision are captivated by the models. Builders who don't have models probably won't hear from most of the people who are interested in building in the subdivision. Even if a builder has signs on lots offering to custom-build houses, the contractors with models capture most of the business. Some buyers want a particular building lot, and they might deal with the builder who owns the lot they want, whether a model exists or not, but by far, a majority of the sales are made from the model homes.

DISADVANTAGES OF MODEL HOMES

The disadvantages of model homes might outweigh the advantages. One distinct disadvantage is the cost of building the model and the cost of paying interest on the money borrowed until the home is ultimately sold. The carry cost on even a modest home can easily exceed $600 a month, even when it is an interest-only loan. To cover this type of overhead, some sales have to be made.

Model homes are not extremely effective unless an attendant is on duty to show the house to prospective purchasers. You could put your office in a model home and do your office work while standing duty as a salesperson. But if you get much traffic through the model, you won't get much of your general contracting work done. Hiring a model attendant on an hourly basis gets expensive and is usually not a very cost-effective way to do business. Putting a commissioned salesperson in the model makes a lot of sense, but you have to be prepared to build in volume if you do this. Salespeople expect to make good livings when selling from model homes, and they can only do this by selling a lot of houses. If you can't build them as fast as your salespeople can sell them, you'll lose your sales force.

Reality vs. dreams

Reality is not always as nice as visions in your dreams. This is true of houses, too. A skillful salesperson can get prospect pumped up on a house just by using line drawings and words. The right salesperson can have potential customers foaming at the mouth to sign a contract for what they know will be their true dream house. If these same customers were to see a tangible example of what they are buying, they might not be so anxious to sign on the dotted line. It is usually easier to close a sale when a physical house is in existence to sell, but some salespeople do just as well, or better, without the benefit of a completed home.

Money

Money is the biggest drawback to building a model home. The house has to be paid for in some way. Most builders arrange either long-term construction financing or actually put a regular mortgage on their models. An interest-only, construction-type loan is more typical. For a builder to be able to arrange financing for a model home, the lender must feel comfortable with the builder's ability to perform well on the loan obligation. If you're new to the business and have not yet developed a successful track record, getting a bank to front you the money for a model home can be very difficult.

Even if you are able to arrange financing for a model home, you must be prepared to make monthly payments on the property. These payments might run as little as $600 or as much as $1200 or more. Can you afford this type of expense for the next year? You don't have to sell many houses to justify the carry cost on a model home, but you do need to sell enough to pay all of your bills.

Unfortunately, not all builders are successful. If you pick a blueprint to sell from and find that no one wants to buy that particular house, you are out the cost of the plans and advertising, but you should survive. Building a model home and then discovering that the public is not at all interested in it can sink your ship. You must weigh this type of risk when deciding whether or not to build a model home.

SUBDIVISIONS

If you are going to be building in subdivisions where you control numerous building lots, a model home can be very beneficial. Some builders construct a model house for every style of home they plan to build in a subdivision. I've done this before, but it gets awfully expensive. To compete with the other builders in the subdivision, I felt that I had no choice but to take the gamble. My luck was good, and I went through the deal with a profit. Some builders didn't.

In my opinion, you must be planning to build in volume to justify a model home. The one exception to this is if you will be using the model home as your residence. When you live in your model, you are killing two birds with one stone, so to speak. Your carry cost and your cost of housing are one, so the overall expense is livable. I wouldn't consider putting up a model home in a subdivision unless I controlled a minimum of 10 building lots.

HERE AND THERE

If you will be building houses here and there, without any particular concentration in one area, I wouldn't invest in a model home. When I was building in Virginia, I would option 10 or 20 lots in a subdivision in the blink of an eye. Typically, I had houses going in three or four subdivisions at once. This gave me an inventory of between 30 and 60 building lots. With this type of volume, model homes were not only viable, they were nearly mandatory.

In Maine, there are very few large subdivisions. The building business in Maine is much different from what it was in Virginia. I don't have a model home at this time, and I cannot see any likely circumstances that would encourage me to build one. You have to match your business strategy to what you do and where you do it.

Builders who are seeking custom-home buyers who already own their own land could benefit from model homes, but they probably can't justify the cost. Model homes work best in subdivisions where builders have a prevalent exposure and a number of building lots. If your business will build only a few homes a year, you can't justify a model home, unless you are going to live in it.

MY PERSONAL OPINION

My personal opinion on model homes is that they are excellent sales tools, but a builder must be able to justify the full costs associated with a model before one is built. I know that model homes are not needed to sell to-be-built houses. In truth, I've sold more houses from blueprints and line drawings than I've sold with models. It is my opinion that a builder should not invest in a model home during the early stages of building a business. I believe the money can be put to better use, such as in the purchase of advertising.

Model homes are a wonderful asset to have when selling, but the homes don't generate a lot of business unless they are in prominent locations in popular subdivisions. You need customers before you need model homes, and advertising is the most successful way to attract customers. If I were you, I'd spend my money on advertising and let the model home wait for a while.

8
Courting bankers

Establishing credit is a task most business owners must tackle. Setting up credit accounts for a new business can be tough for anyone, but for people with poor credit histories, the job can seem almost impossible. Vendors don't want to establish credit for you or your company unless you have a good track record. How can you develop a good track record if people won't give you a chance? The dilemma is something of a revolving door—no matter where you enter it, you seem to go around in circles. This chapter is going to help you to understand the methods used to establish credit.

ESTABLISHING GOOD CREDIT

Good credit is crucial to a growing business. Without good credit, expansion, even survival, can be very difficult. Most businesses need credit accounts, whether they are used on a monthly basis to buy supplies or on a semi-annual basis, as operating capital. If you have an existing business without any credit blemishes, setting up new accounts is not too difficult. If, on the other hand, you have a poor credit rating, getting new credit lines can be an uphill battle. For a new business, establishing credit can be a slow process, but it can be done.

What types of credit do you need? As your business grows, so will your desire for various types of credit. Let's look at some of the types of credit you might need.

Start-up money

The first need you have for credit could be to obtain start-up money. Start-up money is the money you use to get your business started.

It is best to rely on money you have saved for start-up capital. Borrowing money to start your business puts you in a hole right from the start. The burden of repaying a loan used for start-up money only makes establishing your business more difficult. However, many successful business are started with borrowed money.

When I started my first business, I had to borrow money just to buy tools. That first business was started about 18 years ago, and I had to borrow $500 to fill my toolbox. Today, $500 doesn't seem like much money, but back then, it felt like a big risk to take. I took the risk and it paid off. While I can't recommend going into debt to start your business, I did it and it worked.

A loan for start-up capital can be one of the most difficult to obtain. Lenders know that many new businesses don't survive their first year of operation. If you don't have home equity or some other type of acceptable collateral, banks are reluctant to help you get your business started.

If you wait until you quit your job to apply for a start-up loan, your chances of having the loan approved drop considerably. Most savvy entrepreneurs arrange a personal loan while they have their regular job and use the funds for their business start-up. This procedure results in more loan approvals.

Some contractors start their businesses on a part-time basis. By doing this, they are able to obtain the loans they need while they are still employed by someone else and can build up cash reserves. The work they do on the side produces money to repay their loans and can accumulate into a healthy stockpile of ready cash.

The stress and hours of working full-time and running a part-time business can be tiresome, but the results are often worth the struggle.

Operating capital

Once your business is open, you probably need some operating capital. This money bridges the gap between job payments and keeps the business running. If you don't already have a supply of money set aside for this purpose, you might want to apply for an operating-capital loan.

Lenders are a little more willing to make these loans if you have an established track record. However, don't get your hopes up on being approved for any loan as a self-employed person until you can provide two years of tax returns for income verification.

Since you probably need operating capital to survive your first two years, make arrangements for your financing before you jump into business.

Operating-capital loans can be provided from several types of financing. Some business owners borrow against their home equity to generate operating capital. Lines of credit are often set up for the contractor to pull from as money is needed. With this type of financing, you pay only interest on the money you are using. Short-term personal loans are another way of financing your operating capital. These loans are frequently set up with interest-only payments until the note matures, at which time the total owed becomes due.

Advertising accounts

Advertising credit accounts are convenient. These accounts allow you to charge your advertising and pay for it at the end of the month. This eliminates the need for cutting a check with every ad you place. It also gives your advertising a chance to generate income to pay for itself. However, be careful; if you abuse these accounts, you can get into deep debt quickly.

Advertising often pays for itself, but sometimes it doesn't. You cannot afford to charge thousands of dollars worth of advertising that doesn't bring in paying customers. If you do, your business is crippled before it starts. You will be faced with large payments on advertising that was a dud. Use your credit accounts prudently.

Supplier accounts

You need supplier accounts for convenience and to float your material costs until the customers pay their bills. By establishing credit with material suppliers, you don't have to take your checkbook with you every time you need to pick up supplies.

Since you probably have to supply and install materials before your customers pay for them, supplier accounts buy you some float time, generally up to 30 days. You can get the materials you need without paying for them until the first of the next month. This gives you time to install the materials and collect from your customers before paying for the materials.

Again, you must be cautious. If your customers don't pay their bills, you won't be able to pay yours. You must keep the reins tight on collecting your accounts receivable.

Office supplies

Many times you can get better pricing on your office supplies by purchasing them from mail-order distributors. Having a credit account with these distributors makes your life easier. You won't have to place COD orders and be concerned about waiting around to give the delivery driver a check. With a business charge account, you won't have to put business expenses on your personal credit cards. This type of account makes your accounting easier and your business less troublesome to keep up with.

Fuel

Paying cash for fuel to run your trucks and equipment can be a real pain when you have employees. Keeping up with the cash and the cash receipts is time-consuming. By establishing a credit account with your fuel provider, your employees can charge fuel for your company vehicles and equipment. At the end of the month, your account statement is easy to transfer into your bookkeeping system.

Vehicles and equipment

As your business grows, you might need more vehicles and equipment. Most business owners can't afford to pay cash for these large expenses. Financing or leasing vehicles and equipment requires a decent credit rating.

As you can see, you need to use credit in your business dealings on a multitude of occasions. At times credit might make a difference between the survival and failure of your business. Develop a good credit rating and work hard to maintain it.

PICKING YOUR LENDING INSTITUTIONS

One of the first steps in establishing credit could be picking your lending institutions. Not all lenders are alike. Some lenders prefer to make home mortgage loans. Others make car loans and others, secured loans. Some lenders make loans for any good purpose if they feel the loan is safe. Finding a bank that is willing to make an unsecured signature loan can be troublesome.

Before you can set out to find a lender, you must decide what type of loan you will request. If you do not have a well-established and financially sound business, be prepared to sign the loan personally. Most lenders won't make a business loan without a personal endorsement. Are you looking for a secured or unsecured loan? A secured loan is a loan where something of value, collateral, is pledged to the lender for security against nonpayment of the loan. An unsecured loan is a loan that is secured by a signature, but not by a specific piece of collateral. Most lenders much prefer a secured loan.

If you plan to request a secured loan, you must decide what you have to place as collateral. The amount of money you want to borrow has bearing on the type and amount of collateral that is acceptable to the lender. For example, if you want to borrow $50,000 for operating capital, putting up the title to a $10,000 truck is probably not sufficient collateral.

When you are looking to borrow large sums of money, real estate is the type of collateral most desired by lenders. If you have equity in your home and are willing to risk it, you should find that getting a loan is relatively easy. However, home-equity loans can get confusing. Many people don't understand how much they can borrow against their equity. Let me show you two examples of how you might rate the loan value of the equity in your home.

Let's say your home is worth $100,000 and you owe $80,000 on the mortgage. Equity is the difference between the home's appraised value and the amount still owed on it. In this case the equity amount is $20,000. Does this mean you can go to a bank and borrow $20,000 against the equity in your home? No, not likely—most lenders would consider such a loan a high risk. Some finance companies might lend $10,000 on this deal, but they would be few and far between. In most cases you would not be able to borrow any money against your equity.

Lenders want borrowers to have a strong interest in repaying loans. If a lender loaned you $20,000 on your equity, your house would be 100% financed. Basically, if you defaulted on the loan, you wouldn't lose any money, only your credit rating. To avoid this type of problem, lenders require that you maintain an equity level in your house that is above and beyond the combination of your first mortgage and the home-equity loan. Most banks want you to have between 20 and 30 percent of the home's value remaining in equity, even after getting a home-equity loan.

Let's say you have the same $100,000 house, but you only owe $50,000 on it. A conservative lender would allow you to borrow $20,000 in a home-equity loan. This brings your combined loan balance up to $70,000, but you still have $30,000 of equity in the home. If you default on this loan, you lose not only your credit rating, but $30,000 to boot. A liberal lender might allow you to borrow $30,000, keeping an equity position of $20,000.

If you don't have real estate for collateral, you can use personal property. Personal property might include vehicles, equipment, accounts receivable, certificates of deposit, or whatever. Different lenders have different policies, so shop around until you find a lender you like doing business with.

Banks are an obvious choice when applying for a loan. Commercial banks often make all types of loans. But banks are not your only option. Savings-and-loans normally deal in real estate loans, but they are worth investigating. If you belong to a credit union, check its loan policies and rates. Finance companies are usually aggressive lenders, but their interest rates might be high. Private investors are always looking for viable projects to invest in or loan money to. Mortgage brokers are yet another possibility for your loan. A quick look through the phone directory and the ads in the local newspaper can reveal many potential loan sources.

ESTABLISHING CREDIT WHEN YOU HAVE NONE

This section is going to teach you how to establish credit when you have none. It is better to start with no credit than to be starting with bad credit. If you have never used credit cards or accounts, you might find it difficult to get even the smallest of loans from some lenders. Another problem you encounter is the fact that you are self-employed. Most lenders don't want to loan money to self-employed individuals until the individuals can produce at least two years' tax returns. With this in mind, you might be wise to set up your accounts and credit lines before quitting your job.

Since most people don't set up business accounts until they are in business, I'll give you advice on how to get credit without having a regular job. However, if you have the opportunity to establish your business accounts before quitting your job, do it.

Supplier accounts

Supplier accounts are one of the best places to begin establishing your new credit. When you go into business, suppliers want to supply you with needed materials. Suppliers are cautious about granting credit accounts, but they are more liberal than the average bank.

As a new business, you need credit, materials, and customers. As ongoing and competitive businesses, suppliers need new customers. In effect, you need each other; this is your edge. If you handle yourself professionally, you have a good chance of getting a small charge account opened. Now, let's look at how you should go about opening your new supplier accounts.

You can request credit applications by mail or you can go into the stores and pick them up from the credit department. Once you have the credit applications, you need some detailed information to complete them. Most applications ask for personal references, credit references, your name, address, phone number, social security number, bank balances and account numbers, business name, and much more. The application asks how much credit you are applying for. Don't write in that you are applying for as much as possible. Pick a figure, a realistic figure that is a little higher than what you really want. Setting the higher figure gives you some negotiating room.

Most supplier credit applications are similar. Once you fill out the first application, make copies of it for future reference. Not only will the photocopy serve to refresh your memory if the supplier has questions on your application, you can use the copy as a template in filling out other credit applications.

When you have completed the applications, return them to the credit department. You might want to hand-deliver the applications so you have a chance to make a personal impression on the credit officers. Making good impressions on the credit managers can help get you over the hump.

If you get a notice rejecting your credit request, don't give up. Call the credit manager and arrange a personal interview. Meet with the credit manager and negotiate for an open account. When all else fails, ask for a smaller credit line. Almost any supplier will give you credit for $500. It might not sound like much, but it's a start, and a start is what you need.

Take whatever credit you can get. After you establish the account, use it. Having an open account is not enough. You must use the credit to gain a good credit rating. If the credit is not used, it does you no good. You should use the credit account frequently and pay your bills promptly. When the supplier offers a discount for early payments, pay your bills early. By paying early and taking advantage of the discounts, you save money and improve your credit rating.

Suppliers offer the easiest access to opening new accounts and starting to build a good credit rating. After these accounts are used for a few months, your credit rating grows. By keeping active and current accounts with suppliers you are building a good background for bank financing.

Other vendor accounts

Other vendor accounts are also valuable. When you open your business, you need office supplies. You need the services of a printer. Newspaper advertising is normally a part of every new business. All these miscellaneous vendors offer the opportunity for creating credit. It can be easy to set up accounts with small businesses. If you are a local resident of the community, small businesses might not even ask you to fill out credit applications. These opportunities are too good to overlook.

Major lenders

Major lenders are a little more difficult to get credit from. They don't seem to feel the need to encourage business from new companies. Unless you have a strong credit rating, tangible assets, and a solid business plan, many banks are not interested in loaning substantial sums of money. But don't despair, there are ways to work with banks.

Banks, like suppliers, should be willing to make a small loan to you. I know $500 does not buy much in today's business environment, but it is a worthwhile start to building a credit rating. There is another way to build your credit rating and get noticed by your banker.

Bankers like to have collateral for loans. What better collateral could you give a banker than cash? I know, you are thinking that if you had cash for collateral, you wouldn't need a loan. But that is not always true. When you are establishing credit, any good credit is an advantage. Let me tell you how to get a guaranteed loan.

Set up an appointment to talk with an officer of your bank. Tell the banker you want to make a cash deposit in the form of a certificate of deposit (CD), but that after you have the CD on deposit, you want to borrow against it. Many lenders allow you to borrow up to 90 percent of the value of your CD. For example, if you put $1000 in a CD, you should be able to borrow about $900 against it. You are essentially borrowing your own money and paying the bank interest for the privilege. This concept might sound ludicrous, but it works to build your credit rating.

Banks report the activity on their loans to credit bureaus. Even though you are borrowing your own money, the credit reporting agency shows the loan as an active, secured loan. As long as you make the payments on time, you get a good credit rating. This technique is often used by people repairing damaged credit, but it works for anyone.

Building a good credit rating can take time. The sooner you start the process, the quicker you enjoy the benefits of a solid credit history. The road to building a good credit rating can be rocky and tiresome, but it's worthwhile.

OVERCOMING A POOR CREDIT RATING

If you are starting with a bad credit rating, this section gives you options on how to overcome it. There is no question that setting up credit accounts is harder if you have a poor credit history, but you can do it. If you are battling a bad credit report, plan on spending some time in cleaning up the existing report and building new credit. This journey is not easy, pleasurable, or quick, but the results should make you happy.

Secured credit cards

Secured credit cards are one way for people with damaged credit to begin the rebuilding process. Secured credit cards are similar to the procedure described for CD loans. A person deposits a set sum of money with a bank or credit card company. Then a credit card is issued to the depositor. As the card holder, a person can use the credit card with a credit limit equal to the amount of the cash deposit or slightly more. These people are basically borrowing their own money, but they are also rebuilding their credit.

CD loans

I have already talked about CD loans, so I won't go into detail on them again. If you need to rebuild your credit, CD loans are a good way to do it. By depositing and borrowing your own money, you are able to build a good credit rating without risk.

Erroneous reports

Erroneous reports on your credit history are not impossible. If you are turned down for credit, you are entitled to a copy of the credit report information used in making the decision to deny your credit request. If you are denied credit, you should immediately request a copy of your credit report. Credit reporting bureaus are not perfect; they make mistakes. Let me tell you a quick story about how my credit report was maligned when I applied for my first house loan.

When I applied for my first house loan, my request for the loan was denied. The reason I was given for the denial was a delinquent credit history. I knew my credit was impeccable, and I challenged the decision. The loan officer talked with me and soon realized something was wrong. My wife's name is Kimberley and at the time of this credit request I didn't have any children. The credit report showed my wife as having a different name and it showed me having several children. Obviously, the report was inaccurate.

Upon further investigation, we discovered that the credit bureau had issued the wrong credit report to my bank. My first and last names were the same as the person who had the poor credit history. However, my middle initial was different, my wife's name was different, and I didn't have any children. It happened that I lived on the same road as this other fellow. It was certainly a strange coincidence, but if I had not questioned the credit report, I would not have been able to build my first home.

I know from firsthand experience that credit reports can be wrong. I have seen various situations in the past when my clients and customers fell victim to incorrect credit reports. If you are turned down for credit, get a copy of your credit report and investigate any discrepancies you discover.

Explanation letters

If your poor credit rating is due to extenuating circumstances, letters of explanation might help solve your credit problems. If there was a good reason for your credit problems, a letter that details the circumstances might be all it takes to sway a lender in your direction. Let me give you a true example of how a letter of explanation made a difference to one of my customers.

A young couple wanted me to build a house for them. During the loan application process it came out that the man had allowed his vehicle to be repossessed. On the surface it appeared to be a deal-stopping problem. I talked with the young man and learned the details behind the repossession. At my suggestion, the man wrote a letter to the loan processor. In less than a week, the matter was resolved and the couple was approved for their new home loan. How did a simple letter change their lives? Well, the man had been a victim of bad advice. Let me explain:

This man had a new truck with high monthly payments. When he decided to get married, he knew he couldn't afford the payments. My customer went to his banker and explained his situation. The loan officer told the man to return the truck to the bank, and the payments would be forgiven. However, the banker never told the man that this act would show up as a repossession on his credit report. My customer returned the truck with the best of intentions and acting on the advice of a bank employee.

When the problem cropped up, and the bank employee was contacted, he confirmed my customer's story. The mortgage lender for the house evaluated the circumstances and decided the man was not an irresponsible person. It was decided that he had acted on the advice of a banking professional. Under the circumstances, my customer's loan was approved. If you have strong reasons for your credit problems, let your loan officer know about them. Once-in-a-lifetime medical problems could force you into bankruptcy, but they might be forgiven in your loan request. If you provide a detailed accounting that describes your reasons for poor credit, you might find that your loan request gets approved.

ENSURING CREDIT SUCCESS

I am about to give you seven techniques to ensure credit success. These seven methods are not the only ways to establish credit, but they are proven winners. Let's take a closer look at how you can make your credit desires a reality.

1. Get a copy of your credit report

In most cases, with a written request you can get a copy of your credit report. This is a wise step to take in establishing new credit. By reviewing your credit report, you can straighten out any incorrect entries before applying for credit. It is better to clean up your credit before a credit manager sees the problems. Even if the report contains false information, the credit manager might form a bad impression of you.

2. Prepare a credit package

If you prepare a credit package before applying for credit, the chances of having your credit request approved increase. What should go into your credit package? If you own an existing business, your package should include financial statements (FIG. 8-1), tax returns, your business plan, and all the normal credit information that is typically requested. If you are a new company, provide a strong business plan and the normal credit information. A copy of your personal budget (FIG. 8-2) might also be helpful.

3. Pick the right lender

Picking the right lender is a key step in acquiring new credit. Do some homework and find lenders that make the type of loans you want. Once you have your target lenders, take aim and close the deal.

4. Don't be afraid to start small

Don't be afraid to start small in your quest for credit. Any open account you can get helps you. Even if you are only given a credit line of $250, that's better than no credit line at all.

5. Use it or lose it

When you get a credit account, use it or lose it. Open accounts that are not used will be closed. In addition, an open account that doesn't report activity does you no good in building your credit rating.

6. Pay your bills on time

Always pay your bills on time. Having no credit is better than having bad credit. If you have accounts that fall into the past-due category, your credit history suffers, and you will be plagued by phone calls from people trying to collect your overdue account.

7. Never stop

Never stop building your credit standing. The more successful you become, the easier it is to increase your credit lines. However, don't fall into the trap of getting in over your head. If you abuse your credit privileges, it will not be long before you are in deep financial trouble.

FINDING CREDIT FOR YOUR CUSTOMERS

Have you thought about helping to find credit for your customers? A lot of builders don't take any type of active interest in helping their customers find financing. Many builders leave loan issues to real estate brokers and the individuals who are looking to have a house built. You can take this approach, but if you are willing to participate in setting up easy financing for your customers, you should get more business.

I've worked with financing for a long time, both as a builder and as a broker. It is usually financing that makes or breaks a house deal. Having people who want you to build them a house is not worth much if there is no money available for them to pay you with. It's not your responsibility to find financing for customers, but if you do, you're likely to build more houses.

Financial Statement

Your Company Name
Your Company Address
Your Company Phone Number

Date of statement: _____

Statement prepared by: _____

Assets

Cash on hand	$ 8,543.89	
Securities	$ 0.00	
Equipment		
1992 Ford F-250 pick-up truck	$14,523.00	
Pipe rack for truck	$ 250.00	
40' Extension ladders (2)	$ 375.00	
Hand tools	$ 800.00	
Real estate	$ 0.00	
Accounts receivable	$ 5,349:36	
Total assets		$29,841.25

Liabilities

Equipment		
1992 Ford F-250 truck, note payoff	$11,687.92	
Accounts payable	$ 1,249.56	
Total liabilities		$12,937.48
Net worth		$16,903.77

8-1 Financial statement.

```
PERSONAL FINANCES          Mar 1993

Summary                                    Over(Under)
                       Actual   Budgeted    Budget
Total Income          3140.43   2874.52     265.91
Total Expenses        2772.82   2749.87      22.95
Balance                367.61    124.65     242.96


Detail                                     Over(Under)
                       Actual   Budgeted    Budget
Income
 Salary               2874.52   2874.52
 Other                 265.91      0.00  Garage Sale
Expenses
 Withholdings
  Federal Income Tax   115.22    115.23
  State Income Tax       0.00      0.00
  FICA                  90.88     90.88
  Medical               30.21     30.21
  Dental                22.90     22.90
  Other                  0.00     20.00  Investment Fee
   Total Withholdings  259.21    279.22    (20.01)
   Percent of Budget    9.02%     9.71%    -0.70%
 Finance Payments
  Credit Cards          33.80     33.80
  Auto Loan            238.50    238.50
  Home Mortgage        566.81    566.81
  Personal Loan        125.89    125.89
   Total Finance       965.00    965.00      0.00
   Percent of Budget   33.57%    33.57%      0.00%
 Fixed Expenses
  Child Care             0.00     20.00
  Property Tax          21.90     21.90
  Home Insurance        15.21     15.21
  Auto Insurance        80.44     80.44
  Life Insurance       102.55    102.55
  Contributions        310.00    286.00
  Vacation Savings     100.00    100.00
   Total Fixed         630.10    626.10      4.00
   Percent of Budget   21.92%    21.78%      0.14%
 Variable Expenses
  Household             54.80     50.00
  Groceries            395.66    425.00
  Auto Upkeep and Gas   37.10     45.00
  Furniture              0.00     15.00
  Clothing              35.90     30.00
  School                22.44     20.00
  Medical/Drugs         31.30     30.00
  Entertainment         36.88     45.00
  Memberships           10.00     10.00
  Cable TV              19.55     19.55
  Dining Out            38.40     30.00
  Gifts                100.00     50.00
  Pet Care               0.00     10.00
  Other                136.48    100.00
   Total Variable      918.51    879.55     38.96
   Percent of Budget   31.95%    30.60%      1.36%
```

8-2 Personal budget.

Permanent mortgages

Permanent mortgages are available in so many forms that an entire book could be written on them. Traditional 30-year fixed-rate loans and adjustable-rate loans (ARMs) are the two most common types of financing used for homes. Even within these two categories, there can be many differences in terms and conditions. Federally assisted financing is available, as are many other types of loans.

As a builder, you should not start construction on a custom home until you know that your customer has a firm commitment for a permanent loan. This type of loan might come from a bank, a savings-and-loan, a credit union, a mortgage banker, or some other source. Your customers are not going to be aware of the many types of loan programs that are available. If the customers are working with a good real estate broker, the broker can introduce the customers to lenders and loan options. When customers come to you directly without the aid of a broker, you should be able to guide them to some quality lenders. To do this, you must make some inquiries and establish a rapport with lenders before they are needed. Also, find out what your customers are required to bring to a loan application (FIG. 8-3) before the need arises.

Checklist of loan application needs

❑ Home address for the last five years
❑ Divorce agreements
❑ Child support agreements
❑ Social security numbers
❑ Two years of tax returns, if self-employed
❑ Paycheck stubs, if available
❑ Employee's tax statements (i.e., W-2, W-4)
❑ Gross income amount of household
❑ All bank account numbers, balances, names, and addresses
❑ All credit card numbers, balances, and monthly payments
❑ Employment history for last four years
❑ Information on all stocks or bonds owned
❑ Life insurance face amount and cash value
❑ Details of all real estate owned
❑ Rental income and expenses of investment property owned
❑ List of credit references with account numbers
❑ Financial statement of net worth
❑ Checkbook for loan application fees

8-3 Loan application needs.

Construction loans

As you interview lenders, try to find ones who will approve your customers for permanent financing and a construction loan. Your risk is reduced if the customers sign for the construction loan in their own names. When a builder borrows money on a construction loan, there must be two closings when the house is built. The builder has to close on the land and put the property in the name of the building business. When the house is completed, it is sold to the customer and another closing is needed. This increases the cost of the home. By having your customers set the construction loan up in their names, you reduce your risk and their cost.

While you are shopping around for sources of construction loans, you should ask about terms, conditions, and draw disbursements. Some construction loans are active for six months, others run for nine months, and some go for a full year. Make sure that the term of the loan is adequate for you to complete the house.

What is the loan-to-value ratio of the construction loan? A 70-percent loan is common. Loans with a 75-percent loan to value can frequently be found, and some lenders still loan up to 80 percent of a property's appraised value. The higher the loan-to-value ratio is, the better off you are.

Does the construction loan provide a land acquisition disbursement? Most construction lenders advance money from a construction loan to buy a building lot. This can become a big issue. If a lender won't advance money for land acquisition, you or your customer must come up with a lot of cash on your own. Find a bank that will front the money for land.

The interest rate on a construction loan is normally higher than rates quoted for permanent financing. This is because construction loans have short terms and are a higher risk than long-term mortgages. You need to know what the interest rate will be so that you can calculate the cost of financing into your house price, unless your customers are paying the interest payments out of their own funds.

The cost of points and closing costs can add thousands of dollars to the cost of a house. Each point charge is equal to one percent of the loan amount. In other words, a construction loan that charges one point for a $100,000 loan will cost $1000 plus closing costs. Determine what all the financing costs are before you price out a job, and decide how the money is going to be paid. Are you going to pay it and add the expense onto the price of a house, or will your customers pay all financing fees privately and then pay you for the construction work? Construction loans are a key element in building most houses, so spend some time talking to lenders, and get the facts on what types of financing are available for you and your customers.

Bridge loans

Bridge loans can be used to overcome a common problem. Many people have homes that they must sell before they can buy a new one. Sometimes home buyers have sold their houses by contract and are waiting for the closing to take place. If either of these types of people want you to build a new house for them, they might think that they have to wait until their deals close to start construction. This isn't always true. Sometimes a bridge loan can be used to get the ball rolling on a new house while an old house is being sold or closed.

If your customers are strong enough financially, they should be able to arrange a bridge loan that allows them to have a new house built while they are making arrangements to leave their existing home. If you and your customers are not aware of this option, you could lose months of production time waiting for one deal to close so that you can start a new deal. Check into bridge loans when you talk with lenders and be prepared to share your knowledge with customers. It could get you more work.

CREATING AN EDGE

Builders who are able to help customers set up financing have an edge on their competition. Knowing what customers can do and where they can do it makes you a more formidable adversary to other builders. When customers are shopping builders, the contractor who can make the job and the financing easy and comfortable is likely to win the bid. Don't underestimate the importance of financing. Once you become acquainted with loans and lenders, you should find that sales are easier to make.

9

Selling from the hood of your truck

Selling houses from the hood of your truck is a skill that takes some time to perfect. Many people are reluctant to buy a house that has not yet been built. But a good number of people relish the thought of arranging the construction of a truly custom home. These people are the ones you want to talk to. If you can capture their attention, you can make some quick sales without the aid of a model home.

I've had a lot of experience in selling homes, both as a builder and as a broker. When I started selling houses with nothing but a set of blueprints, the job was referred to as selling houses in the dirt. The reason was simple. What was being sold was a building lot and a dream. I struggled through my first several attempts to sell a house from plans, and I was not successful. However, I didn't give up, and eventually I made my first sale.

As my sales career grew along with my building business, I made mental notes of what worked and what didn't when I was in a selling mode. By keeping records of my activity, I was able to review what I'd done and improve on it. Pretty soon, I was quite proficient in the art of selling a set of blueprints for major money.

To presell houses on a regular basis, you need to invest some time to learn how to do it. Basic sales skills are needed, but they are not enough to get the job done. Many salespeople can sell well, but few of them can consistently sell dream homes from paper. If you show enough people blueprints and present enough proposals, you eventually will make some sales, but you will waste a lot of time and money in the process. You need to make every sales meeting count.

LEARN THE BASICS

A multitude of books are available that teach the basics of salesmanship. You should read many of them. Knowing what to say, how to say it, and when to say it is a big part of the sales game. More important, sometimes, is knowing when to keep your mouth shut and your ears open. I'll talk more about that in a moment.

I don't have the space in this book to go into a full-blown tutorial on sales skills. And truthfully, I'm probably not enough of an expert in general sales to teach you all that you could learn. Get your hands on five books that deal with nothing but sales skills and read them. Make notes as you go along. Compare what the different authors are telling you. You should start to see a pattern develop as you read the books. The principles behind successful selling run true whether you are selling insurance, home improvements, or houses.

Once you have gained an awareness of how to use traditional sales skills to better your business, you can refine the techniques and customize them to suit your personal needs and desires. In other words, you can take generic sales methods and shape them around the sale of new houses. Some of your customization comes from trial-and-error experiences. Each time you miss a sale, you should learn from the experience. And, every time you make a sale, you should recount the strategy that made you successful. To get you started in selling houses from blueprints, I can share with you what I've learned over the past several years.

SHUT UP AND SELL

When I'm training new salespeople, I often tell them to shut up and sell more. It's not that I'm trying to be rude. Many sales professionals talk too much. They don't hear what their customers want, and they lose sales. Think about the last time you bought a car or truck. Did the salesperson descend on you and start chatting about insignificant matters? How long did the salesperson monopolize the conversation before letting you get a word in edgewise? Was there a point during the meeting that you wished the talkative vulture would just go away?

Maybe you haven't experienced a talkative salesperson, but I sure have. As a matter of fact, I just bought a new Jeep and the salesman almost lost the sale because he talked so much. When I went to the car lot, I knew I was going to buy a new Cherokee. There was no doubt in my mind what I wanted. As I was looking at the selection of vehicles, a salesman approached me and started talking. One of the first things he did was to try to steer me from a new Cherokee to a used one. I didn't want a used one, and I told the man this. But he kept pushing the old Jeep on me. It wouldn't have taken much to make me move onto another dealer. However, I simply told him that I was going to buy a new vehicle and that I'd appreciate it if he would let me shop in peace. He went away, and I bought the Jeep.

Over the years, I've trained a lot of salespeople. Some of them have been groomed to sell home improvements for my remodeling business. Others have been taught to sell new houses, both built and in the dirt. As part of my training course, I sit in on sales meetings with the new recruits when they are first getting started. It never ceases to amaze me how often the people blow deals by talking when they should be listening.

Of course, you need to talk to customers, but you have to know when to talk and when to listen. A little chit-chat to break the ice is needed. You must be well versed in your product knowledge and be able to weave in interesting and pertinent facts as your sales meeting progresses. But in many cases, you have to shut up and let your customers explain what it is that they want.

If you steamroll customers with a canned pitch like a lot of salespeople do, you might alienate the people and lose your sale. Sitting at a table telling a young couple how great an expandable Cape Cod is for a starter home could come across as an insult. For all you know, the couple might be wealthy and looking for a six-bedroom colonial to fill with children. If you hammer them with your cheap Cape, the meeting could turn sour quickly.

LISTEN TO PEOPLE

Most people like to talk about what they want in a house. Let them. Take notes of what they say. If one person mentions hardwood floors, write it down. Pay close attention to what you're being told. As your sales skills improve, you learn to read between the lines. It won't take long to determine which half of a couple has the most clout in the decision-making process. You must learn to assess customers quickly to appeal to them. While you might not become lifelong friends, your sales can go much more smoothly if you develop a friendly business relationship.

Don't be in too big of a hurry to start talking about the technical aspects of the home your customers want to build. Let time be on your side. Warm up to the customers. Seed the conversation with facts about you and your company that build credibility and confidence. It's much more effective to weave this type of information into a friendly conversation than it is to run down a mental checklist, spewing out rehearsed lines, one after another.

If your customers want to get right down to business, don't stall them to maintain some prearranged sales strategy. Great salespeople are versatile. Not everyone is going to want to discuss dogs, children, or other aspects of his or her personal life. Some people want to cut to the chase and jump right into the specifications for their new home. When these people come along, let them talk. You have to make your customers comfortable in whatever way is necessary.

Your first goal during a sales meeting is to put your clients at ease. Get them to relax. How to relax them varies with different types of people. One way to break the ice is to give the customers books of house plans to look through. While they are skimming the pages, you can be talking in the background. Don't be answering phone calls during your meeting. Avoid having your office staff interrupt you when you're selling. Get those customers comfortable.

One of the most effective ways to keep people from tensing up is to meet with them in their home. People are always more at ease when they are in their comfort zone. If customers come to your office, they might feel trapped. They won't harbor this feeling in their own home. Another advantage to meeting people in their home is that you can see how they live and what's important to them. Pictures on the walls can tell you a lot about your prospects. Even the furniture in the home can give you a hint about the personality of your customers. Use every advantage you can to gain a sale.

GET DOWN TO BUSINESS

Getting down to business with customers seeking a custom home is usually easy. Start by asking what type of house they like. Inquire about the number of bedrooms and bathrooms that they desire. Don't talk about money, references, or workmanship in the beginning. Get the ball rolling with conversation that is fun for your customers. Let them turn on their dream machine and then you can adjust the focus as the meeting goes on.

By the time your customers have finished describing their dream home, they should be all pumped up and receptive to what you have to say. This is when you begin to go into details. If you have stock plans that you normally offer customers, get out the ones that are closest to what the customer wants. Start talking about putting quarry tile in the foyer and wallpaper and a chair rail in the dining room. Play up the features of your homes.

When you are describing what makes your homes different and better, use photographs as visual aids. Pictures can do a lot to gain a person's trust. Even if you have never built a house, you can create a photo album that can help you sell your services. How?

Consumers tend to be better informed about issues that pertain to them than they once were. Many people want to know the nuts and bolts of how their houses will be built. Gone are the days when you could give a broad-brush description of a home and escape without getting into the finer points. Actually, this change is to your advantage, as long as you are prepared for it.

If you set up a photo album and a prop box, you can impress customers very quickly, and it's safe to assume that most builders will not go to the same lengths to secure a sale. Here's how I do it.

When I get to the point of discussing details, I bring out the photo album. I show them pictures of houses that I've built, but I concentrate on the construction components. It is the close-up photos that capture the attention of most serious buyers. My photo album contains an assortment of close-up shots. For example, I show customers a picture of how my carpenters frame a header. You can build a small wall section with a header in it quickly. You can build it in your garage or yard and photograph it. People will assume it's part of a house you've built, whether it is or not.

Product suppliers and manufacturers are happy to provide builders with brochures and photos. Show your customers a detailed close-up of the cabinet construction for their kitchen cabinets. Point out the metal guides and glides that make the drawers work smoothly. Concentrate on as many details of quality construction as time allows.

While you have your customers looking at things, bring out your prop box. Hold up a metal joist hanger and explain what the item is, how it works, and why it makes for more solid construction. Next, pull out an electrical outlet and show the customers how your electrician uses only copper wire and how the wires are always installed under the screws for a better and safer connection. Show them how some contractors stab their wires into the back of an outlet to save time, and explain how this can be bad, since the wires sometimes work loose if they are not under the screws. Continue this type of show-and-tell until you are finished or until you sense that your customers are tired of it.

Have plenty of catalogs and brochures on hand for your customers to look at. Generating ideas can be an important part of a sales meeting when you are selling a house that doesn't yet exist. Looking at catalogs is a great way to maintain a level of enthusiasm in your customers.

Once you have convinced the customers that you are the best builder in the world, hand them an easy-to-read, bulleted checklist of the most important features incorporated in your homes. List the energy efficiency, the extended warranty (if you offer one, and you should), and any other special features. Never stop selling.

When you feel sure that you have your customer's confidence, move closer toward your closing pitch. Maybe you will give your customers a pencil and allow them to mark up a set of plans so that you can see their modifications. If you have a CAD system on your computer, use it to help your customers get the layout they want. Put some paper in their hands. All you have to sell is your words and some paper, so make the most of them.

If your meeting is going well, you should be getting close to the point of talking money and financing. You might have to arrange a second meeting to get into financial details. But, if you're selling stock plans that you've already worked up prices on and options that you've already priced out, seize the moment and keep your momentum rolling. Your goal, under these circumstances, is to get a signature on a contract before you part company with your customers.

Is it really possible to sell an unbuilt house to people in just one meeting? It certainly is. I've done it on numerous occasions. Some people want to talk with their parents or attorney before signing anything. A good number of people want to think it over. But a percentage of your customers will be ready to sign a contract right on the spot. Never be afraid to ask for a sale. You won't make many sales without asking people to buy from you.

REVIEW THE KEY ELEMENTS

Let's go over the key elements in selling from blueprints. First, get your customers comfortable. Gain their confidence. Get them excited about the house they have in their minds. Use visual aids. Remember to listen carefully to what your customers have to say, and by all means, let them talk. Weave your sales pitch into normal conversation. Don't use a canned pitch that makes you sound like a telemarketing computer. Stress your strong points and don't demean any of your competitors by name. It's okay to say that some builders do this or that, but don't name names. It makes you look bad and perhaps desperate. Maintain control of your meeting at all times, but do it in a tactful way. Last, but far from least, ask the customers to make a deal with you before they leave.

I can't begin to remember how many houses I've sold from the hood of my truck and from various dining room tables. It has been a lot of houses, I know that much. Usually, if I can keep customers in a sales meetings for at least two hours, I get a sale. Not always on the spot, but before the game is over, I usually win. It often takes three or more meetings to hammer out enough details to close a sale, so don't be too impatient. On the other hand, never assume that you have to go through an obligatory series of meetings to get a sale. Sell softly from the start and never stop selling yourself.

It takes time to perfect a personal style that is comfortable for you. Once you get the hang of selling houses in the dirt, you'll have a chance to be one of the most successful builders in your area. If you just can't bring yourself to become a salesperson, hire someone who can and will sell for you. Without sales, you have no business.

10
Opening your own office

Office space can be a major factor in the successful operation of your business. For some businesses, location is vital to success. However, most contracting businesses can function from a low-profile location. This is not to say that office space is not needed or is not important. Whether you work from your home or a penthouse suite, your office has to be functional and efficient if you want to make more money.

DECIDING WHERE YOU SHOULD WORK

Should you work from home or in a commercial space? This question can be very difficult to answer. Most people have a good feel of what their needs for a commercial office are. I have worked from home and from commercial offices. My experience has shown that the decision to rent commercial space is dictated by your self-discipline and the type of business you are running. Let's explore the factors you should consider when thinking about where to set up shop.

Self-discipline

Self-discipline is paramount to your success when working from home. If you are not able to make yourself stick to your work, you'll find yourself out of business. It is easy to get caught up in sitting around the breakfast table too long or taking a stroll around your farm. Working from home is very enjoyable, but you do have to set rules and stick to them.

Storefront requirements

Some businesses have storefront requirements. Builders rarely have such needs. Other trades, such as plumbers or electricians, might need to display the fixtures they sell. Most builders have no need to do this. They can send their customers to suppliers of building supplies to view showrooms. If you want or need a storefront window, you probably can't work from home. A storefront might improve your business, but it is not mandatory. Having a physical location for your customers to visit helps to build credibility, which is an important element for success. But can you justify the expense? Storefront space is expensive. If you don't need it, don't pay for it.

Home office

A home office is a dream of many people. Putting your office in your home is a good way to save money if it doesn't cost you more than you save. Home offices can have a detrimental effect on your business. Some people assume that if you work from home you are not well established and might be a risky choice as a contractor. Of course, working from home doesn't mean your business is having financial trouble, but some customers are not comfortable with a company that doesn't have a commercial office space.

I work from home now, and I have worked from home at different times for nearly 20 years. I love it. I am also very disciplined in my work ethic, and even though my office is in the home, it is set up in professional style. When clients come to my home office, it is obvious that I am a professional. I talk more about setting up a home office a little later, but take your home office seriously.

Commercial image

A commercial office can give you a commercial image. This image can do a world of good for your business. However, the cost of commercial offices can be a heavy weight to carry. Before you jump into an expensive office suite, consider all aspects of your decision. I'll talk more about the pros and cons of commercial offices later in the chapter.

ASSESSING YOUR OFFICE NEEDS

Before you can decide on where to put your office, you need to assess your office needs. This part of your business planning is instrumental in the success of your business. Can you imagine opening your business in a fancy storefront and then, say six months later, having to move out of the expensive rental space? Not only would that situation be potentially embarrassing, it would be bad for business. Once people get to know your business and where it is, they expect it to stay there or to move up. A downward move would alarm present customers and scare off a percentage of future customers. This is only one reason why office selection and location are important.

How much space do you need?

One of the first considerations in choosing an office is how much space you need. If you are the only person in the business, you might not need a lot of space in the office. When you consider your space requirements, take the time to sketch out your proposed office space. It helps if you make the drawing to scale. How many people will you meet with at any given time? How many desks will be in the office? I'm now a one-man business, not counting subcontractors, but I have two desks and a sorting table in my primary office. In addition, I have another room designated as my library and meeting room, another space set up as a darkroom, and a photography studio in my basement. My barn stores my tools, equipment, and supplies. So, you see, even a small business can have the need for large spaces.

When you are designing your office, consider all your needs. Desks and chairs are only the beginning. Will you have a separate computer work station? Do you need a conference table? Where will your filing cabinets go? Where and how will you store your office supplies? How many electrical outlets will you need? The more questions you ask and answer before making an office commitment, the better your chances are of making a good decision.

Do you need commercial visibility?

Do you need commercial visibility? Almost any business can do better with commercial visibility, but the benefits of this visibility might not warrant the extra cost. If you are out in the field working every day, and you don't have an employee in the office, what good will it do you to have a storefront? If your business allows you to remain in the office most of the time, a storefront might be beneficial. You would get some walk-in business that you wouldn't get working out of your home. In general, if you can't be there to mind the store, you don't need the store.

Do you need warehouse space?

Do you need warehouse space? If you have to keep large quantities of supplies on hand or deal in bulky items, warehouse storage might be essential. A popular solution for office and storage space is a unit that combines both under the same roof. These office/warehouse spaces are efficient, professional, and normally not outrageously expensive.

If you don't need fast access to the materials you put into storage, renting a space at any private storage facility can be the best financial solution to your needs. At one time, I ran my company from a small office and a private storage facility. This arrangement was not convenient, but it was cost effective, and it worked.

The list of possible considerations could go on and on, but I'm sure you have the idea of what to look for in your office needs. Once you have evaluated your needs, you can consider where to put your office.

MAKING A DIFFERENCE WITH LOCATION

In business, location can make a difference. If you cater to customers in the city, living in the country can be inconvenient, and it can cause you to lose customers. Having an office where you are allowed to display a large sign is excellent advertising and builds name recognition for your company. If your office is in a remote section of the city, people might not want to come into your office. If you work from your home, and your home happens to be out in the boonies, customers might not be able to find you, even if they are willing to try. Location can make a difference in the profitability of your business.

How does your office location affect your public image? The public can be a strange group. Public opinion is fickle but important. If your business is perceived to be successful, it probably will become successful. On the other hand, if the public sees your business as a loser, look out. It is unfortunate that we sometimes have to make decisions and take actions just to create an image the public wants to see, but there are times when we must.

What does the location of your office have to do with the quality of your business? It probably has nothing to do with it, but the public thinks it does. For this reason, you must cater to the people you hope will become your customers. If you are dealing with a business in which a downtown office is expected, you should plan on working your way into a downtown office. If you don't, a time will come when your customer base will peak and stagnate. Only so many people will deal with you when your business is unconventional.

Aside from the prestige angle of office location, you must consider the convenience of your customers. If your office is at the top of six flights of steps and no elevators are available, people might not want to do business with you. If adequate parking is not available in the immediate vicinity of your office, you could lose potential customers. All of these location factors play a part in your public image and success.

DETERMINING HOW MUCH OFFICE YOU CAN AFFORD

How much office can you afford? This question appears to be a simple one, but answering it might be more difficult than you think. When you look at your budget for office expenses, you must consider all the costs related to the office. These costs might include heat, electricity, cleaning, parking, snow removal, and other similar expenses. These incidental expenses can add up to more than the cost to rent the office.

If you rent an office in the summer, you might not think to ask about heating expenses. In Maine, the cost of heating an office can easily exceed the monthly rent on the space. When you lay out your office budget, take all related expenses into account. Come up with a budget number you are comfortable with, and make sure you keep your office rent and related expenses within your budget.

When you begin shopping for an office outside of your home, ask questions, and lots of them. Who pays for trash removal? Who pays the water and sewer bill? Who pays the taxes on the building? These might seem like stupid questions, but some leases require you, the lessee, to pay the property taxes. Who pays the heating expenses? Who pays for electricity? Who pays for cleaning the office? Does the receptionist in the lower level of the building cost extra? If office equipment is available in a common area, like a copier, for example, what does it cost to use the equipment? Your list of questions could go on and on.

Ask all the questions and get answers. If you are required to pay for routine expenses like heat or electricity, ask to see the bills for the last year. These bills can give you an idea of what your additional office expenses might be.

Before you rent an office, consider ups and downs in your business cycle. If you are in a business that drops off in the winter, will you still be able to afford the office? Do you have to sign a long-term lease or will you be on a month-to-month basis? It usually costs more to be on a month-to-month basis, but for a new business, a long-term lease can spell trouble. If you sign a long lease and default on it, your credit rating can become scarred. What happens if your business booms and you need to add office help? If you are in a tiny office with a long-term lease, you've got a problem. If you do opt for a long-term lease, negotiate for a sublease clause that allows you to rent the office to someone else if you have to move.

It can be easy to fantasize about how a new office will bring you more business. It's fine to enjoy this thought, but don't put yourself in a trap. When you project your office budget, base your forecast on your present workload. Better yet, if you've been in business a while, base the projections on your worst quarter for the last year. If you can afford the office space in the bad times, you can afford it. If you can only afford it during the summer boom, you're probably better off without the office.

Many new business owners get carried away with their offices. They look for space with marble columns, fancy floors, wet bars, and all the glitter depicted in offices on television. Well, unless you are independently wealthy, these lavish work spaces can rob you of your profits. Not only will the cost of the office drain your cash flow, you might lose business because of your expensive taste. That's right — you might lose business by renting a great office.

Consumers aren't stupid. It doesn't take long to figure out that if a company has high office overhead, the customers are paying for it. The flashy office might be fun, and it might be impressive, but it can also be bad for business. Of course, this depends on the type of business you have and the clientele you cater to, but don't assume that an expensive office is going to get you higher net earnings.

DECIDING BETWEEN ANSWERING SERVICES AND ANSWERING MACHINES

When the pros and cons of answering services are compared to answering machines, you'll find many different opinions. Most people prefer to talk a live person rather than a cold, electronic machine. However, as our lives become more automated, the public is slowly accepting the use of electronic message storage and retrieval.

When you shop for services, which do you prefer, an answering service or an answering machine? Do most of your competitors use machines or people to answer their phones? This is easy to research—just call your competitors and see how the phone is answered. It's pretty well accepted that the use of an answering machine can cause you to lose business, but that doesn't necessarily mean that you shouldn't consider using a machine.

Answering machines are relatively inexpensive. Most machines are dependable. These two points give the answering machine an advantage over an answering service. Answering services are not cheap, and they are not always dependable. More callers will leave a message with an answering service than on a machine. This point goes in favor of the answering service. Answering services can page you to give you important and time-sensitive messages; answering machines can't. Another point for answering services. Are you confused yet? Don't worry, we'll sort it all out.

To determine which type of phone-answering method you should use, make a list of the advantages and disadvantages of each. Once you have your list, it should be easy to arrive at a first impression on which option you should choose. It might be necessary to change your decision later, but at least you'll be off to a reasonable start.

What should you look for in an answering machine?

There are several qualities you should look for in an answering machine. These qualities should include the ability to check your messages remotely. Most modern answering machines can be checked for messages from any phone. Choose a machine that allows the caller to leave a long message. Many machines allow the callers to talk for as long as they want. These machines are voice activated and cut off only when the caller stops speaking. Pick a machine that allows you to record and use a personal outgoing message. Some of the answering machines are set up with a standard message, one that you can't change. It is beneficial to customize your outgoing message. If you buy an answering machine that meets all of these criteria, you should be satisfied with its performance.

What should you look for in an answering service?

What should you look for in an answering service? Price is always a consideration, but don't be too cheap; you might get what you pay for. Find an answering service that answers the phone and takes messages in a professional manner. You want a dependable service, one that sees that you get all of your messages. Ask if you can provide a script for the operators to use when answering your phone. Some services answer all the phones with the same greeting, but many will answer your line any way you like.

Ask what hours of the day the service provides coverage. Most services provide 24-hour service, but it generally costs extra. Ask if the service will page you for time-sensitive calls. Most will. Determine if your bill will be a flat rate or if it will fluctuate depending on the number of calls you get. Inquire about the length of time for which you must make a commitment. Some answering services allow you to go on a month-to-month basis, and others want a long-term commitment.

If you decide to use an answering service, check on its performance periodically. Most services provide a special number for you to call for picking up your messages, especially if they base your bill on the number of calls taken on your phone line. Even if you have to pay for calling in on your own line, do it every now and then. When you call in on your business number, you can get firsthand proof of how they handle your calls. Have friends call and leave messages. The operators won't recognize the voices of these people and will treat them like any other customer. These are the best ways to check the performance of human answering services.

Which way should you go?

Which way should you go in your decision between machines and humans? I think the business you lose with an answering machine is more valuable than the money you save. If you can hire a human answering service, I think you should. I have tried having my phones answered each way, and I am convinced that human answering services are the best way to go.

Pulling together all of the components of a functional office can take time. Don't expect to make all of the best decisions on your first attempt. You might have to experiment a little to find out what works best for you. Remember—don't jump into a long lease that can run your business into the ground. You are better off to grow slowly rather than not grow at all.

11
Finding the best building lots

Finding the best building lots is the cornerstone of success for home builders. You can have a long line of willing home buyers standing outside your office and never make a penny if you can't find and secure building lots that suit their needs. You might think that finding lots is no problem, but it can be a big problem. This is especially true if you are working in a rural area with first-time buyers who need cheap land.

I've worked as a builder in different types of markets. I started out near a big city. Lots were readily available to builders who could commit to buying them in high volume, but spot lots (individual lots in or out of subdivisions) were hard to come by, and they were very expensive.

When I moved my building operation to a less-populated area, lots of all sizes, shapes, and prices were abundant. Finding a lot for any type of buyer was easy. Now that I've moved to Maine, I'm faced with difficulty in finding lots for affordable housing. There is plenty of land in Maine, but most of it is in large parcels and much of it, large or small, carries a hefty price tag.

I recall one occasion when my advertising was working particularly well. In one week, I had 63 prospects respond to my ads. All of these people were looking for starter homes. It was no problem for me to build the houses at prices they could afford, but I could find only two lots that were priced in a range that would work with affordable housing. It pained me greatly to watch 61 disappointed prospects leave my office. Not all of them were qualified to buy a house, and I'm sure many of them were just curious about home ownership, but the lack of lots undoubtedly cost me a bundle of money.

Not all building lots are created equally. Some are clearly better than others. In Maine, full basements are very popular among home buyers. A majority of houses are built on full, buried basements. But not all building lots are suitable for homes with basements. Much of the land in my region is full of bedrock, or what Mainers call ledge. To build a basement on a lot full of ledge requires blasting, and that gets very expensive. An inexperienced builder could walk into a world of trouble if a price was given to a customer for a home on a basement and then ledge was discovered a few feet down.

Maine has ledge, and Virginia has soggy soil and other building hazards. I'm sure that other parts of the country present their own special types of obstacles for builders. Buying a lot that has hidden expenses associated with it is a sure way to give up your profit on a job. You have to be knowledgeable of what to look for, and you have to be judicious in the lots you select.

Not all hazards are created by nature. Some are related directly to human beings. For example, you might buy a lot that looks great today and find out halfway through the construction of a spec house that an airport is being built nearby and that the flight pattern will be right over your new spec house. Something like this can make the value of your property plummet.

Before you buy any land, you have to go through a series of checks. Routine real estate requirements, such as checking to see that the land can be transferred with a clear title, are to be expected. You should check zoning requirements and all other aspects of risk associated with buying real estate. Any good real estate lawyer can walk you through the typical real estate concerns. But, as a builder, you have to look for more than just the typical legal problems. You must assess the land for its potential to be built on.

UTILITY HOOK-UPS

One of the first things that a builder must take into consideration is the availability of utility hook-ups. This issue is not always a simple one, but it is always an important one. You might be surprised at how often building lots are affected by utility problems.

Water

Houses require water. If you are building in cities and towns, you probably expect a water main to be within easy reach of a building lot. When an existing water main runs along a property line, it makes life simple for builders. A phone call to the proper authorities can tell you right away how much it is going to cost to get water to the house you are building. But there can be some hidden expenses, and it's up to you to avoid costly surprises.

Under normal conditions with a municipal water main, a branch pipe is run from the main to a location about 5 feet inside the property line of a building lot. The placement of this feed pipe might be in the front or back of the building lot or even on one of the sides. Most water mains are installed under roads, so it is logical to assume that the water hook-up will be located in proximity to the road, but this is not always the case.

When a building lot is purchased that will be served by a municipal water main, a tap fee is usually required. The tap fee is not part of the price of the building lot. It is a fee that is required for the privilege of connecting to the city water system. An important fact to remember is that the tap fee doesn't always guarantee that a water connection will be placed on a lot for the price of the fee. Sometimes a contractor is required to make the actual connection, with the fee covering only the privilege to tie into the main.

Tap fees vary from city to city. While living in Maine for the last eight years, I have not built a house that was served by town water. The last tap fees I paid were in Virginia, nearly a decade ago, and they cost close to $2500 back then. As you can see, the price you have to pay to gain access to potable water can be steep. I expect the fees are higher in some locations and lower in others, but it is definitely an expense that you must factor into your job cost. Remember, the tap fee doesn't always mean that a connection will be made and placed on a lot for you. It's very possible that you will have to pay a private contractor to make the actual water tap.

If you are responsible for cutting a road surface and repairing it to gain access to a water main, be prepared for a big expense. Opening a road up is no big deal, but repairing it to town, city, or state standards can cost a small fortune. It's entirely possible for builders to lose most of their profits if they neglect to figure in the cost of getting water to a building lot.

Sewer connections

Sewer connections in urban areas are much the same as water connections. Tap fees are generally charged for the permission to tap into a sewer main. The cost of a sewer connection is likely to be as much, if not more, than the cost of a water tap. Don't overlook this type of soft cost when you are working up a building budget.

Electrical connections

A modern home without electrical connections isn't worth much. Builders who work in cities tend to take their electrical connections for granted. Rural builders know better than to assume that electrical service is available. They also know that the cost of getting an electrical hook-up can be substantial.

You might have to deal with two types of electrical service. You know that regular electrical service is needed. You also might need temporary power to use while building a home. Getting temporary power can cost several hundred dollars and take a few weeks to arrange. If you will use generators on your jobs for power, you can skip the temporary setup. Once the house you are building is in dried-in condition, meaning that the interior of the home is protected completely from weather, you can have permanent power installed and use it to complete your construction.

If you are used to working in urban areas, you probably won't give much thought to getting electrical service to a house. Your biggest concern would be whether to have the power lines brought in overhead or buried underground. Builders who work out in the boonies, like I now do, have to make sure that power is available and what the cost is for getting it to a building lot.

When I built my most recent personal home, I was required to sign an agreement with the power company that said I would pay a $75 fee every month for more than two years to cover the cost of having power lines extended to my land. I live about half a mile down a private road, so I'm being charged for the cost of installing poles and wires to my home. This kind of an expense can sneak up on you if you're not thinking about utility costs.

If you or your customers have to pay thousands of dollars to obtain electrical power, phone service, or cable television, a cheap building lot might not be as good a deal as you think it is.

Wells

Building lots in rural areas are often required to have their own wells as private water sources. The good news is that there is no tap fee to be paid to a city and the homeowner won't receive monthly water bills. The bad news is that wells and pump systems are not cheap. A typical well and pump setup can run anywhere from about $2000 to more than $4500. This type of expense has to be factored into the cost of a building lot.

Septic systems

Septic systems, like wells, are common in rural locations. A simple septic system might cost less than $4000, but a complex system could cost more than $10,000. When you compare these figures to the cost of a sewer hook-up fee, it's easy to see that country lots that seem inexpensive might just be nightmares waiting to happen. You have to be diligent in discovering all costs associated with a lot and the construction of a house on a lot before you can determine if the property is a good value.

LAY OF THE LAND

The lay of the land on a building lot can affect the cost of building a house. A lot that slopes off in any direction requires more foundation work than a level lot. In some cases, the slope can result in a builder having to spend considerably more for a foundation. Keep this in mind the next time you are walking a building lot and thinking about buying it.

Soggy ground

Soggy ground can give builders problems during and after construction. Installing a foundation on land that is soggy is tough. You must dig down to a depth where solid, dry ground can be found. If you have to go very deep, you've increased the cost of your footings and your foundation walls. After a house is built on this type of land, moisture problems in or under the home can plague your customers, who will, in turn, hound you. Be very careful if you are buying soft ground.

Rock

Buying a building lot that is hiding rock just below grade can cause you all sorts of problems if you are required to install a septic system or a basement. Blasting bedrock is an expensive proposition. If you work in an area where rock is likely to be a problem, do some probing in the ground with a steel drive rod or have a backhoe operator dig a few test pits for you. Don't agree to build a basement or install a septic system before you know if rock is going to stop you. Even putting in normal sewer and water service can be very difficult if rock is encountered. Look before you leap.

Flood zones

Buying building lots that are located within established flood zones is very risky. The land might be buildable, but financing can be very difficult to arrange for properties located in floodplains and flood zones. If a property is situated in a flood hazard area and financing is needed, the lender will almost certainly require flood insurance. This type of insurance is expensive. Also, buyers might not be interested in owning a home where a flood might wipe out everything they own. I'd stay away from flood risks.

Trees

Trees on a piece of land increase the cost of building a home. This cost is usually offset by the desirability of a lot with trees, but you should still be aware that extra costs will be incurred. Having trees cleared can tack several hundred dollars onto the price of a house. I've known builders who looked for lots without trees to save this expense. Personally, I usually buy wooded lots and pay to have them cleared, because I feel that a wooded setting helps to sell most houses faster.

ACCESS

Access to a building lot might not seem like something that you would have to worry about, but it can be. Suppose you bought a lot that came with a 25-foot driveway easement for access—would it be a good deal at a low price? I doubt it. Most lenders require a property that is accessed by a right-of-way to have an easement that is at least 50 feet wide. Buying a lot with a narrow right-of-way might mean that you or your customers will not be able to obtain conventional financing for construction.

Maintenance agreements

Road maintenance agreements are not uncommon in many areas. If several building lots are situated along a private road, it is customary for the land developer to create a road maintenance agreement. Basically, these agreements say that each property owner is responsible for an equal portion of road repairs and maintenance. This might not sound like much when you are looking at land, but the cost of keeping up a road can be steep. Just having my road plowed when it snows costs me $50 a storm. Adding layers of stone is very expensive, and the cost of paving can be astronomical. For this reason, some road maintenance agreements can be such red flags that you should run away and look for a lot in some other location.

Dues and assessments

Some housing developments have a structured plan in which homeowners must pay dues and assessments for maintenance and amenities. This practice is not at all uncommon in planned developments. The homeowner dues can go up year after year. Building in subdivisions in which homeowner dues are required might hurt your chances of selling a spec house. Certainly, a lot of people are willing to assume financial responsibilities in return for good roads, playgrounds, parks, swimming pools, and similar amenities, but you should at least be aware that the costs of a property-owner association could cost you some sales.

RESTRICTIONS

Many subdivisions contain lots that are controlled by covenants and restrictions. Your lawyer should look for these restrictions when assessing the deed and title of a piece of land. As a builder, you must be aware of what the restrictions are. For example, many subdivisions require any house to contain a minimum amount of living space. It could be 1500 square feet, 1800 square feet, or whatever. If you are planning to build a ranch-style home, you might learn that the subdivision does not allow any one-story homes. The restrictions can be very detailed.

I've built in subdivisions in which roof colors and types were regulated, as was the color and type of siding that could be used. Even the foundation materials are sometimes specified in the restrictions. These restrictions help maintain a quality subdivision, but they can be a shock to an unsuspecting builder. Make sure that there are no prohibitions on the type of house you plan to build before you agree to buy a building lot.

THE CREAM OF THE CROP

When you're looking for building lots, finding the cream of the crop can take a lot of work. How you go about locating lots depends somewhat on the type of builder you are. High-volume builders usually make deals with land developers to acquire control over many lots on a graduated take-down schedule. I'll discuss this practice extensively in the next chapter.

Some builders buy raw land and develop their own lots. Done properly, this can be a very profitable venture. However, it can also be the downfall of a good construction company. Chapter 13 addresses the issue of developing your own lots.

If you are going to build only a few homes a year, you most likely will find your lots in a more traditional manner. Advertisements in your local newspaper are a good place to start. Talking to real estate professionals can lead to some good lot acquisitions. Chapter 15 discusses the many pros and cons of working with brokers.

Buying lots from ads in the paper or from signs that you see in various locations is a common way of doing business as an average builder. But it is not always the best way to find and secure ideal building lots. Sometimes you have to dig deep to turn up lots that are not advertised for sale. In fact, many of the lots you go after might not even be for sale when you start making offers to buy them. Let me expand on this idea a bit.

Unadvertised specials

Some builders do an excellent job of finding unadvertised specials. If you learn to become one of these builders, you can enjoy better lots, better prices, and less competition. When a building lot is advertised in a newspaper or placed in a multiple listing service, the demand for the lot can be great. Sellers who are seeing a lot of activity might hold out for top dollar, and they might not be willing to negotiate special purchase or option terms. If you can locate suitable lots that are not for sale but might be buyable or lots that are getting very little exposure on the open market, you're in a better buying posture.

Developers of subdivisions frequently have some lots that simply don't sell during the major sales effort of a subdivision. These lots become leftovers, and they can be an excellent value for a small-volume builder. I have built houses on numerous leftover lots, and with the exception of one lot, I've always done well with the ventures. The one lot that I didn't do so well on had underground water problems that I didn't anticipate, which ran up the cost of construction and ate into my profits.

Finding leftover lots is easy. Get in your truck and ride around in some established subdivisions. Take note of any lots that have not been built on. Determine their addresses or legal descriptions and make a trip to city hall to find out who the owners are. If lots are sitting empty in a mature subdivision, there's a very good chance that you can buy them outright, especially if they are still owned by the developer.

By checking public records in the local tax office or court house, you can see who owns what and what they paid for it when they purchased it. A little more detective work can reveal an address and maybe a phone number. Once you know how to make contact, you simply call or write the owners and let them know of your interest in buying their land. This gets the ball rolling, and the worst that can happen is that you will receive no response or will be told that the land is not for sale. In the best case, you might get a quality lot at a bargain price.

Many times people sell their land if they are asked to, but they have not yet gone to the trouble to find buyers. Developers sometimes lose interest in trying to sell lots that just don't seem to sell, so they stop advertising and move on to new projects. In either case, sellers might be willing to do business with you on discount terms. And speaking of terms, you might be able to arrange some attractive purchase options, since the sellers are not in a big hurry to make a sale. This can help your cash flow and still give you control over lots you can built on.

Small publications

Small publications often allow people to place free or nearly free ads to sell their possessions. This type of publication has existed in every state that I've lived in. Some real estate brokers advertise in these little papers and booklets, but most brokers spend their time and money with more aggressive publications that have larger circulations. The deals in these little buy-and-sell papers are frequently average at best, but there are times when a really wonderful opportunity shows up. I bought the land for my personal home from an ad in such a publication, and I got a great deal on it.

Many builders look in their local papers and deal with real estate brokers who are members of multiple listing services. Some builders dig deeper, but many don't. By watching obscure publications, you just might stumble across some excellent opportunities. Don't expect to find them fast. It might take months, but if you are diligent in your search, something worth buying will show up eventually. It always has for me.

Brokers

As I said earlier, I talk extensively about working with real estate brokers in Chapter 15. For this reason, I don't want to dwell on brokers in this section, but I think

I should mention them. If you develop a quality relationship with a few outstanding brokers, you might enjoy some fabulous land deals. Many builders shun brokers for various reasons. I'm both a builder and a broker, so I know the game from both sides of the table. A lot of inadequate brokers and agents offer their services in the real estate industry. Don't judge all real estate professionals on the actions of a few. I tell you in Chapter 15 how to find and identify the types of brokers you should probably be working with, but keep an open mind. The right brokers can work wonders for you, both in acquiring first-class lots and in making sales for your new homes.

Do it yourself

We've all heard the old saying, "If you want something done right, you have to do it yourself." This statement has a lot of truth to it. If you're capable of doing a job yourself, it's difficult to find someone who you feel can do the work better, whether it be painting, carpentry work, or land development. Becoming your own developer is a big and often risky decision, but it can pay big dividends. As an alternative, you might do well to work with some land developers to obtain exclusive rights to good lots. Let's turn to the next chapter to see how.

12
Controlling desirable lots

Controlling desirable lots in subdivisions without buying them is a great way to eliminate completion while maintaining a good cash flow. Doing this is easier than you might think, but it is not without risk. You must approach this method of lot acquisition with the same respect as you would when buying lots straight out.

As a volume builder, I've often worked out deals with developers so that my company controlled dozens of building lots with very little actual cash being tied up. When I was developing subdivisions on my own, I worked out deals with other builders to allow them the opportunity to control lots without buying them. I've made the deals as both a developer and as a builder, so you have the advantage of hearing how options and take-downs work from both perspectives.

When land developers create subdivisions, they need builders to come in and buy the lots that have been created. Few developers depend on the general public to buy their lots, although some developers will sell to anyone who has enough money. In most cases, developers contact established builders and try to sell them all of the available lots. One builder usually does not take over an entire subdivision, but it does happen from time to time. Usually three or four builders are solicited to work in building out a subdivision. Wise developers pick their builders carefully.

A developer likes to see builders working together to create a harmonious housing development. Wise developers look for builders who complement each other without providing direct competition. Some developers have little concern over who builds what, as long as the lots sell. There is no tried-and-true rule of how the sell-out of a subdivision works, but it is a pretty consistent fact that developers seek out builders to buy a majority of all lots created. This is where you come in.

You can wait for a developer to contact you, or you can go directly to the developers and solicit them. If you don't have a track record as a builder, some developers might be apprehensive about doing business with you. Developers want builders who they feel can uphold their take-down schedules and build out a subdivision in a timely fashion.

BUYING WITH TAKE-DOWN SCHEDULES

If you're not familiar with what take-down schedules are, let me explain them to you. They're really quite simple. Let's say that I'm a developer and you are a builder. As part of my routine, I normally won't sell lots in less than a 20-lot package. My building lots sell for $30,000 each. To buy all of the lots at one time, you need $600,000. Few builders can handle this type of up-front expense, and not many lenders will take such a big gamble on a new subdivision. This leaves me, the developer, with a problem. I've got 80 lots to sell, and I can't find builders who are flush enough to buy 20 at a time. What am I going to do?

In this situation, I'm going to offer you a take-down schedule. You sign a contract that says you will purchase 20 lots for $30,000 each. I get an earnest-money deposit when the contract is signed. The amount of the deposit might be $500 or less, but it could be much more. It's whatever the two of us agree to. Then you buy your first lot right away. That leaves 19 lots to be purchased over some period of time. To determine how many lots you must buy in a given period of time, we create a take-down schedule. You might buy one lot each month, or you might buy a new lot every time the lot you are building on at the present time is sold. The take-down schedule might call for you to buy all of the lots within a 12-month period, regardless of when you buy which lots. The options for terms in a take-down schedule are limited only by the agreements made between us.

Working with take-down schedules is good for both of us. You get control of 20 lots with a minimal amount of out-of-pocket cash. I get a commitment for 20 sales that I can show my banker, which frees me up to keep working with a new line of credit. I don't have to peddle my lots off one at a time to the general public. If you default, I keep the deposit money, maintain ownership of the unpurchased lots, and sue you for your lack of performance. Assuming that the deal works out well, we both win. You have control of exclusive lots that you can buy gradually, and I have sold a chunk of my subdivision.

Take-down schedules are created frequently between builders and developers. They are about the only practical way for developers to make large sales consistently. Builders like using take-down schedules because their customers have 20 lots, or whatever the number, to choose from, even though the builder might actually only own one of the lots. It's a win-win situation for both the builder and the developer.

As I said earlier, take-down schedules do carry some risk. Since little money changes hands when a deal is struck, it's easy for builders to get greedy. They might sign up for more lots than they can reasonably buy in a set period of time. If a builder defaults, a developer is likely to attack with legal proceedings. This is bad news for the builder, both from a public-image point of view and from a financial perspective.

You could go into a new subdivision with a great deal of optimism and secure 20 lots on a take-down schedule, only to find that the public is not enthralled with the housing development. Depending on the terms of your take-down schedule, you might have to buy a lot each month, even though you are unable to sell the houses needed to fill them. This can be a crushing blow. There is a thin line between not committing to enough lots and making a commitment that can sink your building business. Knowing what to do is part luck and part experience. If you have enough experience and invest enough time in researching local demographics, you can create your own luck.

CONTROLLING PROPERTY THROUGH OPTIONS

Land options (FIG. 12-1) are another way to control property without paying for it. Once you have contractual control over a lot, you have an advantage over other builders. If a customer falls in love with a lot that you control, it's impossible for any other builder to build that customer's home unless you are willing to settle for just making a lot sale. The more lots you control, the less competition you have.

Options, like take-down schedules, are not governed by set rules. A buyer and seller can work out any terms that the two of them are comfortable with when setting up a purchase option. Investors and builders often use options to gain control

---

Purchase Option

In consideration of the payment by _____, hereinafter referred to as optionee, in the amount of _____ _____ ($_____) receipt of which is hereby acknowledged by _____, optionor, agrees to grant optionee the option to purchase the real estate described as _____, and commonly known as _____, in the city of _____, for a purchase price of _____ ($_____) under the following terms and conditions:

If option is exercised, said option will place into force the purchase and sale agreement signed by all parties dated _____, which contains a contingency allowing this option to purchase prior to the purchase and sale contract being effective.

If not exercised by _____, this option shall expire and optionor shall be released from all obligations from this agreement and from the purchase and sale agreement. Optionee's rights shall cease and the above-named consideration shall be retained by the optionor.

Time is of the essence in this agreement.

To exercise this option, the optionee shall deliver written notice to the optionor at the address of _____, prior to the expiration of this option. An additional deposit in the amount of _____ ($_____) shall be placed with an escrow agent as detailed in the purchase and sale agreement.

If notice is mailed the date received indicated on the return receipt of the certified mail shall be the date of notification.

_____   _____
Optionor   Date   Optionee   Date

_____   _____
Optionor   Date   Optionee   Date

12-1 Purchase option.

of real estate with minimal cash demands. If you smell a great opportunity but aren't completely sure of yourself, an option is a good choice. The most you stand to lose when you option the purchase of a piece of property is the money that you put up with the option. Let me explain how some types of options might work and why they're advantageous to you.

Short-term options

I describe short-term options as options that must be acted on within 90 days. Three months is not a lot of time, but it can be enough for a builder to make evaluations that could make or break a big deal. Setting up a short-term option is easier than working out the details of some longer options. If the seller of a piece of property is anxious to sell, it's unlikely that any option will be considered too strongly. Once an option is placed on a piece of property, the seller's hands are tied until the person holding the option decides to act. But short-term options can often be placed with developers and investors who are accustomed to playing the real estate game. Typical consumers who are selling their personal land are less receptive to option angles.

When you offer a seller terms for an option, you can make the offer in any form that you wish. The seller isn't obligated to accept your offer and might counter it with a different set of terms. It never hurts to ask for an option. The worst that can happen is that you won't get it. Since, as a builder, you are more likely to be interested in long-term options, let's discuss the details of all basic option plans with the longer term in mind.

Basic options

Basic options for builders usually run for longer than 90 days. A six-month option is not unusual. Four months is very common, and other terms exist. To illustrate how this type of deal is done, let's assume our roles of you as a builder and me as a developer. You want to control three lots in my subdivision, but you're not comfortable with committing to a purchase contract and a take-down schedule until you can test the public opinion for my new subdivision.

As a developer, I plan to sell all of my building lots within a three-year time period. You contact me and offer to option three of my lots for six months. In the offer, you agree to give me $3000 in nonrefundable option money. If you don't buy the lots, I keep the three grand. If you do buy them, the money is applied to the purchase price. I would prefer a contract to buy and a take-down schedule, but I have a lot of property to sell and I have three years to sell it in. Taking your money now might result in three sales. At the worst, I'll get $3000 and still have my lots to sell later. Under the circumstances, I counter your offer with an offer that requires you to put up $6000.

You now have an opportunity to control $90,0000 worth of lots for only $6000. It's your call. If you take the option, advertise the lots with houses built on them and achieve quick sales, you can option more lots and maintain your momentum. You decide to do the deal. Options are as simple as this. The most you stand to lose is your $6000. You're not getting in over your head by committing to a $90,000 deal, but you are gaining contractual control over three very desirable lots. It's not a bad deal, but you decide to make the deal a little sweeter for yourself before you let me know that you would take the deal as it stands if you have to.

You counter back to me that you will put up the $6000, but that for every lot purchased, $2000 of the money is applied to the purchase price. This way, if you buy and sell one lot, the most you can lose is $4000. If you buy two lots (FIG. 12-2), you're at risk for only $2000. Since I have plenty of lots and time, I accept your offer and we cut the deal. You're off and running with very little cash out of pocket and minimal risk. This is the way that smart builders test the water before they jump in with both feet.

Basic options allow both buyers and sellers to arrive at terms that suit each other. There are no real rules to the game. Offers go back and forth until a deal can be struck. If you can find sellers who are not panicked for quick cash, options are very effective. When making offers on leftover lots and lots that are not officially for sale, sellers have very little to lose by accepting options. Keep this in mind the next time you are shopping for lots. If your gut instincts turn out to be wrong, you will have lost your option money but not your shirt.

Exercise of Option

To _____, Optionor, as Optionee, it is my intent to exercise the option agreement dated _____, between the

Purchasers, _____, and the sellers

_____, for the sale of the real estate

commonly known as _____.

The escrow agent for this transaction is:

Name

_____

Address

_____

Phone

_____

The amount deposited in escrow is

_____ ($_____) and this deposit

was made on _____.

| Optionor | Date | Optionee | Date |

| Optionor | Date | Optionee | Date |

12-2 Form used to exercise an option on real estate.

STRETCHING YOUR MONEY

Learning how to stretch your money with options and take-down schedules is very important for spec builders. Tying your money up in lots reduces your ability to handle the carry cost of keeping houses on the market for longer than the period of time you had budgeted. Even if you are building custom homes, it can be quite beneficial to have a diverse stable of building lots available only to you.

When you are first starting out, buying and optioning lots might be too risky. You can advertise your services and hope to find customers who already own their own lots. Some customers do. You can also assume that the customers you take on in the early stages of your business will find suitable lots on the open market. I can't tell you that you will or won't be successful by locking up building lots. The person who controls the land controls the building, so keep this in mind. Weigh your opportunities carefully and don't overextend yourself. Most builders who fail do so because they leverage too much on credit. Building your business slowly might be agonizing, but it is often the best route to take. You have been given a lot of solid information, and you will get a lot more in future chapters, but you have to mold your new knowledge to fit your personal circumstances. What works for me might not work for you.

13
Developing your own building lots

Developing your own building lots can sound like a good way to control your own subdivision and make a lot of extra money at the same time. Developing land can result in these types of advantages, as well as others. On the flip side, getting into land development can be very risky for anyone who is not experienced in the process. Unless you've worked with a developer, a surveyor, or someone else who has shown you the ropes, calling yourself a land developer can be very dangerous financially.

When I first entered the trades, I worked for various developers who were also builders. The first two people of this type that I worked with did large developments. My first experience came with single-family homes and lots. When I changed jobs, I was introduced to a 365-unit townhouse development. Even though my work didn't include development duties, I got to see a lot of what goes on.

When I went into business for myself, I never contemplated becoming a developer. My goal was plumbing and remodeling. Even when I started building houses, I never gave much thought to being a developer. Then I ran into a developer who was aging and looking to sell out his inventory of land. Some of his property was already developed into ready-to-build lots, and some of it was still in the raw stage. This was when I first thought seriously about trying to develop my own lots.

At the time my first development opportunity came along, I was young and ambitious, but I was also nervous about testing the waters of land developing. After hearing horror stories and talking to developers who didn't succeed, I decided not to pursue the land deal. It might have been a mistake, but I believe I did the right thing.

About a year after I declined my first development opportunity, I ran into a developer who used the same survey and engineering firm that I did. We talked on occasion and got to know each other one week at a time. I built a new home for one of the partners in the surveying firm and that got me closer to the inner circle. Before long, I was talking about development deals with the developer I had met. Pretty soon, we were working together. The development projects included commercial space for a shopping center, in-town land for housing, and rural land where we created 10-acre mini-farms.

I've been fortunate with my development deals. So far, I've not gotten in too deep or lost serious money on land deals. Based on what I know of the industry, I'm an exception, rather than the rule. Buying and developing raw land is risky. But the reward is often in direct relationship to the amount of risk taken.

Land development can be as simple as buying a tract of land and cutting it in half. Complex development deals can take years to see results. The approval process that generally is required for major developments can be very lengthy and expensive. If you don't have deep pockets, big development deals are not good ventures to get involved with.

LITTLE DEALS

Little deals are a good place to start as a developer. Buying some land and cutting it into a few building lots is something that most builders can handle if they put their minds to it. The cost for this type of developing is usually low, in comparative terms. Let's talk about some of the potential expenses that you must be prepared to handle if you want to create your own building lots.

Survey and engineering studies

Survey and engineering studies are usually one of the first expenses incurred once a developer has found a piece of land that is intriguing. Surveys are done for obvious reasons. You want to know how much land you're buying and what shape it is in. If the land is in a flood zone or floodplain, a thorough survey will reveal it. Elevations are included on comprehensive surveys, which show the grade of roads that must be built, the drop-off of house foundations, and so forth. A full-blown survey probably costs anywhere from several hundred dollars to a few thousand dollars. The cost depends on past surveys, land size, geographic location, and other factors.

The engineering studies can be extremely expensive, or they can amount to only a few hundred dollars. It depends on what you are looking for. If you just want to establish approved perk sites for septic systems, the cost is not staggering. When you want to know soil consistencies, compaction rates, water-retention details, and so forth, the cost can reach well up into the thousands of dollars.

Engineering firms often represent developers during the approval process of large developments. Paying professional hourly rates for meeting after meeting, month after month, can run into some major money. This expense won't normally be needed on little deals, but it is another potential expense to be aware of.

Acquisition

Once you have had property checked out from north to south, you need money or credit to secure the land. Down payments on land are often required to be at least 20 percent of the land's market value and a 30-percent down payment is common. Depending on the price of the land, this can amount to quite a bit of cash. But there is often a way around this problem.

Sellers of land frequently finance a sale themselves. If you're getting owner financing, the down payment could be 10 percent or less. The last piece of land I bought was conveyed to me with just a $500 deposit and interest-only monthly payments. When you are cutting a deal with a seller who has clear title to land, you can make some very creative deals.

Filing fees

Filing fees, legal fees, and other costs associated with a development deal can run up to a sizable sum. The largest portion of these expenses is normally the legal fees, closing costs, and points. If your deal is a simple one and you are financing the property with the owner, your costs could wind up being less than $500.

Hard costs

The hard costs of land development vary with the type of land and its location. If you must pay to have electrical power, water mains, or sewers extended to the land, you could be looking at thousands of dollars. If utilities are already available, the cost is much less. In the case of rural property, where wells and septic systems will be used, you don't have to worry about water and sewer fees. The builder has to foot the bill for a well and septic system.

Clearing a piece of land is usually left for a builder to do. Developers sometimes have to pay for some clearing to allow the installation of roads, but it is unusual for a developer to prepare lots to a point where house sites are cleared.

Installing a road, even a gravel road, can be very expensive. The cost of a road can be enough to make a development deal go sour. This expense, like all others, should be determined during the time in which you still have an escape clause in your contract. The last thing you want to do is buy a piece of land fast and then find out that the cost of developing it makes the venture too expensive to proceed with. You will be left with a piece of raw land that is worth no more than what you paid for it, and you'll have to bear the burden of selling it. Get estimates on all your projected development costs prior to making an iron-clad commitment to purchase land. Your engineering firm can help you identify the types of costs to consider.

WHEN THINGS GO RIGHT

When things go right, developing raw land into a small number of house lots is not too complicated. Let me give you an example of how simple it can be. I recently bought 18 acres of rural land. If I wanted to cut this acreage into two building lots, say one 8-acre parcel and one 10-acre piece, I could do it for less than $700. The land would have to be perk tested to be sold as building lots, but the tests would cost less than $200. In my case, a formal survey is not needed at the time of development. Using town tax maps and their scale, I can draw my own plot plan. Legal expenses would run me about $250. The land has road frontage and available electrical service, so there's no cost in these categories. Filing fees with the town would be less than $50. It should cost about $500 to do the deal, so I would figure $700 to be safe. My out-of-pocket expense for buying this land with owner financing was a $500 deposit and less than $300 in legal fees. So for between $1300 and $1500, I bought and can develop the land into two building lots. The process would take less than two months to complete, and I would probably triple my investment since I got the land cheap.

If I chose to divide my 18 acres up into more than two lots, the game would change. Creating more that two lots would put me under different rules and regulations. My engineering costs would escalate, as would most of my other expenses. The amount of time needed to gain approval for a larger subdivision of the land could exceed one year. In theory, I should make more money by having an additional lot to sell, but I might not. Between the time lost, the extra money spent, and the fact that the lots would be smaller, I might not make as much money, and I would certainly have to endure a great deal more frustration. By keeping your first few deals simple, you can get a taste of what developing is like, and if you're fortunate, you can make enough money to fund larger deals when you are prepared to do them.

MID-SIZED DEALS

Mid-sized deals are usually what I consider to be a no-man's land. This is one reason why I passed on my first development opportunity. When you do a little deal, your costs are low and your profits tend to be high in terms of percentages. Big deals that work produce fabulous profits, and there is enough profit potential in a big deal to warrant an all-out promotional campaign. Mid-sized deals, such as a 12-lot subdivision, require a lot of money to get going, and they don't produce enough profit to justify major ad campaigns and publicity. This is why I see them as a type of quicksand for developers. You can get sucked into a bad situation quickly when you take on a mid-sized deal.

I can't tell you that you can't or won't enjoy success with medium-sized developments. Some developers do okay with them, but I think the odds are against you. There are a few exceptions to my rule. One of the deals I was involved in consisted of about 200 acres of land. We cut the parcel into 20 10-acre lots, built a spec house on the most visible lot, and promoted the properties as mini-farms. The land was located within a 20-minute drive of a major city, and the deal worked well. I considered it a mid-sized deal, and it paid off. But we had a lot going for us. The location was good, the cost of development was affordable, with the road being our biggest expense, and our package price for house and land was low. This combination led us to victory. Still, I'd be very careful about doing a deal like this one.

BIG DEALS

Big deals can make you dream of early retirement. The amount of money that can be made by developing a shopping center or a large housing development is staggering. Unfortunately, so is the cost of preparing the project for sale. I don't have a lot of firsthand experience as a major developer, so I won't attempt to tell you all the ins and outs. I simply don't know them. As a part of a team, I have worked with big development deals. I saw enough to know that I was not strong enough with either money or experience to tackle such a task alone.

Millions of dollars can be made on big development deals, but a developer needs a tremendous amount of money or backing to pull off the process. And the deal could go bad, causing everyone involved to wind up bankrupt. I'm biased by a lack of knowledge, but I would advise you to avoid big deals unless you can do them like I did and work with someone who knows what to do and when to do it.

It's tempting when you see a parcel of land for sale that you feel would make a good subdivision. You can draw out a plot plan and count the number of lots you would get out of the property. Then you will probably punch numbers into your spreadsheet program to see how much money you would make by doing the deal. This is fun, and the numbers can make your pulse rate climb, but you have to be realistic.

Installing roads eats up buildable land. Some builders don't take this into consideration when mapping out a development plan. Retention ponds, common space, and other zoning requirements can diminish the amount of land left to build on. What looks like a 25-lot subdivision might yield only 15 lots. There's a huge difference in the profit potential between 15 lots and 20 lots. Take your time and make sure your calculations and estimates are correct. It's very easy to make mistakes in the development business that can put you out of business entirely.

14
Building on speculation

Building houses on speculation is riskier than it used to be. There was a time when builders could sell spec houses faster than they could build them. Then the times changed, and builders found themselves holding spec houses for a year, or more, using up all of their profit making interest payments on their construction loans. Opportunities for spec builders are not as good as they once were, but they are better than they have been, and there is some hope that the market will continue to improve.

Builders who concentrate on spec houses avoid a lot of the hassles that custom builders must face. When you build on spec, you call the shots until a buyer comes along. Custom builders have to work under the watchful eye of their customers from start to finish. It is considerably easier to sell a house that has already been built than it is to sell one from blueprints. Spec builders enjoy the luxury of selling tangible products, while custom builders are often selling paper and blue lines. The list of advantages to building on speculation goes on, and I talk more about them as this chapter progresses.

PICKING LOTS AND PLANS

Picking building lots and house plans is one of the first things that a spec builder has to do. If you're going to build on spec, you must have lots to build on. We've talked in a previous chapter about finding desirable house lots. Don't take this part of your job lightly. The location, size, shape, and adjoining and adjacent properties affecting a lot all contribute not only to its market value, but to its appeal to customers.

Speculative building is a gamble in every way. You are buying a lot that you hope can make some home buyer happy. You are building a house that you believe has market appeal. If you're able to sell the house, you make money. When a spec house stagnates and doesn't sell, you've got to buy it yourself or give it back to the bank, which damages your credit and your credibility for a long time to come. You could put $25,000 profit in your pocket, or you could wind up in a bankruptcy court. There is, of course, some middle ground with different options. The bottom line is spec building is a gamble, and that can't be denied.

Since I've talked about lot selection previously, I'll concentrate on house plans in this section. Good spec builders have a feel for their markets. They stay in touch with brokers, other builders, current subdivision projects, and, of course, competitive advertising. The best builders track comparable sales by working with either appraisers or brokers. By doing this a builder can see exactly what types of houses are selling, what price they are selling for, and how long it is taking to sell them. This type of information is invaluable when deciding what types of houses to build on speculation. You can't take the gamble out of spec building, but you can pad the odds in your favor with enough research.

If you have access to a book of comparable sales, and any real estate broker worth a second look has this type of access, you can get a wealth of information to help you in choosing plans for a spec house. A comp book lists the foundation size of a home, the size and number of rooms in the house, the type of electrical service provided, specifications on the heating system, and even the floor coverings. By reading through comp books, you can see exactly what's going on in your real estate market, and the information is generally accurate and dependable.

Watching competitive advertisements is a necessary function, but you have to take the ads with a grain of salt. Many builders and brokers are good at running teaser ads that don't tell you much until you come in for a qualifying meeting. Unless you pose as a prospective buyer, you gather only limited information from ads. Read them. Pay attention to them. Learn from them. But don't rely on them.

Ride around new subdivisions and see what other builders are building. Inspect the construction, if you can, to see how it stacks up to what you will offer your customers. Compile as much data as you can before you commit to building a certain type of house in a particular way. I believe the book of comparable sales is your best tool, but ads and ride-bys are effective in their own right.

Once you have gathered as much information as you can, you have to start making decisions. You can run with the pack, or you can break away and build something completely different. Getting radical is risky. It is normally better to stick with what's working for other builders, but add some special twist of your own. Picking the right plan involves more than simply saying that you are going to build a split-foyer or a ranch. You have to assess and target your market.

TARGETING YOUR MARKET

If you target your market for spec houses, you are much less likely to fail in your attempt to sell your houses quickly. Some markets overlap. Others don't. A lot of builders want to build large, expensive homes, because their profit is usually established in proportion to the home's value. But some builders, like me, deal in less-expensive homes and operate on a volume basis. Both approaches have merit.

Builders of massive homes can boast that they only have to sell one or two a year to make a comfortable living. A builder who specializes in starter homes might have to build two or three times as many homes to make the same money. But if a massive house doesn't sell, the carry cost is substantial, and no income is derived. If I build six small houses and sell four of them, I can afford to carry the other two and still survive. Building one or two big houses and hanging your hat only on those pegs could be disastrous. On the other hand, I do have to sell more houses, and this could be a problem.

Buyers of expensive homes are usually well qualified, but they often have a house to sell. My buyers are coming out of rental property, so I can put deals together quickly that are not contingent on other sales. This, in my opinion is a big advantage. I might have to work a little harder to find a loan for my first-time buyers, but if my connections come through, it's a done deal.

You have to decide what price range you will build in. I can't tell you what will work best for you. First-time homes have done well for me, but I've also built many more expensive houses with success. The key is to know what you want to do in advance. You can't pick a plan and run with it until you have all the details worked out.

HITTING YOUR MARK

Hitting your mark, once you have targeted your market, is easy, if you know what to do. If you have defined your market tightly enough, reaching that market with the tools available to you from today's options is simple. Don't attempt to advertise until you feel quite comfortable that your target audience has been identified in a way that labels it as easy to reach.

Good advertising is a bargain. Poor advertising can be a disaster. Knowing how to maximize your advertising dollar is a crucial aspect of being successful as a spec builder. If you blindly spend hundreds of dollars each week for display advertising in your local paper, you could be wasting your hard-earned profit. Potential home buyers do often read ads in newspapers, but these prospects are not always the best ones to work with. There is often a hidden market that is much richer to mine.

General advertising to the public works, but targeted marketing is often more cost-effective. Running commercials on television is a great way to build name recognition, but it probably isn't an efficient way to sell one or two spec houses. If you are promoting an entire development, television advertising is great, but it costs too much to use for the sale of a single spec house.

When you buy advertising on radio, television, or in the newspaper, the cost of your ad is directly related to the number of people who listen to, watch, or read your chosen media. Your ads could be reaching tens of thousands of people, but how many of those people want what you have to sell? You simply don't know. Oh, you can gauge it to some extent with demographics, such as age groups of readers and listeners, but that's a far cry from a tightly defined audience.

Advertising is discussed later in this book, but a few aspects of it apply especially to spec builders, so I touch on these subjects now. What do you think is the best way to sell one or two spec houses? Most builders would either say that listing the homes with a broker or advertising the houses for sale in the local paper is what works. Either of these methods can work, but I've found great success with other methods. Let me give you a couple of examples.

Let's say that you are building a name for yourself as a builder of log homes. On a percentage basis, the demand for log homes is not too great. But plenty of people will buy log homes. It's just that their numbers are minuscule when compared to the general home-buying public. With this being the case, you would be paying a tremendous amount of money for advertising on a per-lead basis if you advertised in the general media. Your ad bill is based on the total circulation of a newspaper, not on the number of readers who might be interested in a new log home. Is there a better way? Yes, there is.

Direct mail is an awesome way of reaching a target market. By renting name lists, you can mail promotional pieces to people who fit the criteria you are looking for in a buyer. What, for example, would be points of interest in your selection of a name list for log-home buyers? Income would certainly be one qualifier. If the house you are building is small or large, the number of members in a family might be important in culling your name list. A family of five would not be likely to buy your two-bedroom cabin. Have the prospects shown any past interest in log homes? If they subscribe to magazines about log homes, and a mailing list company can probably tell you if they have, these potential customers are a prime target.

When you rent names from reputable list brokers, you can require that all sorts of criteria be met by the names placed on your list. Each set of criteria adds to the cost of a list, but the breakdowns are well worth their cost. The list of names that I'm preparing to order soon is going to cost me six cents per name. My list will arrive on pressure-sensitive labels, and it is sorted by location, income, and the fact that everyone on it lives in rental property. For less than $250, I'm getting a qualified list of 4000 prospects in my area.

I'll use the list I'm ordering to solicit first-time home buyers. That's why I wanted to make sure that everyone on the list lived in rental property. These people are not encumbered by a home to sell, so I should be able to make some quick deals.

Direct mail is not the only way to hit your target. Most towns and cities have publications that are aimed at particular markets. There is a military base located near where I work. The base is a good source on incoming housing prospects. By advertising in the small paper that serves the base, I can target my ads to buyers who have an ability to use Veterans Administration (VA) loans.

If I wanted to hit the out-of-state market that is so lucrative here in Maine, I know of a few national magazines that I can advertise in that have proven successful as real estate vehicles. Vacation homes and camps are very popular in Maine. I could advertise in many types of local outdoor publications to get work building seasonal cottages and cabins.

The point I'm making is that once you know who you want to sell to, you can get your message through to them in many good ways. Don't waste your money on expensive, general-media campaigns when you are selling a single spec house. An ad in the paper is a good idea, but it doesn't have to be a big, expensive ad. The money spent on advertising is one of your largest overhead expenses, so invest it wisely.

CREATING A SAFETY NET

Would you walk a tightrope over a pool of hungry crocodiles without a safety net below you? I wouldn't do it under any conditions, but there are always some daredevils out there. Building houses on speculation is not the same as balancing yourself on a cable over gaping crocodile mouths, but it can be nearly as dangerous. If you hope to survive year after year as a spec builder, you have to plan for the unwanted experiences that are likely to surface from time to time.

I haven't built a spec house in several years. My reasoning is simple: I've got enough contract work to do that I don't have to take the risk of building on speculation. At times I would prefer the fewer hassles that go along with spec building, but it's safer to build contract houses.

At my peak of spec building, I had an unusual but very effective safety net. I established groups of investors who would buy any spec house that I had trouble selling. The sales to investors had to be discounted deeply, but a 5-percent profit was better than a loss.

During my early years as a spec builder, my safety net was building each spec house with the intent to live in it, at least for a few months. If the house didn't sell quickly, I lived in it until it did. Another ploy I used was to arrange permanent financing for my spec houses to be used as rental property if they didn't sell well. If a house was costing me too much money in interest to keep it on the open market, I would close on the 30-year loan and rent the house out.

You don't have to have a contingency plan when you build on speculation, but you should. A lot can go wrong, and it doesn't take long for a builder to lose all profit in a house to the carry costs. I strongly suggest that you have at least one backup plan in mind for every spec house you build. Otherwise, your ticket to fame and fortune could wind up being a one-way trip to financial disaster.

CHOOSING COLORS AND PRODUCTS

The colors and products associated with a spec house often have to be chosen by the builder, which carries a certain risk. Colors that you like might not appeal to potential buyers. A brand of carpeting that appeals to you might not suit a buyer. Avoid making these types of decisions until the last minute. When you must make decisions on an item, make sure that you are taking a neutral route.

The color of a home's roof and siding must be decided early in the construction stage. Interior paint colors might wait until a buyer is found, but they often are chosen by spec builders. Go with noncommittal colors, like white, off-white, cream, and so forth. Don't pick stand-out colors that make definitive statements. By the same token, be careful about letting customers choose their colors before their financing is arranged.

Let's say that your spec house is almost ready for interior paint, and you have a couple who wants to buy the house. The people want one bedroom painted pink, one papered with dinosaurs, and other rooms painted in a variety of special colors. If you go along with their wishes and paint and paper the house while their loan is being processed, you could wind up in trouble. Suppose their loan request is denied? What will you do with a spec house that is haunted with custom colors? It will be more difficult to sell. It's good to let customers make their own choices on colors, floor coverings, plumbing fixtures, and other items, but you have to make sure that the people you are selling to will, in fact, be able to complete their obligation to the sale.

My rule-of-thumb as a spec builder has always been to let people pick their own selections and to stop construction until their loan is approved. This slows down the process, and I've made some exceptions to my rule, but generally, I've stuck with it. Customers won't like the fact that you don't continue construction until their loan is approved, but you have to weigh the risk. If the customers select colors and products that you feel would be acceptable to most prospective buyers, keep the construction rolling. But when you get someone who has unique taste, or, in other words, someone whose taste might be considered weird, hold off on the customization until you know the sale is a done deal.

SELLING YOUR SPEC HOUSES

Selling your spec houses is of paramount importance. Not only do you have to sell them, you need to sell them quickly. Paying interest on a construction loan erodes profit every month that a house sits on the market. Rather than going into great detail on selling in this chapter, I think it would be best to save the topic for the next chapter.

15

Working with real estate brokers

Working with real estate brokers is a subject that draws a lot of fire from builders. Some builders swear by brokers, others swear at them. A good broker can do wonders for your building business. Bad brokers can break your bank account. When it comes to selling houses, whether you are selling in the dirt or selling on spec, you must perfect a plan that works for your business. There's no clear-cut best way to sell houses. Each builder and every selling situation varies, at least to some extent. The variations muddy the waters and force you to define optimum procedures for your company.

Real estate agents and brokers fall into two distinct types. The first group is considered to be working for the seller of a property, which, in this case, is you. Buyer's brokers, the second group, represent buyers. Both types of brokers work with buyers and sellers, but their fiduciary responsibility is to one or the other, not both. After you get past the basic grouping, you'll find several subgroups of real estate professionals. Some specialize in multifamily housing. Others concentrate on commercial properties. A few are generalists: people who sell any type of real estate for a commission. If you look hard enough, you might find a new-construction specialist. They exist in major markets but can be difficult to locate in areas where new construction is not prolific enough to warrant a specialization. The list of possible subgroups goes on, but the types mentioned are enough for our purposes.

As a builder, you have many options in how you sell your houses. If you have the time and the talent, you can make the sales yourself. Some builders set up in-house salespeople to work selling only the homes being constructed by the builder. Many builders put the sales of their homes into the hands of brokers. Which of these three ways appeals to you?

I've been in the building business long enough to experiment with all three methods of selling. My personal sales efforts have been very effective, but when I was building in high volume, I simply didn't have the time to make all the sales myself. Using brokers and agents to sell for me has worked with limited success, but it has proven to be my least effective method. When I created an in-house sales team, I saw adequate results to maintain my volume building. To define your needs, let's look at the three options more closely.

SELLING IT YOURSELF

When you want to build a house, if you can sell it yourself, you have more control over the sale and you should make more money. You won't have to pay a sales commission, so you keep all the money. This works fine for builders who are accomplished in sales skills and who are building a low enough volume of houses to maintain an active presence as both a builder and sales professional. Unfortunately, many builders don't possess good sales skills. These skills can be learned, and I think every builder should develop at least minimum sales ability.

Outside of the money you save, one of the biggest advantages to selling your own homes is that you know what customers are being promised. I've had real estate agents promise customers the moon to get a sale. These promises make it very difficult for the builder who has to live up to the expectations of the customers. If you're pitching the deal, at least you know how far you're reaching to make a sale.

Assuming that you either already know how to sell or are willing to learn, I think a start-up builder can do best when making all sales on a personal level. If you're just getting started, you probably can't handle the volume that an in-house sales staff might produce, and you don't need the trouble that might come from working with independent agents and brokers. This is only my personal opinion, but it is a thought that I'm comfortable with.

As I said from the start, builders share different opinions on the best way to make sales. I can tell you about my personal experiences, and I can share recollections of conversations with other builders, but I can't define what might work best for you. It's your responsibility to assess all the data available to you when making a decision. Basically, if you are not a good salesperson, don't try to sell your own homes.

ESTABLISHING AN IN-HOUSE SALES STAFF

An in-house sales staff is something that most new builders can't handle. The cost of the salespeople and the volume of sales that you hope they will create can cause a start-up builder more trouble than they're worth. Once you have the bugs worked out of your building operation, an in-house salesperson can be very valuable. But most builders have to grow into a position where having someone on payroll to make sales is sensible.

In-house sales associates are sometimes paid an hourly rate, just like a carpenter, but this is rarely the best way to structure a compensation package. Many big-league builders have their personal sales staff on a commission-only basis. If the salespeople don't sell, they don't get paid. This type of arrangement is ideal, but it's not practical for a small-time builder.

Commission-only salespeople need to sell a lot of houses to make a good living. They also need the houses completed and closed quickly so that they can afford to live between paychecks. The sales professionals don't normally get paid until a house is built and the property transfer is complete. Small builders usually don't have the ability to offer enough volume or quick turn-around time to attract top-notch, commission-only sales associates.

This leaves the small builder with the option of paying a draw against commission. The sales staff is technically paid by commission, but they receive regular paychecks to keep them alive until house deals close. This requires a builder to incur the expense of the paychecks until deals close. If you are working with a tight budget, you probably can't afford to put out money for salespeople before you have closed on houses. Here lies the problem.

If you can't sell houses very well yourself, and you can't find or keep great salespeople on staff, what's left? Independent agents and brokers are the remaining solution. Going to a real estate brokerage offers a builder several advantages. One of the largest is the fact that getting a broker to represent your houses won't cost you a cent until a house is sold, built, and closed on. There are, however, some drawbacks to brokers that tarnish the gleam on such a wonderful solution. So, let's get into the pros and cons of listing your homes with a brokerage.

SELLING THROUGH BUYER'S BROKERS

Buyer's brokers represent buyers rather than sellers. You would not list your house for sale with a buyer's broker, but you could very well do business with this type of broker. If a buyer's broker has a client who wants to have a new house built, the broker might contact you for the purpose of providing estimates and proposals for your services. Keep in mind, the broker is working with you but for the buyer.

Buyer's brokers can be paid by either sellers or buyers (FIG. 15-1). This point confuses a lot of people. Many people assume that since buyer's brokers work for buyers, they are paid by buyers. This is not always the case. If the broker's client wants to do business with you, the contract offer to purchase might stipulate as part of the terms that it will be your responsibility to pay the broker. This is not necessarily bad, but it is something that you should be aware of and understand.

Most builders build the cost of a real estate commission into the sales prices of their houses. If a brokerage is used, the fee is paid. Builders who sell homes themselves pocket the extra profit. In my opinion, paying a reasonable commission to a buyer's broker for bringing a deal to the table makes a lot of sense. It's a sale you wouldn't otherwise have. For the sake of this discussion, it's enough that you understand that you might be asked to pay the fee of a buyer's broker, so look for the fine print in any offer that is presented to you by a broker or buyer.

SELLING THROUGH SELLER'S BROKERS

Seller's brokers are the most common type of real estate salespeople. These are the traditional brokers who work for sellers and with buyers. If you've been a builder for a while, you've probably been contacted by seller's brokers who were asking you to list your homes with their brokerages.

Is it a good idea to list your houses with a brokerage? It can be, but it can also put you in a difficult spot. The reasons can get complicated and confusing, so let's look at them one at a time.

If you list a house with a broker who is experienced and aggressive, you might see a contract in just a few days. But if the broker is lazy, inexperienced, or not motivated to sell houses that have to be built before a commission can be made, you could have your house tied up for months with no sales activity. Picking a listing broker is a delicate job that must be done right if you want to avoid all the headaches that can go along with the use of real estate agents and brokers.

Broker Commission Arrangement

If _____ , broker of the

_____ agency procures

an acceptable offer for the purchase of my real estate, commonly known as

_____ , and the property

is successfully sold, the real estate agency shall receive a commission equal to

_____% of the closed sale price. The listed price of this property is

_____

($_____). This commission agreement will remain in effect from

_____ to _____ . Seller agrees

that if the property is sold within six months to anyone the broker has registered

with the seller, as a prospective buyer, the broker shall be entitled to the above

commission. This does not apply if the seller lists the property with a licensed real

estate brokerage on an exclusive basis.

_____   _____
Seller   Date   Broker   Date

_____
Seller   Date

15-1 Broker commission arrangement.

Some real estate salespeople sell by puffing up the truth. They don't usually lie to customers, but they might not hesitate to imply a little of this or a little of that. And they might make things sound a little better than they really are. A broker with these habits can make your job extremely unpleasant. You might find that customers are coming to you complaining that their house wasn't built the way it was supposed to be. It's your job to satisfy the customer or live with the bad reputation they are likely to give you. I've gone through this type of thing, and I know how miserable it can be. In fact, let me give you a few examples of what independent brokers have gotten me into.

I remember one house where the broker who sold it told the customers a dishwasher was included in the asking price. The spec sheet on the house never mentioned a dishwasher except as an option. When the house was completed and a walk-through was done for a punch list, the customer complained to one of my field supervisors about the lack of a dishwasher. After checking into the matter, we learned what had happened. Rather than leave a customer with a bad taste for my company, I paid to have a dishwasher installed. I never listed another house with the broker who gave away the appliance to make the sale. I admit, a dishwasher was a fair trade for a good contract, but I should have been consulted before the deal was made.

I had a broker promise customers that my workers would help them haul in and install their clothes washer and dryer. These were not new appliances that we were including with the house. They were old appliances being moved into the house. All of a sudden, my field supervisors were turned into moving men. Go figure.

My list of stories is a long one, but I won't continue to put you to sleep with them. I will tell you, however, that some salespeople can create a lot of trouble for you when they start giving customers whatever it takes to make a deal.

Listing your homes with a seller's broker can be a good idea and a pleasant experience. The key is in finding the right agent or broker and the best brokerage. To do this, you must understand a little bit about how the real estate game is played.

Listing with big names

Real estate companies with big names have their advantages. National franchises get a lot of exposure, and they often draw large numbers of prospects. These factors could make you think that you are better off to list your house with the brokers who wear gold coats than you would be to list it with a little mom-and-pop office. Is this the case? Not always. In fact, sometimes the opposite is true.

There is no question that being a household name is an advantage in the real estate business. Several franchise names come to mind when I think about recent television commercials. If I were making a long-distance move, I might very well contact one of the big-league brokerages to help orchestrate my move. But the big-name companies don't always do better for builders than a small company can.

When you list a house with a big-name outfit, your home might be only one listing out of 300. Your house is likely to get less personal attention and exposure than it might at a different brokerage. Listing your house with a one-person brokerage could get your home a lot more attention and advertising. It only takes one broker to sell your house. You don't need to list with an army of agents to get results. What you have to do is find the right combination of agent, broker, and brokerage to do business with, which is not an easy task.

Asking questions

Before you agree to list your house with a brokerage, ask a lot of questions. For example, how many new-construction homes has the broker and brokerage sold in the last 90 days? If they haven't sold many, they are probably the wrong team for you. Selling from blueprints and shell homes takes a different type of talent than selling a completed home. You need to list with people who have what it takes to sell what you have to offer. You have an outrageous number of brokers and brokerages to choose from, so keep searching until you find the right one.

How often will your home be advertised? Will it be a spotlighted listing? What media will be used to promote your home? Ask these questions. You deserve answers to them. If the answers are not what you want to hear, keep looking for a good broker to work with.

What type of listing are you being asked to sign? An open listing allows anyone to sell a home. This type is good for you in one way but bad in another. The good thing is that you are free to sell the home yourself without paying a commission. Bad news enters the picture because most brokerages do not advertise open listings, since they might not be the agency getting credit for a sale.

An agency listing is, in my opinion, the best option for a builder. This type of listing gives a brokerage owner an exclusive listing agreement that protects the brokerage. Only the listing brokerage can take credit for a sale, with the exception of you or your in-house sales staff. If you sell the house, you don't have to pay a commission. When some other agency brings a buyer in, the listing agency gets a split of the commission, usually half. This type of listing provides some protection for the brokerage and leaves you free to sell your own property without paying a fee.

An exclusive listing is what most brokers are after. This means that the broker gets a commission regardless of who sells the house, even if you sell it yourself. The good thing about an exclusive listing is that the brokerage should work harder to make a sale, but this doesn't always happen. If I were you, I'd stick with agency listings so that I could sell my own houses without paying a fee.

Regardless of what type of listing you agree to, make sure you understand all of the terms and conditions. For example, how long will the listing run? Most of them go for four to six months. Some are shorter and some are longer. Can you terminate the listing at any time if you are not pleased with the service you are getting? Will you have to pay a termination fee if you stop the listing early? Ask these questions. Get a listing that you can terminate at any time without paying a penalty fee.

Paying commission

The commission you agree to pay a brokerage is strictly up to you and the brokerage. The amount could be 3 percent or 10 percent. New homes often pay a broker 5 percent, but 6 or 7 percent is not uncommon. Higher commissions are rare, and so are lower ones. But there is no set rule as to what a commission has to be. It is a negotiable issue.

Some brokers spend all their time getting listings and rarely sell a house. Other brokers seldom go after listings and spend most of their time selling. When you are checking out a brokerage, see how many of its in-house listings are sold by its own people. A brokerage that sells its own listings should be a good one for you.

Using multiple listing services

Most real estate brokerages belong to some type of multiple listing service (MLS). Being in an MLS isn't mandatory, and a broker can sell effectively without the benefit of MLS services, but I believe that you will be better off if you list with a brokerage who uses an MLS system. When a brokerage puts your house into an MLS system, a lot of other brokerages become aware of your house, which can reduce the house's time on the market. You want quick sales, so look for all the advantages that you can find.

Letting brokers do the dirty work

Brokers do the dirty work for builders. It is the broker who pays to advertise a house. When phone calls come in, the broker handles them. Going out after dark to show houses is a responsibility that good brokers assume. All you do, as the builder, is wait for the broker to bring you a purchase offer (FIG. 15-2). A good broker should prequalify prospects, help with financing, and serve as a buffer between you and the customer. When the system works well, brokers are a favorable asset. Unfortunately, the system doesn't always work, and builders sometimes judge all real estate professionals on the actions of one agent.

I would say that most builders, especially new builders who aren't comfortable making their own sales, should list their homes with brokerages. If you do, you are sure to encounter some problems, but the good should outweigh the bad if you do enough homework to select an ideal broker. Take a moment to become familiar with some of the other types of paperwork brokers can handle for you (FIGS. 15-3 through 15-10).

Contract for Sale of Real Estate

Contract made this _____ day of _____, 19_____, at _____

_____, State of _____, by and between

_____ (Seller)

and _____ (Purchaser). Seller

hereby agrees to sell, and Purchaser hereby agrees to purchase, a certain lot or

parcel of land with any building or improvements thereon (premises) situated in

_____ ,

State of _____, and described as follows: _____

_____

_____

The following items to be included in this sale: _____

_____

_____

_____

Said premises shall be conveyed within _____ days from the date of this contract by

a good and sufficient _____ deed or Seller conveying good and merchantable

title to the same free from all encumbrances, except existing easements, restrictions,

conditions, and covenants of record, existing building and zoning laws, and usual

and customary public utility easements servicing the premises; however, should the

title prove defective, then the Seller shall have a reasonable time after due notice of

such defect or defects too remedy the title, after which time, if such defect or

defects are not corrected so that there is a merchantable title, then the Purchaser

may at their option, be relieved from all obligations hereunder and withdraw

earnest money or deposits, if any.

PAGE 1 OF 4   INITIALS_____

15-2 Contract for the sale of real estate.

And for such deed and conveyance Purchaser shall pay the sum of _____

_____ dollars ($_____), payment to be made as follows:

1. $_____ received of Purchaser as earnest money in part payment on account for said lot or parcel of land with any buildings or improvements thereon and items included, if any.

That _____ shall hold said earnest money on deposit and act as escrow agent until transfer of title; that _____ days will be given for obtaining the Seller's acceptance; and in the event of the Seller's non-acceptance, this earnest money shall be promptly returned to Purchaser.

2. $_____ to be paid at the time of delivery of the transfer deed in cash, or by certified, cashier's, bank, or treasurer's check.

3. For a total purchase price of $_____

($_____).

This contract is subject to following conditions:

_____

_____

_____

_____

_____

Full possession of said premises shall be delivered to Purchaser at the time of the delivery of the transfer deed, said premises to be then in the same condition in which they now are, except in the case of new construction. New construction shall be completed according to attached plans and specifications and approved for occupancy by the local code enforcement officials. Reasonable use and wear of the buildings thereon are the only exception for existing buildings.

PAGE 2 OF 4   INITIALS_____

15-2 Continued.

The following items will be prorated as of the date of the transfer of said deed: Utilities, fuel, rents, real estate taxes for the current taxing period, for the town/city/county of _____.

The risk of loss or damage to said premises by fire or otherwise until the transfer of title hereunder is assumed by the Seller.

All covenants and agreements herein contained shall extend and be obligatory upon the heirs, personal representatives, and assigns of the respective parties.

That in case of failure of Purchaser to make either of the payments, or any part thereof, or to perform any of the covenants on its part made or entered into, this contract shall, at the option of the Seller, be terminated and Purchaser shall forfeit said earnest money; and the same shall be retained by Seller as liquidated damages and the escrow agent is hereby authorized by Purchaser to pay over to Seller the earnest money, if any.

This contract is also subject to a satisfactory water test, by a testing service approved by the State of _____.

The results of said inspection must be conveyed to all parties within _____ days of the final acceptance of this contract. Cost of this test to be paid by

_____.

If a broker is involved, Purchaser acknowledges that _____ _____ represents the Seller and Seller acknowledges that _____ represents the Purchaser.

PAGE 3 OF 4   INITIALS_____

15-2 Continued.

Witness our hands and seals on the day and year first above written.

I/we hereby agree to purchase the above described premises at the price and upon the terms and conditions above set forth.

_____   _____
Witness   Date   Purchaser   Date

_____   _____
Witness   Date   Purchaser   Date

I/we hereby accept the offer and agree to deliver the above described premises at the price and upon the terms and conditions above set forth.

I/we further agree to pay _____ a commission for his/her services herein, _____ percent of the sale price.

_____   _____
Witness   Date   Seller   Date

_____   _____
Witness   Date   Seller   Date

PAGE 4 OF 4

15-2 Continued.

Contingency Release

This contingency release shall become an integral part of the purchase and sale agreement dated _____, between _____, Purchasers and _____, Sellers of the real property commonly known as _____.

The following contingencies are hereby removed from the above mentioned contract:

_____

_____

_____

_____

_____

_____

_____

Purchaser	Date	Seller	Date

Purchaser	Date	Seller	Date

15-3 Contingency release.

Earnest Money Deposit Receipt

This will serve as receipt for the earnest money deposit received of _____

_____, Purchasers of the real estate

commonly known as _____.

The Sellers of this property, _____, will place this

earnest money on deposit with the following lender _____

_____, of _____

Any special arrangements for this deposit will be as follows:

_____

_____

_____

_____

_____

_____

_____	_____	_____	_____
Purchaser	Date	Seller	Date
_____	_____	_____	_____
	Date	Seller	Date

15-4 Earnest money deposit receipt.

Estimate of Purchaser's Closing Costs

SALES PRICE

_____(A)

ESTIMATED COSTS:

Escrow fees _____

Document preparation fees _____

Loan origination fee _____

Legal fees _____

Loan assumption fee _____

Transfer tax _____

Pest control fee _____

Loan application fee _____

Recording fees _____

Points _____

Trustee's fees _____

Notary fees _____

Prorated taxes _____

Interest _____

FHA/MIP (mortgage insurance) _____

Inspection fees _____

Credit report fee _____

Hazard insurance _____

Title insurance _____

Downpayment _____

Other fees _____

Total costs _____ (B)

Credits _____

_____

_____

Total credits _____ (C)

Estimated total cash need for closing equals (B) minus (C)

Total estimated closing costs $_____

15-5 Sample form of a purchaser's closing costs.

Promissory Note

City _____

State _____

Date _____

Face amount of note $_____

For value received and/or services rendered, the undersigned promises to pay

to the order of _____ at _____

the principal sum of $_____.

($_____) with interest thereon at the rate of _____ percent per

annum, said interest to be paid in monthly payments of $_____,

($_____) for _____ months. The balance is due in full and payable

on _____. This note shall be secured by the personal guarantee of the

undersigned and their heirs. Further security for this note shall be as follows:

_____

_____

_____   _____
Debtor   Date   Debtor   Date

_____
Witness   Date

15-6 Promissory note.

Balloon Promissory Note

City _____

State _____

Date _____

Face amount of note

$ _____

For value received and or services rendered, the undersigned promises to pay to

the order of _____ at _____ the principal sum of

$ _____ ($ _____) with interest thereon at the rate

of _____ percent per annum, said principal and interest to be paid in full

on _____. This note shall be secured by the personal guarantee of the

undersigned and their heirs. Further security for this note shall be as follows:

_____

_____

_____

_____

_____   _____
Debtor   Date   Debtor   Date

_____
Witness   Date

15-7 Balloon promissory note.

Counteroffer

This counteroffer is in response to the purchase and sale agreement dated

_____ , between the Purchasers,

_____ , and the Sellers,

_____ , for the sale of the real

estate commonly known as _____ .

All other terms shall remain the same. Seller retains the right to accept any other

offer prior to written acceptance and delivery of this counteroffer back to the

Seller. This counteroffer shall expire at _____ o'clock

AM/PM on _____ , unless an executed accepted copy is

returned to the Seller prior to the above deadline. The following counteroffer is

submitted for your review:

_____

_____

_____

_____

_____

_____

_____

_____

_____

_____

| _____ | _____ | _____ | _____ |
| Seller | Date | Purchaser | Date |

| _____ | _____ | _____ | _____ |
| Seller | Date | Purchaser | Date |

15-8 Counteroffer.

Contract Extension

The time for performance of the purchase and sale agreement dated _____

_____, between _____,

Purchaser and _____, Seller

for the sale of the real estate commonly known as _____

_____, is

hereby extended until _____.

Witness our hands and seals this _____ day

of _____, 19_____.

_____   _____
Purchaser   Seller

_____   _____
Purchaser   Seller

15-9 Contract extension.

Seller's Disclosure Form

Owner _____

Owner's address _____

_____

Property address _____

_____

Age of structure _____.

How long has the seller owned the property? _____.

Water Supply Information

Public—Yes/no   Private—Yes/no   Drilled/dug/artesian

Other (describe) _____

_____

Location _____ Date installed _____.

Installed by whom? _____

Have any problems ever been experienced with the following:

Water quality _____ Quantity _____ Pump _____

Discoloration _____ Other _____

Has the water ever been tested?   Yes/no   Date of test _____

Are test results available?   Yes/no   Have any test results ever been unsatisfactory

or satisfactory with notation?   Yes/no   If yes, what steps were taken to remedy

the problems? _____

Waste Disposal System

Public—Yes/no   Private—Yes/no   Quasi-public—Yes/no

Have there been any problems with waste disposal?   Yes/no

If yes, explain _____

If system is private, circle the appropriate type of system:

Septic   Leach   Holding tank   Other _____

Tank size _____   Tank installation date _____

Type of tank: Concrete   Metal   Other _____

15-10 Seller's disclosure form.

Waste Disposal System *(continued)*

Tank location _____

Company providing service _____

Have you experienced any malfunctions? Yes/no   If yes, explain _____

_____

Does system comply with current code requirements? Yes/no   If no, explain.

_____

Comments _____

_____

Insulation

	Type/amount	None	Unknown
Attic/Capwalls	_____	_____	_____
Floors	_____	_____	_____
Other	_____	_____	_____

Was insulation installed during your ownership?   Yes/no   If yes, by whom?

_____

Comments _____

_____

Hazardous Materials

Do you have knowledge of current, or previously existing, known hazardous
materials on or in the property, such as:

Toxic materials	Yes	No	Unknown
Landfill	Yes	No	Unknown
Radioactive material	Yes	No	Unknown
Other	Yes	No	Unknown

Comments _____

_____

_____

15-10 Continued.

Hazardous Materials *(continued)*

Asbestos

On heating pipes	Yes	No	Unknown
In the siding	Yes	No	Unknown
In floor coverings	Yes	No	Unknown
In roof shingles	Yes	No	Unknown
Other locations	Yes	No	Unknown

Comments _____

_____

Radon

Has the property been tested for radon?   Yes/no   If yes, what types of test

were conducted?   Air/water   Test date _____

Are results available? Yes/no   Have any steps been taken to reduce radon? Yes/no

Comments _____

_____

Lead-Based Paint

Does the property contain lead-based paint?   Yes/no/unknown

Are you aware of any cracking, peeling, or flaking paint?   Yes/no

Comments _____

_____

Underground Storage Tanks

Are there now, or have there ever been, any underground storage tanks on your

property?   Yes/no unknown   If yes, are tanks in current use?   Yes/no

What materials are, or were, stored in the tanks? _____

_____

Age of tanks _____   Size of tanks _____

Location _____

_____

15-10 Continued.

Underground Storage Tanks *(continued)*

Have you ever experienced any problems such as leakage?   Yes/no

Are tanks registered with the authorities?   Yes/no

If tanks are no longer in use, have tanks been abandoned according to the local authority's regulations?   Yes/no

Additional items of disclosure are provided on the disclosure addendum, if any. Are there any?   Yes/no

The Purchaser is encouraged to seek information from professionals regarding specific issues of concern involving hazardous materials or other concerns arising from this property.

_____		_____	
Seller	Date	Purchaser	Date
_____		_____	
Seller	Date	Purchaser	Date

15-10 Continued.

16

Making the most of your time and money

Are you good at managing your money? Do you think you will be good at making the money in your business stretch to its maximum potential? What about your time? Do you possess good time management skills? Do you understand the principles behind time management? Well, if you are weak in either of these areas, you are going to have some trouble in your business. This chapter is going to help you strengthen your skills in time and money management. Knowing how to make the most of your time lights the path to making more money.

It is a cliché, but time is money. Whether you bill your time on an hourly basis or a contract basis, lost time translates into lost money. To get the most out of your business, you have to get the most out of your time and money. How many times have you said you don't have time for this or that? Do you find yourself rushing around to get everything done, only to be frustrated that you didn't accomplish your goals? Poor time management is usually a factor in these circumstances.

There is a difference between not having time to accomplish a task and not taking the time to complete the job. Let's look at a quick example. Consider yourself as a busy business owner. Your morning starts early, and you work until the phone stops ringing at night. If you have ever done this before, you know your day can include 12 to 14 hours of work. Obviously, time is a coveted commodity in this situation. Even with all this work, you have started putting on weight since you got successful. Getting out of the field and into the office has caused you to go a little soft. You want to work out, but when could you possibly manage to find the time?

This scenario is not unusual. Business owners are often obsessed with the advancement of their business. This obsession can ruin marriages, damage relationships with children, and drive the business owner to extreme behavioral swings. If you are going to survive in the fast lane, you have to learn how to make pit stops. Otherwise, you are going to burn out. I'm not a great one to preach slowing down, since I always seem to be in high gear, but it is true that you need to allow some time for yourself.

Over the last 12 years I have changed my work habits considerably. I no longer go from 6 a.m. to 11 p.m. I spend time with my wife and children. I pursue some hobbies, and I'm going to start working out real soon, really I am. Okay, so some nights I write until 3:00 a.m., but I'm not missing time with my family—they're asleep. I wouldn't consider myself a workaholic, but I am very aggressive. In fact, if you want to win the business battle, you have to be aggressive. If you move too slowly in the rat race, the faster rats will run over you.

When you have something you want to do or something you should do, you can probably make time to do it. In the case of making time to work out, you can work out before work, at lunch, or after work. When you consider that most authorities say you only need 20 minutes of exercise every other day, that's not much time. You probably spend that much time reading the paper, drinking coffee, or thinking about what you are going to do on your day off, if you ever take one. Time management is an individual act. Every individual adjusts to changes in his or her schedule differently. Let's see how you can budget your time better.

BUDGET YOUR TIME

Do you budget your time? Most people don't; they react rather than act. This is a major mistake for a business owner. If you think about a football game or boxing, the advantage of acting rather than reacting is easy to see. The team that gets off the ball first or the man who throws the first punch has the advantage. The other side has to react to this aggressive action. The opponent of the aggressive team or fighter doesn't have the luxury of planning the attack, because they are too busy trying to thwart the offensive action. The same is true in business. Let me show you what I mean.

Consider this example. You are a building contractor. You do much of your own work. Your strongest competitor is also a builder, but she has an outside sales staff and uses subcontractors for her jobs. Your competitor has more time to spend on business projections and evaluations. Both of you make about the same amount of money.

You try to be on the job by 7:30 a.m., and you rarely leave before 4 p.m. When you get off from working with your hands, you take care of estimates and paperwork. By 6:30 p.m., you're home and having supper with your family. Then, from 8 p.m. to 10 p.m., you are returning phone calls, setting up subcontractors, and paying bills, among other things. By midnight, you're in bed. It's a tough schedule to keep, but it isn't an uncommon one for small contracting companies.

While you are thrashing around and working yourself into exhaustion, your competitor seems to be gliding through life. She is in her office by 9 a.m., a full hour and a half after you are on the job. She handles her office work during the day while you are sweating it out on the job. She tends to customer service and supervision, leaving subcontractors to do the heavy work.

Commissioned salespeople make the estimate calls and sales. At 5 p.m., your competitor goes home. An answering service picks up her business calls and transfers them to the appropriate salesperson or worker. From 5 p.m. on, your competitor has a personal life. What are you doing wrong?

If you're happy, you're not doing anything wrong. However, if you resent your competitor, you need to work smarter, not harder. Both of you are operating profitable businesses, but you are working much harder. Your competitor has outmaneuvered you in the business arena. There are pros and cons to the lives each of you live, but she seems to have the better life.

Doing business your way, you probably have less on-the-job problems, because you are on the job. You are not paying out money to commissioned salespeople. You are in total control of your business, as much as anyone ever is. You are basically married to your business. You have chosen to create a working life; your competitor has created a business. Which would you prefer to have? Your competitor has her business running smoothly because she has made good management decisions. Part of this management is time management.

I started out by having a working life; now I have a business. This is not to say I don't work. I probably work more hours than most business owners, but I enjoy most of what I do. For me, work is not a drudgery. I use subcontractors, sell my own work, do some of my own field work, do my own photography, writing, consulting, and real estate sales. As you can tell, my business is diversified. I have variety in my work life. This variety allows for a change of pace and a more even keel. However, to accomplish this goal, I've had to perfect my time-management skills. If you want to shape your business, you also need to learn how to use time management.

Know when you are wasting your time

For effective time management, you must know when you are wasting your time. It is easy to get caught up in the heat of the battle and lose your objectivity. If this happens, you won't realize you are working harder and losing ground. To run a good business, you must be able to step back and look at the business operation from an objective point of view.

Taking an unbiased look at your operational procedures can be troublesome. Many of my clients engage me to troubleshoot their businesses for just this reason. They are too close to the fire to see the individual flames. However, you don't have to hire an outside consultant to evaluate your business. You can do it yourself.

To determine how much time you are wasting, you might have to waste a little time. I realize this may seem redundant, but it's true. One of the best ways to pinpoint your wasted time is to spend time making a time log (FIG. 16-1). This log can expose how and where you are losing time.

Your log can be written or taped on a tape recorder, whichever you are most comfortable with. As soon as you wake up, start your log. Keep track of everything you do from the time you wake up until you retire for the evening. The log should include not only your business activity, but also your personal functions.

I know it might be inconvenient, but you must discipline yourself to make entries in your log with every activity you undertake. Whether it is brushing your teeth, walking the dog, going to the mailbox, or making business calls, enter your actions in the log. If you don't take the log seriously, this experiment won't work.

Keep your log for at least two weeks. At the end of that time, review the log, page by page. Scrutinize your entries for possible wasted time. For instance, if you find you talk for more than five minutes in your business calls, take a close look at what you are talking about. Certainly, sometimes business calls deserve 30 minutes or more, but most calls can be accomplished in five minutes or less.

Look for little aspects of your daily life that could be changed. Do you sit at the breakfast table and read the paper? How long do you spend reading the paper? Does this reading help you in your business? If your reading doesn't pertain to your business, maybe you should consider curtailing the time you spend at the table. If

```
Appointments and Events for:    Feb 16, 1993

   6 AM :
   7    :
   8    :    Meet with Mike about insurance
   9    :    9:25 car inspection
  10    :
  11    :
  12 PM :
   1    :
   2    :    Decision needed on blueprints for Walker job
   3    :
   4    :    Write memo about blueprint decision
   5    :
   6    :    Tennis
  EVE   :    No appointments
```

16-1 Time log.

you derive pleasure from reading the paper, it's not a bad thing to do. However, if you are caught up in a habit of reading the paper for half an hour and wouldn't feel deprived without this time, you have just found time for your workout.

You will probably find many red flags as you go down your time log. Almost everyone has little habits that rob them of time. Most of these routine activities bear little importance in daily life, but they go on because they are habits. Before you can change your bad habits, you must identify them. A time log helps expose your wasted time.

Control long-winded gab sessions with employees

Part of your time-management routine should include controlling long-winded gab sessions with employees. Employees are supposed to make money for you, but they can drain you of your profits. If you spend too much time talking with employees, you are losing money in two ways. You are not free to do your job, and your employees are not working. Let me give you a case history that drives this point home.

During one of my consulting assignments, I was working with a service company to see how the company could improve its efficiency. The business owner was smart. He knew that if I was brought in and introduced as a consultant, the employees would not act as they normally did. For this reason, I was introduced as just another employee.

I acted like an employee, dressed like an employee, and got to know the employees. The first week I was inside the company, I found a major loss of income. The income loss was the result of upper management talking with employees.

When the crews would come into the office for their daily assignments, it was common for them to hang around for half an hour, talking to the managers. The talk was not business-related. There were 10 employees, an operations manager, and an office manager involved in the conversations. This company was charging $35 per hour for its labor rate. Based on billable time, the 10 employees wasting 30 minutes a day were costing the business owner $175 a day. The office manager and the operations manager were making a combined income in excess of $45,000 a year. When you tally up the cost to the business owner for these morning talks, the annual total of lost income was in the neighborhood of $45,500. What seemed like a simple, friendly morning talk was actually a business-threatening cash loss.

SET YOUR APPOINTMENTS FOR MAXIMUM EFFICIENCY

If you learn to set your appointments for maximum efficiency, you will see increased time in your day. You can convert this extra time into money. If you prefer, you can spend the time you save following your hobbies and being with your friends or family. In any event, setting efficient appointments gives you more time to use for the purpose of your choice.

How do you set efficient appointments? You set efficient appointments by arranging the meetings in a logical order. It is also beneficial to schedule meetings to be held in your office instead of a customer's home or office. By meeting in your office, you save time. However, if you are having a sales meeting, you might be better off to sacrifice some time and meet with the potential customer on his or her home turf. People are more comfortable in their own home or office. When you are in a sales posture, you want the customer to be as comfortable as possible. But, for now, we are dealing with time management, so let's concentrate on why you want the meetings to convene in your office.

By setting appointments in your office, you save time and gain control. How do you save time? You save time because you can work until the person you are meeting arrives. How many times have you gone on appointments, only to have the other party be late or never show up? Have you ever thought of how much time these tardy or broken appointments cost you? Schedule appointments for your office whenever possible. If your client is late, you can continue working. If the appointment is broken, you haven't lost any time from work.

There is a side benefit to meeting people in your office. You will be more at ease, and the people you are meeting will feel at a disadvantage. This can be detrimental in a sales meeting, but typically, it works to your advantage. Under these circumstances, you have the control. How much you use or abuse it is up to you.

Did you know that, in general, people feel more intimidated when a desk is between them and the person they are meeting with? Are you aware of the signals you are sending when you put your hands behind your head and rear back in your chair? Body language says a lot. If you investigate sales techniques, you'll discover how your body language can influence your meetings. It is true that keeping a desk between yourself and the person you are meeting with is intimidating. In fact, you are often in a more advantageous position if you move out in front of your desk. People are more receptive and less intimidated. Rearing back in your chair, with your hands behind your head, signals that you are in control of the conversation. Body language is a powerful sales tool, and good salespeople know how to read the signals people are sending with their physical movements.

At any rate, scheduling meetings in your office is much more efficient than leaving your work space. You can work until the meeting starts. If the person never shows up, you haven't lost much. You are sitting beside your phone and won't miss calls. All in all, try to schedule meetings in your office.

REDUCE LOST TIME IN THE OFFICE

To maximize your time, you must reduce lost time in the office. It is common to think that if you are in the office, you're working; this theory doesn't always wash. A lot of people sit in the office, thinking they are working, without accomplishing any productive work. These people often lose more than time; they lose money.

Where do you lose time in the office? Do you sit in the office for hours at a time, waiting for the phone to ring? If you do, get a cellular phone or an answering service, and get out there looking for business. If your phone isn't bringing you business, you must go out after it.

Can you type? Do you take an hour to type a single proposal? If you do, it might be worth your while to find a typist who is an independent contractor. Hire the typist to transcribe your voice-tapes into neatly typed pages. This eliminates work you are not good at and allows you to do what you do best.

Assess your office skills and rate them. If filing is not your strong suit, find someone to do your filing for you. If you have an aversion to talking on the telephone, hire someone with an excellent phone presence to tend to your calls. You can use a checklist to rate your in-office performance. After referring to your checklist, concentrate on what you need to work on.

REDUCE LOST TIME IN THE FIELD

If you reduce lost time in the field, you can make more money. Reducing lost time in the field is similar to reducing time in the office. You can use the same type of checklist to appraise your performance. However, the ways for improving the quality of the time you spend in the field require some different tactics.

Whether you are doing your own work or supervising subcontractors, you can spend most of your time in the field. There is nothing wrong with being out of the office as long as you are being productive. But how productive are you?

There are many ways to reduce the time you lose in the field. A mobile phone can make a huge difference in your in-field production. Working with a tape recorder also gives you some advantages. Setting up a filing system in your work vehicle helps you stay organized and saves time. Let's take a look at the wonders a tape recorder can work for you.

Use a tape recorder to improve efficiency

You might be amazed at the results of using a tape recorder to improve efficiency. Tape recorders can boost your productivity into a whole new category. Many contractors spend an enormous amount of time driving. Tape recorders can turn this previously wasted driving time into productive time. Letters can be dictated, notes can be made, marketing ideas can be recorded, and myriad other opportunities await the users of tape recorders.

Tape recorders can improve efficiency in and out of the office. For an investment of less than $50, you can convert wasted time into productive time, which should result in extra money. Tape recorders are definitely worth strong consideration.

Use a mobile phone

Should you have a mobile phone? The answer to this question depends on you and your business. Cellular phones can be a boon to your business. Almost any owner of a service business can benefit from a mobile phone. While it is true that cellular phones are expensive on a cost-per-minute basis, they can be a bargain. If a $5 call results in a $10,000 job, you've made a wise decision.

What benefits can you derive from a cellular phone? The potential benefits of being able to communicate on the move are many. A general contractor can call from a remote job site and cancel a concrete delivery. If a job has failed its footing inspection, the concrete must be stopped. If the delivery is not stopped in time, the contractor has to pay for the concrete, even though it's not used. This can amount to a savings of at least $400.

A service technician can call ahead to make sure the next homeowner on the schedule is home and ready for service. If the homeowner has forgotten the appointment, the service technician can move up the schedule to the next service call, saving at least an hour's labor charge. When you add up this type of savings over a year's time, the amount can be substantial.

If you have a mobile phone and get stuck in traffic, you can call ahead and let someone you have an appointment with know you are on your way and you're going to be late. By keeping your customers and business associates aware of your schedule, you'll have fewer broken appointments and disappointed clients.

If you have on-the-road communication, you can always be reached for emergencies or highly sensitive issues. When customers know they can reach you at any time, they are more comfortable doing business with you.

Can you afford to operate without a mobile phone? With the progress in modern technology and the competitive nature of service businesses, I think cellular phones are nearly a necessity.

Time truly is money. You might benefit from sitting in on one of the many seminars available on the subject of time management. Learning how to budget your time is just as important as knowing how to budget your money. After all, your time is what you're selling. By honing your time-management skills, you can become a better builder, and your odds of survival are much higher.

17
Getting computerized

Computers and software have made a monumental impact on the way modern business is conducted. With the use of computers and their associated peripherals, a business can do almost any office task faster and, in most cases, better. Modems can be used to send documents over the phone lines from one computer to another. E-mail is as common today as a phone call was a few years ago. Credit checks can be flashed on a monitor in a matter of moments. Bookkeeping can be done by people who would never attempt such a task without the aid of computers and software. Scanners can put hand-drawn sketches into a computer. Businesses can sell stock, conduct banking functions, and exchange information over their phone lines with the use of computers. Estimating programs can take the guesswork out of job quotes, and worldwide communication is at the fingertips of computer users. Computers are truly remarkable.

As good as computers are, they are only as good as the individuals using them. If the person operating the machine doesn't know how to use the tool, the job cannot be done correctly. Many people are intimidated by computers. This group of people would prefer to muddle along the old-fashioned way than learn new skills. They present arguments for the time they might lose to become computer literate. Money is another defense. They come up with excuses for not being able to afford a computer system, even though many systems can be purchased for less than $1000. Some people might never learn to benefit from modern technology. When you compete against this group of people, you can gain a competitive edge by using computers.

Whether you love them or hate them, computers are a way of life. Most businesses that have used computers for any length of time wouldn't know how to operate without them. Once you are bitten by the computer bug, the attraction can be consuming.

It starts with a word-processing software package. You see how much easier it is to prepare your contracts and correspondence on the computer. You no longer need correction tape for the old typewriter. Forms saved on your computer make it fast and easy to turn out proposals, letters, and much more.

The next step is spreadsheet software. You start playing the what-if game and get excited about the potential of your company. Suddenly, forecasting the future and tracking your budget is fun. You are starting to get hooked.

Then, you experiment with database programs. Mailing lists were never so easy to accumulate and use. Your marketing is much easier with your new, computerized customer base. Checking inventory is a snap, and storing historical data is a breeze.

With a little more time, you get into accounting programs and automated payroll. With a little playful study, you find you can do what you've been paying other people to do for you. For the one-time cost of your software, you have eliminated a routine overhead expense.

Once you go online, you download bid-sheet information, scan job opportunities, and exchange comments with other computer users. Sooner or later, you give in to games. Playing golf on the computer might not be the same as kicking the dew off grass at the club, but it's not a bad way to relieve stress.

By this time, the computer bug has you in it's grip. You are addicted and would rather give up your easy chair than your computer. If you don't like the idea of using computers, you might find this scenario hard to believe, but don't be surprised to find yourself in a similar situation.

I used to be a computer-hater. I didn't believe in them and wanted nothing to do with them. My wife, on the other hand, was fascinated with computers. When she wanted to buy one, I didn't object, but I showed no interest in the mechanical monster.

Kimberley went about her business with the computer. She often tried, unsuccessfully, to get me to use it. When Kimberley wanted to computerize our business, I was adamant it would never happen. Well, I was wrong.

We did put our business on the computer, and after a while, I started to see the benefits of the change. The office requirements were being met much more quickly than before the use of the computer. At first, Kimberley handled all of the computerized tasks. In time, I started to play with the system. I never liked having a system, even one for simple filing, that I couldn't understand. At first, I was frequently frustrated, but after a short time, I started getting into the new technology. Soon, I insisted that we upgrade our system to a more modern and more powerful level. The rest is history; I now know and love computers.

IMPROVING YOUR BUSINESS

How can a computer help your business? A computer can help your business in myriad ways. You can perform almost any clerical or financial function on a computer. You can draw blueprints on a computer. Marketing and advertising are easier when you have the help of a computer. Inventory control and customer billing almost take care of themselves on a computer. When you take the time to look, you can see that your business can benefit from the use of computers in dozens of ways. Let's take a closer look at exactly how you can derive benefits from computerizing your office.

Marketing your services

You can do marketing tasks more efficiently with a computer. You can build a database file of all your customers and potential customers. Then you can print mailing labels for all of these names. If you use telemarketing, you can use computers to conduct your cold-calling surveys.

Drawing programs enable you to create logos and flashy flyers. By using the creative options available through your drawing program, you can attract more attention and cut down on your outside printing costs.

Following historical data is easy with a computer. With just a few keystrokes you can see what decks were selling for two years ago. A few more taps on the keys can display the type of work that makes you the most money. Computers can definitely make your marketing more effective and faster.

Preparing payroll

If you have employees on the payroll, you will enjoy the automated features available in payroll software. You enter a minimal amount of data, and the computer runs your payroll for you. These programs can be real timesavers for companies with several employees.

Costing jobs

Job-costing results can be printed out of your computer files in minutes. What used to take hours to do can now be done in minutes. Not only do computer-generated job-costing numbers come about more quickly, they are probably more accurate.

Tracking your budget

Tracking your budget (FIG. 17-1) is no problem with the right software. You can boot up the computer and see where you stand at any time during the year. This accessibility is not only convenient, it can keep you from busting your budget.

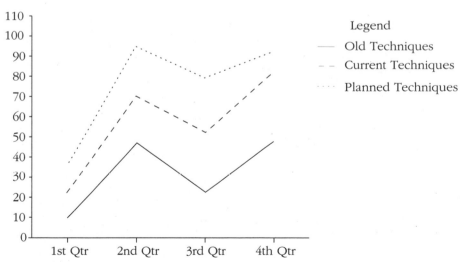

Sample Tracking Chart

17-1 Tracking chart created on computer.

Projecting tax liabilities

Projecting tax liabilities is another function a computer can perform for you. If you are wondering how much money you will need at tax time, just ask the computer. The computer can give you the information you need to plan for your tax deposits.

Estimating

Computerized estimating can make your life easier. Systems are available that almost do the estimates for you. Of course, you have to give the machine a little help, but not much. By moving a light pen across the blueprints, you can have take-offs done before you could get started manually.

Word processing

Word processing makes all of your written work easier. With built-in spell-checkers and thesauruses, today's word-processing software takes the drudgery out of writing a letter. If your grammar skills are rough around the edges, you can incorporate grammar software to correct your writing.

Employing databases

Databases can be used to store information on any subject. Whether you want to know the cost of a ton of top soil or when your trucks are due for service, a database can do the job. You can design the database file to include as much or as little information as you like. Sorting these electronic files is much easier than digging through the old filing cabinet.

Writing checks and checking balances

If you don't like writing checks or keeping track of your bank balance (FIG. 17-2), let the computer do it for you. Dozens of programs and forms are available for paying your bills with a computer. As a side benefit, the computer can make adjustments to your bookkeeping records as it goes.

Improving customer service

Customer service can be improved with the use of computers. If you receive a warranty call, you can quickly determine if the job is still covered under its warranty. When your customers are having a birthday, your computer can remind you to send out a birthday card. Whatever reason you have for wanting to stay in touch with your customers, a computer allows you to do it faster and easier.

Controlling inventory

Inventory control is made faster and more efficient with the use of computers. As your stock is depleted, the computer can tell you what to order. At the end of the

Checks and Deposits

		Starting Balance			$1,599.44
323	1/2	Office Rent	$566.81		$1,032.63
324	1/4	Service Station	$ 9.50		$1,023.13
325	1/5	Stationery Store	$ 14.90		$1,008.23
	1/6	Kile Job		$ 100.00	$1,108.23
326	1/10	Electric Company	$112.88		$ 995.35
327	1/10	Natural Gas Company	$ 66.81		$ 928.54
328	1/10	Telephone Company	$ 56.88		$ 871.66
	1/11	Baker Job		$ 345.00	$1,216.66
329	1/14	Plumbing Supplier	$ 99.50		$1,117.16
	1/15	Salary		$1,388.41	$2,505.57
330	1/19	Nails	$ 55.90		$2,449.67
331	1/21	Permits for Dean Job	$123.45		$2,326.22
332	1/22	Tarp	$ 36.22		$2,290.00
333	1/27	County Tax Department	$ 44.90		$2,245.10
	1/30	Salary		$1,388.41	$3,633.51
334	2/2	Office Rent	$566.81		$3,066.70
335	2/4	Hardware Store	$ 44.20		$3,022.50
336	2/5	Electrician-Drill Job	$588.40		$2,434.10
	2/6	Apex Job		$ 345.00	$2,779.10
337	2/9	Electric Company	$105.33		$2,673.77
338	2/9	Natural Gas Company	$ 63.55		$2,610.22
339	2/9	Telephone Company	$ 65.73		$2,544.49
340	2/11	Post Office	$ 20.00		$2,524.49
341	2/11	Office Supplies	$ 19.44		$2,505.05
	2/15	Salary		$1,388.41	$3,893.46
342	2/20	Shipping Expense	$ 15.00		$3,878.46
343	2/20	Blueprint Copies	$ 40.00		$3,838.46
344	2/23	Gas Station	$ 10.00		$3,828.46
345	2/27	Bill's Restaurant	$ 45.90		$3,782.56
	2/28	Salary		$1,388.41	$5,170.97

17-2 Financial report generated on computer.

year, you won't have to spend hours in the back room counting your inventory. All you have to do is ask the computer to print a report, and in moments, your inventory is done.

Having a personal secretary

With a memory-resident program, your computer becomes a personal secretary. The machine can beep to let you know it is time for your next appointment. You can store phone numbers in the computer and have it dial your calls for you via a modem. You can even send a fax with a computer. The right software all but replaces your old appointment book. With the touch of a hot key, you can scan all of your appointments as far in advance as you like.

The fact is, computers and software can tackle about any job you throw at them. I'm not suggesting that people should be replaced with machines, but there is definitely a place for computers in your business.

BUILDING CREDIBILITY

Have you ever considered building credibility with your customers through the use of computers? Computers can lend an air of distinction to your company. In some circumstances, credibility is gained as a result of customers having direct contact with the computer. In other cases, the confidence is built by what the computer allows you to do. Let's see how these two different approaches work.

Direct-contact approach

The direct-contact approach is the method most often recognized. An example could be having customers come into your office and comment on your computer system. There is no doubt that the visual effect of the hardware has impressed the customer.

Another means of building credibility through direct contact could come from demonstrating the power of your computerized system. For example, assume you have a young couple come to your office to discuss house plans. The couple has a rough sketch of what they want, but it is not drawn to scale. Other builders they have visited have made photocopies of the sketch and let it go at that. But you are going to be different; you are going to make a lasting impression on this young couple.

When the couple produces their well-worn line drawing, you look it over. Then you get up and put it on you flat-bed scanner. The couple watches, at first thinking you are making a normal copy of the drawing. When they see their sketch come up on your computer screen, they are amazed. But the show's not over yet.

Now you take your mouse and begin drawing lines on the monitor. In less than 15 minutes, you have converted the rough sketch into a scale drawing of the couple's dream home. As you talk to the potential customers, you make adjustments in the on-screen drawing, moving the kitchen sink down the counter, adding an island work space, drawing in a fireplace, and so on. In less than an hour, you have a viable plan drawn for the anxious home buyers.

Now tell me, which contractor would you hire to build your house, the one who made a photocopy or the one who took the time and had the technology to produce a professional drawing? I think the answer is obvious. Even if the computerized contractor wanted more money for the same job, you would probably assume he was worth it. As you can see, with a little thought, you can use computers as sales tools.

Indirect approach

The indirect approach to building credibility with a computer can be equally effective. The indirect approach relies not on what the system looks like but what you do with it.

Assume for a moment that you are a prospective homeowner. You are soliciting bids for a new house. During your search, you have narrowed the field of general contractors down to two. However, the two contractors are evenly matched in their prices, references, and apparent knowledge of building. Whom do you choose? You must look deeper to find what separates these two contractors. You start your reconsideration by going back over the bid packages.

The first contractor, Harry's Homes, Inc., has submitted a bid package similar to the other general contractors. You received a fill-in-the-blank proposal form, the type available in office supply stores. The proposal had been typed, but it was obvious the preparer was not a typist. Harry's references checked out, even if they were handwritten on a piece of yellow legal paper. The plans submitted were drawn in pencil. The graph paper was stained, probably from coffee. The spec sheet in the package was vague. In general, the bid package was complete, but it wasn't very professional.

Harry's competitor, Cornerstone Builders, Ltd., submitted essentially the same information, but the method of presentation was considerably different. The contract from Cornerstone was clean and neatly printed with a laser printer. The paper the contract was printed on was a heavy, high-quality stock. The accompanying specification sheet was very thorough. It listed every item and specified the items in complete detail.

The reference list supplied by Cornerstone was printed with the same quality as the contract, and the references were all satisfied customers. The plans in this bid package were extraordinary. These plans were drafted on a computer and showed elevations, floor plans, and cross-sections.

The bid package from Cornerstone Builders, Ltd., was compiled in an attractive binder. This bid package exuded professionalism. Since all factors, except the bid packages, seem equal, you decide to give the job to Cornerstone.

If this was a true story, which contractor would you have chosen? I believe you would have picked Cornerstone. Sometimes all it takes to win the job is a good presentation. Computers can help you make fantastic presentations.

CHOOSING SPREADSHEETS, DATABASES, AND WORD-PROCESSING PROGRAMS

Spreadsheets, databases, and word-processing software applications are what the majority of business owners use the most. You can buy each of these programs as stand-alone software, or you can buy a software package that incorporates all of the features into one package. Most of these combo packages are called integrated packages.

Should you buy an integrated package or stand-alone programs? Both types of software can be good, but the answer to the question depends on your needs and desires. If you are new to computers, an integrated package might be your best choice. Many stand-alone programs are more complex than the individual modules in integrated packages. Let's take a look at some of the pros and cons of each type of software.

Spreadsheets

Spreadsheets serve many business functions. Choosing the best spreadsheet for your business needs might require some research. Dozens of programs can get the job done. Some of these programs cost hundreds of dollars and others can be had for about $50.

You don't have to spend a fortune for software. There are many good deals available. Once you get into the computer scene, you'll find a flood of inexpensive software. Some of this bargain-basement software is great, and some of it leaves a lot to be desired. However, for less than $5, you can test drive the software with a demo disk.

By shopping, you can find almost any software combination you want. If you opt for a major-brand software, you might need some time to get acquainted with your purchase, but some excellent reference books are available to help you learn the software applications. If you go for the inexpensive, generic software, you might have more trouble finding after-market resource directories.

Most spreadsheets do approximately the same functions. Some are easier to learn than others, but the end result is about the same. When you are ready to buy a spreadsheet program, do your homework. Many good programs are available.

Databases

Databases are electronic files. These computerized files are generally more efficient than manual files. You can think of database software as your opportunity to file and retrieve any information you want.

Databases provide you with nearly unlimited chances to store and retrieve information. Sorting and retrieving information in a database is easy. Printing reports and mailing labels are no problem when you have a database program. If you would keep it in a filing cabinet, you can keep it in a database.

Word processing

Word processing is probably one of the most-used types of software available. It's great! You can write and store form letters. You can merge names into the form letters to achieve a personal mailing. If you make a mistake in your typing, you don't need correction tape or correction fluid. With a word processor, you simply type over your mistakes. Word-processing programs can be used for all your correspondence needs.

Word-processing programs don't have to be complicated. However, some of them are. What type of word processor do you need? You need a word processor that does what you want it to do. Again, you can buy a stand-alone program or an integrated system. Some software allows you to produce newsletters, books, graphics, and just about anything else you would like to.

If you don't want to spend the money for a top-notch word processor, you can invest in software with less extensive features. Your decision to go with less expensive software shouldn't affect the results of your most basic business needs.

In recent years, it has become common for new computers to be sold with a host of software already loaded on their hard drives. The type of software that comes with a computer purchase can range from a spreadsheet to an encyclopedia. In many cases, you might not need any software, except possibly some estimating and CAD (computer-aided drafting) programs, other than what comes loaded on the computer you buy.

Integrated programs

Integrated programs can fulfill all of your basic needs. These programs compile spreadsheet, database, and word-processing software into a single piece of software.

These combination programs are efficient and relatively inexpensive. While most integrated programs are not full-featured software, they can satisfy the needs of many small- to average-sized businesses.

I have used a few integrated packages. For the most part, I use stand-alone programs for my business needs, but integrated packages are capable of giving good results. My first integrated package did a great job. I still use an integrated program for my database needs. With my writing requirements, I do use a stand-alone, major-market word processor. My spreadsheet applications are done on a very inexpensive clone of a major spreadsheet program.

So, what should you do? If you are new to computer operations and need multiple features, as most businesses do, an integrated software package can serve your needs well. Unless you are willing to spend extensive time learning what you need to know, integrated packages should provide the easiest learning curve. Don't get the wrong idea. Integrated software is not a compromise; it can be an excellent choice.

USING A COMPUTER-AIDED DRAFTING PROGRAM

If you design your own home plans, you need to learn how computer-aided drafting (CAD) programs can change your life. CAD programs allow people without the natural ability to draw freehand to draw professional plans and diagrams. I'm a good example of the type of person who can benefit from a CAD program.

My wife is a fine graphic artist. Kimberley can envision and draw about anything she wants. I, on the other hand, have trouble drawing any detailed, freehand drawing. Sure, I can draft a floor plan, a building elevation, or similar drawings, but without my computer, I'm at a loss. When I bought my first computer drafting program, I had my doubts.

I purchased the software to offset my frustration with freelance artists. You see, I bought the software to illustrate my books. This book doesn't contain the type of illustrations that require a CAD program, but many of my other books do.

My first attempt at CAD was a miserable disappointment, but my further attempts produced excellent results. After I got into computer drafting, I saw a way to improve my contracting business. I gave you an example earlier about how a contractor captures business with a CAD system. This procedure has proved successful for me.

The results you can achieve with a computer drafting program are unlimited. Once you get the hang of putting drawings on the computer, you are likely to enjoy it. You might even find computer drafting to be therapeutic.

SELECTING YOUR HARDWARE

Selecting your hardware is not a playful task. While some people buy computer equipment to satisfy their curiosity or their got-to-have-it impulses, selecting hardware is serious business. What should you look for in computer hardware? Well, let's find out.

Desktop computers

Desktop computers, or PCs, are far and away the most popular computers among business owners. However, desktop computers are not what they used to be — they're better. Desktop units are available in many configurations and speeds. No matter what your needs are, there is a desktop computer available to meet them.

Desktop computers can be purchased for less than $1000, or you can spend upwards of $10,000. How can there be such a disparity in prices? The difference in price is determined by many factors. One factor is name recognition; famous names command more money. Other factors include speed, memory, features, and more. As efficient as desktop computers are, laptop and notebook computers are taking the market by storm.

Laptop computers

Laptop computers are very popular. These mini-powerhouses are capable of the same tasks as many desktop units. However, the compact size and portability of laptops make them especially desirable. For busy people on the move, a laptop can be the ideal computer.

You might think you have to give up features for portability, but you don't. It is very feasible to get as many features in a laptop as you find in the average desktop unit. It is not uncommon to find laptops with huge hard drives and blazing clock speeds. Couple these features with a VGA screen and the freedom of battery-operated power, and you've got a winner. However, notebook computers are seizing much of the market that laptop units once owned.

Notebook computers

Notebook computers resemble laptop computers. Notebooks are typically smaller and lighter than laptops. For this size advantage, you have to sacrifice some features. For example, the keyboard on a notebook computer is smaller than that on a laptop. Smaller keys and screens are about the only noticeable drawbacks to a notebook computer.

I have two desktop computers and a notebook computer. The notebook replaced my previous laptops. The light weight and portability of the notebook is ideal for my needs. In fact, I find myself using the notebook much more often than I do the desktop units.

I don't recommend buying a notebook computer as your only computer, but having a computer that fits neatly into a briefcase with a lot of room left over is nice. You could most likely perform any function you needed to on a notebook, but a desktop model makes a better one-and-only computer.

Computer memory

How much computer memory do you need? Older computers operated with 512 kilobytes of random access memory (RAM). There are still some old computers available with 512K of memory, but this amount of memory is no longer sufficient if you plan to work with modern software programs.

If you don't want to be restricted in software, buy a newer computer with more memory. Computers change so quickly that it's hard to predict what amount of memory will be needed next year. Your best bet is to buy a unit with the most memory that you can get and afford. Eight megabytes of RAM is currently a fairly standard amount of memory in new computers. If you are looking at used computers, you might find many models with much less memory. Buying an older, cheaper computer is no bargain if you can't do the work you want to with it.

If you will be doing a lot of CAD work, you probably want as much memory as possible. When a computer has limited memory, complex drawings cannot load from the files into the computer's memory. You won't be able to get the drawing on your monitor.

Computer speed

Computer speed is another issue to consider. How fast is fast enough? Some applications are painfully slow with older, slower computers. For example, if you are doing CAD work, a slow computer can drive you crazy. In the time it takes a slow computer to redraw your drafting project, you can get a cup of coffee, open your mail, and pace around the computer. If you plan to use CAD programs, get the fastest machine you can afford.

If word processing is your primary use for the computer, speed and memory are not as important. Almost any computer allows you to use word-processing software.

Database work is more enjoyable with a fast computer, but high speed is not essential. Memory requirements depend on the type of database software you buy, but most programs work with minimal memory.

How do you judge the speed of a computer? The first way is by name. You can get an idea of a computer's speed by its number designation. A 486 machine is extremely fast, a 386 machine is very fast, and a 286 machine is fast. A Pentium, which could also be called a 586, is the fastest computer currently available.

To determine the exact speed of individual computers, you also must look at their megahertz (MHz) ratings. The higher the number in the megahertz rating, the faster the machine is. For example, not all 486 machines run at the same speed. You might find one with a speed of 33 MHz and another with a speed of 66 MHz. The machine with the 66-MHz rating is the faster of the two machines.

Based on current technology, you should buy at least a 486 machine. Anything less than a 386 is questionable when using modern software packages.

Hard drives

Hard drives or hard disks are storage devices found inside a computer. Not all computers have a hard disk. Some computers rely on floppy disk drives to operate. Computers without hard drives are basically a thing of the past. Modern computers use floppy disk drives in conjunction with hard drives. I would not recommend buying any type of computer that does not have a hard drive.

Many software programs cannot be used without a hard drive. Due to the size of the files in the software, a floppy disk drive is not capable of running it. Hard drives come in many sizes. Their size determines how much data can be stored. For example, a 20-megabyte (20MB) hard drive holds only half the data that a 40MB drive holds. What size hard disk do you need? As a minimum, you should have a 40MB hard drive. If you are planning to store a high volume of graphic files and CAD files, a much larger hard drive is in order.

My first hard-drive computer had a 20MB hard drive. When I replaced that computer, I got a 40MB hard drive. From there, storage capacity blasted into the hundreds of megabytes. I'm looking at a computer magazine right now that offers a desktop computer with an 850MB hard drive for less than $1300. You might not be able to imagine needing anywhere near that much storage space, but it doesn't take long to fill a hard drive up. You can do most builder-type work with a 40MB hard drive, but the larger hard drives allow you more options.

Floppy disk drives

Floppy disk drives can be found mounted in the housing of a computer or they can be independent units that connect to the computer with a cable. Not long ago, many computers depended on floppy disk drives for their operation. Today, hard drives have taken over. Even when computers have hard drives, they usually have floppy drives, as well. The common floppy disk drives come in two sizes: 5¼ inch and 3½ inch. The most popular size is the 3½-inch floppy disk drive. In fact, finding a 5¼-inch floppy disk drive on a modern computer would be rare, indeed.

CD-ROMs

CD-ROMs have opened up what is known as multimedia. One CD can store a huge amount of information. If you have children, a multimedia system is worth the extra cost just to provide the learning tools possible from such a system. There is no true need for a multimedia system in a building business, but the cost of stepping up to one is not too great, and the benefits are wonderful. If you're into playing games, you can't beat what's available with a multimedia system.

Monitors

CGA

CGA monitors are color monitors. They are at the low end of the price range for color monitors. If you will only be using your computer for short periods of time, a CGA monitor will do fine. If, however, you will be spending hours staring at the screen, you owe it to your eyes to get a better monitor. Also, if you will be doing graphic work, you'll need a more expensive monitor. CGA monitors are not much more common than dual-disk floppy drive computers. CGA monitors are still available in the used market, but I wouldn't recommend buying one. Before you buy any monitor, make sure it is compatible with your computer. Not all monitors work with all computers.

EGA

EGA monitors are a step up from CGA monitors. However, if you want more than a CGA monitor, I recommend moving up to a VGA monitor. EGA monitors, like CGA monitors, are nearly extinct.

VGA or SVGA

VGA monitors are much easier on your eyes than CGA or EGA monitors. If you are doing graphic work, you almost must have at least a VGA monitor. When you spend your whole day in front of the monitor, a VGA is much easier on your eyes. SVGAs (the S stands for super) are even better, and they are commonly sold as part of a modern computer system.

Printers

Laser printers

Laser printers are considered to be the best printers you can buy, and their prices reflect it. Laser printers do turn out beautiful work, and they offer creative options not available with other printers. Prices have come down quite a bit over the last few years, and laser printers are now affordable. However, most contracting businesses can get by without a laser printer.

If you plan to do graphics and CAD work, a laser is well worth the expense, but for letters and reports, a dot-matrix printer is okay, and a inkjet printer is fine. Both are less expensive than lasers.

Dot-matrix printers

Dot-matrix printers can serve most of your needs. These impact printers are fast. When you want to check your job cost, you can run off a report in draft mode before you can walk across the office. If you want your text to be dark and pretty, you can switch to the near-letter-quality (NLQ) mode. In this mode, the printer is about twice as slow as in draft mode.

The quality of the text produced with a dot-matrix printer can't compare with that of a laser or inkjet printer, but it is fine for most business applications. I still have a dot-matrix printer that I use for rough drafts, but all of my finish work is done on a laser printer.

Other printers

There are, of course, other types of printers. There are thermal printers and printers that spray ink on the paper. With a little shopping, you can see all of the available printers. Inkjet printers are extremely popular and quite affordable. You can even by an inkjet color printer for less than $300. Shop around for your computer equipment and see the various types of printers and equipment in action. This is the best way to decide what you want and need.

Modems

Modems are devices that allow computers to talk to each other over the telephone lines. Some modems are internal to the computer, and others, called externals, sit outside the computer and attach with a cable. You can add a modem at any time, and you probably don't need one at first. If you want to get into e-mail, online services, or things of that nature, you need a modem.

Mouse

A mouse, in computer terms, is a device used for drawing on the computer and for clicking on commands. If you will be doing graphic work, you most likely need a mouse. For everyday office work, you should be able to get by without a mouse.

However, most modern software is geared towards the use of a mouse. A mouse is inexpensive and easy to use, so don't rule the device out.

Package deals

Package deals on computers are abundant. You can find them in stores and mail-order catalogs. When you buy a computer system in a package deal, you normally get everything you need all at one time. Let me give you a couple of examples of what I'm talking about.

The dealer that I buy from sends me a new catalog every few months. The catalog always offers some enticing deals. For example, you can get an inexpensive desktop (PC) system for under $1300. What do you get for your money? It starts with 8MB of RAM, an 850MB hard drive, 100MHz speed, an SVGA monitor, a keyboard, a mouse, loads of do-it-all software, and much more. The only thing left to buy is a modem, fax, and printer.

Maybe you would prefer a notebook computer package. Well, my catalog lists them at prices less than $1800. You get a color VGA screen, 4MB of RAM expandable up to 36MB, built-in stereo speakers and microphone, a 340MB hard drive, a speed of 75MHz, a lot of software, and other neat stuff. If you want to add a fax-modem with voicemail, the cost is $145. A built-in pointing device is included, as is a three-year warranty.

CHOOSING THE BEST SOFTWARE FOR YOUR BUSINESS

Choosing the best software for your business can get confusing. There are hundreds of types of software to choose from. Many of the various types of software perform the same functions with only slight differences. The costs for these software packages can range from less than $20 to well over $700. So how will you ever decide on the right software?

Well, finding software that does what you want it to do isn't too hard. But finding software that does what you want it to do, in a way that you want it to do it, can be more difficult. When it is time to buy software, you need to do some research.

The first step in acquiring the proper software is knowing what tasks you want the program to handle. If you know you want a word processor that has a built-in spell-checker, you know part of what to ask for. However, the list of software packages that have an on-board spell-checker is very long. So you have to narrow your search parameters.

One of the best ways to choose your software is in person. By going to a store that lets you work with the software, you can get a feel for the features offered in the package. You can also assess the difficulty of the learning curve for the software.

Some software sellers offer demo disks. These demo disks allow you to work with the software on your own computer at your own speed. Of course, the demo disks do not have the full features of the program, but you can gain enough insight to make a buying decision.

Don't buy software until you are sure you want it. Most computer software is not returnable unless it is defective. Since computer disks can be copied, the disks cannot be opened and returned.

If you are new to computers, a good integrated-software package is probably the best place to start. By purchasing a do-it-all package, you can experiment and see what type of work you do the most. If you outgrow the integrated package, you can step up to a stand-alone program.

Why do you need a computer? Well, you probably don't *need* a computer, but if you have a computer, you have many advantages. A computer allows you to save time. Your work is less tedious once you learn to use a computer. Chances are good that your business will be better organized and more efficient with a computer. There are lots of reasons to buy a computer and very few reasons not to.

Computers can give you an edge in marketing your services. Once your computer system is online, you can perform marketing ploys you have never done before. With the automated features offered by your computer, you can experiment and find new ways of getting business. Creating mailing lists is simple. Designing your advertising can be fun. Tracking past successes (FIG. 17-3) and failures allows you to refine your marketing plan. In general, computers are a great aid in building your business.

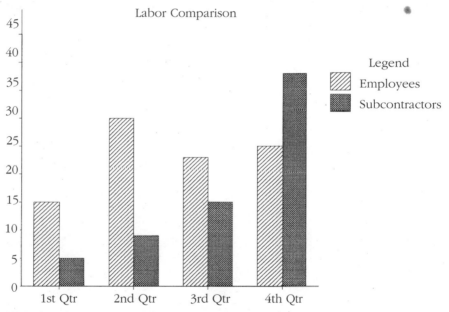

17-3 Comparison chart created on computer.

18

Keeping track of your cash

Keeping track of your cash might be more difficult than you think. This is especially true if you're a builder who also offers remodeling and repair services, as many builders do. Extending credit to customers is risky business, but it is often a necessary evil when acting as a contractor.

As a pure home builder, most of your accounts will be paid on time. Construction loans and draw disbursements make this possible. However, not all customers rely on construction loans. Although they are few and far between, some people pay with their own savings to have homes built. As soon as you let people run up a tab with your business, you're at risk. You can, however, employ many methods to keep your business out of red ink.

KEEPING YOUR CREDIT ACCOUNTS UNDER CONTROL

Keeping your credit accounts under control is very important. If you don't learn to manage your money, the amount of money you make will have little bearing on your success. It is not a matter of how much money you make, but how much money you retain.

Credit accounts can be the downfall of any business. If a business gets behind in its credit accounts, the business is likely to spiral downward. First come late notices. Then come threats of turning the account over for collection. If the situation is not rectified, lawsuits and judgments follow. In the end, the business will be penniless and stuck with a bad credit rating. If you think starting a new business is a challenge, you'll be staggered by the difficulty in bringing a wounded business up out of the ground.

Keep your accounts current (FIG. 18-1). If you are unable to pay your bills, talk with your creditors. Credit managers are not ogres, but they do have a job to do. If you are honest and open with your suppliers, most of them will work with you. If you try to ignore the problem, it will only get worse. A healthy business needs a good credit history. If you have good credit, cherish it; if you don't, work to get a good rating.

PROTECTING YOUR CASH FLOW

Cash flow is paramount to any business's success. Paper profits are nice, but they don't pay the bills. Have you ever heard about a person who is land-rich and cash-poor? Well, it's true; I've had tremendous financial statements and nearly no cash.

Accounts Payable

Vendor	Job	Amount Due	Date Due	Date Paid

18-1 Accounts payable form used to keep track of money owed.

Having a business with a high net worth is not worth much if you don't have enough money to pay the bills. Cash flow is very important to a healthy business.

I've seen a large number of businesses forced to the brink of bankruptcy even though they had significant assets. These businesses held valuable assets but couldn't convert the assets into cash. If a business becomes cash-poor, it is handicapped. A business without cash is like an army without ammunition. The cash might be on the way, but if the enemy attacks before it arrives, the business cannot defend itself. Having tanks behind you doesn't do any good if they are inoperable. Regardless of your assets and business strength, if you don't have cash, you are in trouble. As the old saying goes, "You've got to pay to play."

One of the biggest traps you can avoid is staying away from bad jobs. Jobs that result in slow pay or no pay can be your undoing. Don't get greedy; greed is a major contributor to business failure. Most business owners are anxious to reach their goals quickly. It is better to take a slow approach and reach your goals than it is to run full out only to fail.

LOOKING AHEAD TO FINANCIAL OBSTACLES

If you are looking ahead to financial obstacles, you are on the right track. Even the best business encounters setbacks. If you are prepared for these drops in business, you have a better chance of surviving them. Business owners that learn to project the future are the ones that are around in the future.

What should you look out for? You must anticipate cycles in your business. The economy historically cycles in upward and downward swings. If you can't cope with these changes, your business won't be around for the next change. Some unfortunate businesses get caught in the middle of a cycle. These businesses don't usually last to see the good times.

Sometimes I seem to have terrible timing. At one time, I had a building business that was building 60 homes a year. I enjoyed about 18 months of a very prosperous business, but then the tax laws changed. What did tax laws have to do with my building business? About half of the homes I was building were for partnerships. These homes were supposed to be tax shelters. I had devised a plan, with the help of tax attorneys and accountants, that was nearly perfect. The program was working without a flaw until the tax laws changed. When the tax advantages were removed from passive-income investments, my business suffered a great deal. This is only one example of how the economy has affected my businesses. I went through the recession of the early 1980s and I've done okay in the economic downturn of the early '90s. I'm a survivor, and you can be too.

How can you overcome severe financial obstacles? With good planning and forecasting, you have a better-than-average chance of seeing light at the other end of the tunnel. By researching historical data, you can begin to see an economic pattern forming. The economy moves up and down; these cycles can be identified and charted. While there is no guarantee the cycles will repeat themselves, history indicates they will.

Performing future projections can be fun, but don't get too infatuated with them; they might not come true. However, successful businesses seem to share the traits of effective projections. You must look ahead and plan. Having backup plans can make the difference between survival and failure.

Without money management, you won't have a business for very long. Money management is crucial to the successful operation of your business. If you are not experienced in money management, take some business courses at a local community college, read more business books, go to seminars, and do what it takes to get yourself educated in the intricate skill of managing money.

GETTING CONTRACT DEPOSITS

Contract deposits strengthen cash flow. Getting money from customers before starting jobs is not always easy, but it does ease the cash-flow burden. It is not unusual for contractors serving homeowners to receive cash deposits. A typical scenario finds the contractor getting one-third of the contract amount when the contract is signed, one-third of the contract amount at the halfway point of the job, and the balance upon completion. This, of course, is not the case with builders who are constructing new homes. Any deposit given on a new home is likely to worth less than 10 percent of the contract amount. In fact, it is not unusual for home builders to begin work without any significant deposit.

When advance deposits are given, such as in the case of remodeling jobs, contractors are working with money provided by the customers. Without deposits, contractors must work with their own money and credit, which can put a serious strain on a builder's bank account.

As consumers become more aware of contracting rip-offs, they become more wary of giving deposits. The public is very aware of the risks involved with giving money to someone before the work is done. This trend in eliminating deposits puts more stress on contractors. When contractors receive deposits, they have money to work with. If the contractors don't get deposits, they are more exposed and must use their credit lines and cash to maintain their businesses. The refusal of deposit payments by consumers seems to be a fact of life. With this in mind, make arrangements to work with your own money and credit. In addition, you have to be more selective in your customers. If you are footing the bill up front, you have to be sure the customer will pay you. If you don't get deposits, you are at risk.

Cash deposits (FIG. 18-2) enable you to do more business. However, it is important that you use the deposit money for the job the deposit was made on. All too often, contractors use deposit money to make truck payments, rent payments, or even to buy materials for other jobs. These situations are potentially explosive. You should never take one person's deposit and use it on someone else's job. You also should never use deposit money for ordinary operating expenses. If you get into this habit, you might be out of business in less than a year.

ELIMINATING SUBCONTRACTOR DEPOSITS

Eliminating subcontractor deposits is another way to manage your money wisely. Just as homeowners are reluctant to give deposits to contractors, you should be selective in giving deposits to subcontractors. Once a sub has your deposit money, you can have a hard time getting it back if the subcontractor doesn't perform as expected.

When subcontractors are dependent on your deposit to do jobs, they might be in financial trouble. If you give into these deposit requests, your money is at risk. Make it a rule to never give anyone money they haven't earned.

Cash Receipt

Date _____

Time _____

Received of _____

Address _____

Account number _____

Amount received _____

Payment for _____

Form of payment _____

Signed _____

18-2 Cash receipt used to document money received.

STRETCHING YOUR MONEY

If you learn to stretch your money, you can do more business. The more business you do, the more money you make, hopefully. How can you stretch your money? Well, there are many ways to make your dollars go further. Let's see how you can increase the power of your cash.

- Rent expensive tools until you know you need to buy them.
- Don't give subcontractors advance deposits.
- Collect job deposits from homeowners whenever possible.
- Make the best use of your time.
- Forecast financial budgets and stick to them.
- Buy in bulk whenever feasible.
- Pay your supply bills early and take the discount.
- Put your operating capital in an interest-earning account.
- File extensions and pay your taxes late in the year.
- Don't overstock on inventory items.
- Keep employees working, not talking.
- Consider leasing big-ticket items.

These are only a few of the ways you can manage your money for more potency. If you study your business, you can find other ways to maximize you money.

EXTENDING CREDIT TO CUSTOMERS

Should you extend credit to your customers? Not unless you have to. Although it is common for service companies to allow customers 30 days to pay their bills, this policy crunches your cash flow. Further, if you don't collect what's due you on the spot, you might never see your hard-earned money. People seem to have little problem in neglecting to pay for services they receive and put on credit accounts. However, if you make an attempt to collect as soon as the job is done, most people will pay you.

You might find that you must allow credit purchases to get business. But remember, working and supplying materials that you don't get paid for is worse than not working. If you allow credit purchases, be selective. Ideally, you should run credit checks on all customers who wish to establish a credit account. After all, do banks loan you money without checking your credit history?

COLLECTING PAST-DUE ACCOUNTS

Collecting past-due accounts is a job few people enjoy. As a business owner, you no doubt will have occasions to collect old money (FIG. 18-3). This part of your job can get very frustrating. When you call to collect an old debt, you are likely to hear some very creative excuses.

People who are behind in paying their bills often lie. When you are listening to excuses for why you can't get the money owed to you, don't get soft-hearted. Remember, if customers don't pay you, you cannot pay your bills.

If you start to acquire a long list of accounts receivable, be advised, you are heading for deep trouble. Never let debts get too old. As soon as a customer is in default, take action. Consult a local attorney for your options in collecting bad debts.

USING CONSTRUCTION LOANS

Constructions loans can be very different in their payment structure. I've worked with loans on which I could submit invoices each week and receive checks. Some lenders limit the disbursement process to once a month, which still isn't bad. Many construction loans pay out a percentage of the loan amount based on what work has been completed. Some loans are set up on an even payout basis, such as one-third when a house is under roof, one-third when drywall has been hung and taped, and one-third upon completion.

As a builder, you have to manage your money carefully. Part of doing this is knowing how and when you'll be paid. I should clarify that we are not talking only about your profit payments. Many contractors pay their subcontractors out of their own funds and then get reimbursed by construction loans. This is fine as long as everything is planned and goes as planned. But, suppose you pay out $15,000 to subs and then find out that a draw disbursement is being held up due to some type of inspection? How do you replenish your bank account? It's easy to get into financial trouble as a builder.

Keeping tabs on your money, such as petty cash (FIG. 18-4), and the money that is due you can be a time-consuming process. If you're not good with budgets, schedules, and money, you need to give serious thought to hiring someone who is. Money is the oil that keeps a building machine going. Without cash, your business will crash. When you hang out your sign as a builder, you must have your financial ducks in a row.

Accounts Receivable

Date	Account Description	Amount Due	Date Due	Date Received
	Total Due			

18-3 Accounts receivable form used to keep track of money owed to you.

Petty-Cash Record

Month _____

Year _____

Vendor	Amount	Item	Date	Job

18-4 Form used to maintain records on petty-cash expenses.

Taxes, accounting, legal considerations, and the paperwork that goes along with them are all serious aspects of any business. These areas of business confuse and intimidate many business owners. Some business owners try to take an out-of-sight-out-of-mind attitude towards these less-than-desirable business responsibilities. However, this approach doesn't work, at least not for long.

If you don't tend to your tax requirements, you can regret it for a long time to come. The same can be true about the legal aspects of running your business. A poorly written contract can cause months of frustration, not to mention lost money.

Professional help is not inexpensive, but it is better to pay an attorney to draft good legal documents than it is to pay to have improper documents defended in court. The same can be said about accountants. A good certified public accountant (CPA) can save you money, even after the professional fees you are charged.

Using and maintaining the proper paperwork can make all aspects of your business better. But when it comes to taxes and legal issues, organized paperwork is invaluable. If you ever have to go to court, you'll learn the value of well-documented notes and agreements. An IRS audit will prove the importance of keeping good records. Now, let's examine these issues and how they might affect your business.

UNDERSTANDING YOUR TAXES

Taxes are a fact in any business. The tax laws are complex and can be confusing. For the average business owner, many tax advantages go unnoticed. These business owners know their businesses, but they don't know all they should about income taxes.

Unless you are a tax expert, you should consult someone who is. If you broke your leg, would you attempt to set the bone and apply a cast? Would you ever consider filling a cavity in your own tooth? These are medical questions, but the same questions should be asked of your tax-filing habits. If you prepare your own taxes, you might be giving the government much more money than you are obligated to. Tax experts can save you money on the taxes you pay. When you talk with a CPA, you are talking to an expert. These professionals know the tax requirements like you know your business.

Do you ever get upset at do-it-yourselfers who try to handle their own home needs and wind up calling you after they have made a mess of the job? Well, if you don't consult a tax expert, you might be putting yourself in the same position as the do-it-yourselfer who calls you for help. If you have been in business, you know that wading into the middle of a job that has been attempted by an amateur is worse than taking on the job from the start. The same principle can be applied to your taxes.

If you wait until a month before tax time to meet with a tax specialist, there might not be much the expert can do for you, short of filing your return. However, if you consult with tax experts early in the year, you can manage your business to minimize the tax bite. Early consultations can result in significant savings for you and your business.

If you make money, you have to pay taxes, but you can reduce the amount of money you pay in taxes. How can you reduce your tax spending? A logical step is to meet with tax experts early and allow them to set a path for you to follow. Accountants do more than bookkeeping. A good CPA can provide business planning that results in lower tax consequences.

What type of planning can reduce your tax liabilities? Well, I'm not a tax expert, but I have found many ways, with the help of tax professionals, to reduce the money I spend in income taxes. If you take the time to meet with tax experts, you can find the best ways to run your business for maximum tax advantages.

Tax manipulation is an art, and CPAs are the artists. When you meet with these professionals, you might find numerous ways to save on your taxes. You might be told to lease vehicles instead of buying them. You might learn to keep a mileage log for your vehicle to maximize your tax deductions. A CPA might recommend a different type of structure for your business. For example, if you are operating as a standard corporation, the accountant might suggest that you switch to a subchapter S corporation to avoid double taxation.

If you work from home, your tax expert can show you how to deduct the area of your home that is used solely for business. A tax specialist can show you how to defer your tax payments to a time when your tax rate may be lower. Investment strategies can be planned to make the most of your investment dollars. You can coordinate a viable retirement plan by talking with an authority on tax issues.

Perhaps one of the most important aspects a tax specialist can educate you on is what is not a legitimate deduction. I have seen numerous businesses in which the business owners were required to pay back taxes. Most of these business owners had no idea they owed taxes until they got a notice to pay them. These people find that deductions that they thought were correct are not allowable. It can be a major burden catching up on taxes that were due last year or years before. This type of expenditure is never planned for, and it can drive a business into financial hardship.

The best way to avoid unexpected tax bills is to make sure your taxes are filed and paid properly. The most effective way to ensure that your taxes are done properly is to hire a professional to do them for you. If you resent paying someone to do a job that you think you can do, you are not alone, but you might find that by paying a little now, you save a lot later.

SURVIVING AN IRS AUDIT

If you are concerned about surviving an IRS audit, don't worry—I did it, and you can do it. For years, an audit was one of my biggest fears. Even though I knew, or at least thought I knew, that my tax filings were in order, I worried about the day I would be audited. It was a lot like dreading the semiannual trip to the dentist. Then one day, it happened; I was notified that my tax records were going to be audited. My worst nightmare was becoming a reality.

Before the audit, I scrambled to gather old records and went over my tax returns for anything that might have been in error. I couldn't find any obvious problems. I went to my CPA and had him go over the information I would present in the audit. There were no blazing red flags to call attention to my tax return.

When I called the auditor, I was told that my return had been chosen at random. The person went on to say that there probably was nothing wrong with my return and that a percentage of returns are picked each year for audit. Gaining this information made me feel a little better, but not a lot.

On the day of the audit, my CPA represented me. I was not required to attend the meeting. Even though I wasn't at the meeting, my mental state was miserable that day. When my CPA called, he told me the meeting had gone well, but that I had to provide further documentation for some of my deductions. These deductions were primarily travel expenses and books.

After digging through my records, I found most of the needed documentation. However, some receipts I couldn't document. For example, I had written on receipts that they were for book purchases, but I hadn't listed the title of the book. The IRS wanted to know what books I had purchased. I couldn't remember what the titles were, so I put notes on the receipts to explain that I could not document the book titles.

The travel expenses were easier to document since most of them had been paid for with credit cards. By finding the old receipts, I was able to remember where I had gone and why. After finding as much documentation as I could, I gave the package back to my CPA.

The auditor and my accountant had another meeting. I was expecting the worst, and thought I would have to pay back taxes on the items I couldn't identify properly. But to my surprise, I was given a clean bill of health. The auditor accepted my I-can't-remember receipts, since most of my receipts were documented and I hadn't written fictitious names on the receipts.

I always thought an audit would be horrible. However, I didn't have to attend the audit personally, my CPA did a great job, and I didn't have to pay any serious tax penalties. My audit was over without much pain. I can't say that all audits are this easy, but mine was.

Going through that audit convinced me of what I had believed for years; accurate records are the key to staying out of tax trouble. If my records had been misplaced or substantially incomplete, I might have been in a serious bind. However, my good business principles enabled me to survive the audit with relative ease.

If you are afraid of being audited, keep detailed records. Before taking a questionable deduction, check with a tax expert to confirm the viability of the deduction. Don't cheat on your taxes. Not only is it wrong, you never know when you will be caught. Honesty is the best policy.

HANDLING THE LEGAL SIDE

The legal side of being in business can be perplexing. There are so many laws to abide by that the average person has trouble keeping up with them. Some laws require that certain posters be displayed for employees. Contract law is an issue any contractor has to deal with. Laws pertaining to legal collection procedures for past-due accounts can affect your business. Discrimination laws come into play when you deal with employees. The list of laws that might affect your business could fill a small library. Most of them can affect how much money you make or keep.

It is your responsibility as a business owner to comply with all laws. Ignorance of the law is not a suitable defense. The penalties for breaching some of these laws are extensive and might involve imprisonment and cash fines.

Business law is a vast subject, much too broad to cover in a single chapter. If you are not well versed in business law, spend some time studying the topic. You might find it necessary to go to seminars or college classes to gain the knowledge you need. Books that specialize on business law might be all the help you need. If you have specific questions, you can always confer with an attorney who concentrates on business law.

CHOOSING AN ATTORNEY AND AN ACCOUNTANT

Choosing an attorney and an accountant is not always easy. Finding a professional with the experience, knowledge, and skill that you require can take some time. These two professional fields incorporate an enormous amount of facts and requirements under two simple names. An individual attorney cannot possibly be fluent in all the laws. Accountants cannot be expected to know every aspect of financial law and practice. For these reasons, you must look for professionals who specialize in the type of service you require.

Finding a specialist is easy. Most professionals list their specialties in the advertising they do. For example, an attorney who specializes in criminal law often makes this point clear in advertising. CPAs who concentrate on corporate accounts direct their advertising to this form of service. However, finding professionals who advertise certain specialties is only the beginning.

Once you narrow the field to professionals working within the realms of your needs, you must further separate the crowd. Finding 10 attorneys who work with business law is not enough. You must look through those 10 lawyers and find the one that is best suited to work with your business.

How will you know when you've found the right professional? You can begin your process of elimination with technical considerations. Make a list of what your known and expected needs are. Ask the selected professionals how they can help you meet these needs. Inquire about the past performances and clients of the professionals. As you go down your list of questions, you can begin weeding out the crowd.

Once you have covered all the technical questions, ask yourself some questions. How do you feel about the individual professional? Would you be comfortable going into a tax audit with this CPA? Are you willing to bet your business on the knowledge and courtroom prowess of this attorney? Are you comfortable talking with the individual?

How you feel toward the professional as a person is important. You will very likely expose your deepest business secrets to your accountant and your attorney. If you are not comfortable with the professional you are working with, you won't get the most out of your business relationship.

Another key factor to assess in choosing professionals is the ease with which you can understand them. The subject matter you are discussing is complex and possibly foreign to you; if it wasn't, you wouldn't need the professionals. You need professionals who can decipher the cryptic information that confuses you and present it in an easy-to-understand manner.

DOCUMENTING YOUR BUSINESS ACTIVITY

Documenting your business activity is an absolute must. Good documentation is essential to running a successful business. The documentation can be used to track sales (FIG. 18-5), to keep up with changes in the market, to forecast the future, to defend yourself in court, or to substantiate your tax filing, just to name a few uses.

There are many good reasons and ways for documenting your business activity. A carbon-copy phone-message book is a simple, yet effective, way to log all of your phone activity. Written contracts are instrumental in documenting your job duties and payment arrangements. Change orders and addendums are the best way to document your actions when the actions cause you to deviate from the original agreement. Letters can be used to confirm phone conversations, creating a paper record and avoiding confusion. Tape recorders can help you remember your daily duties. The ways and reasons for documenting your business activity are many.

Contracts, change orders, and related essential paperwork help keep your business on the right track. While most people abhor doing paperwork, it is recognized as a necessary part of doing business. Computers have helped reduce the amount of paper used and have made the task of doing paperwork easier, but even computers cannot eliminate the need for accurate paperwork. To understand the need for so much clerical work, let's take a closer look at the various needs of business owners.

Contracts

Contracts (FIGS. 18-6 and 18-7) are a way of life for contractors. The contracts that a contractor enters into are the lifeblood of the business. There are two types of contracts—oral and written. Oral contracts are legal, but they are essentially unenforceable. Written contracts are the other choice, and these agreements are enforceable.

Cash Receipts

Date	Account Description	Amount Paid	Date Received

18-5 Form used to track all cash receipts.

Your Company Name
Your Company Address
Your Company Phone Number

PROPOSAL

Date: _____

Customer name: _____

Address: _____

Phone number: _____

Job location: _____

Description of Work

Your Company Name will supply, and/or coordinate, all labor and material for the above referenced job as follows:

_____

_____

_____

Payment Schedule

Price: _____($_____),

Payments to be made as follows:

_____

_____

_____

All payments shall be made in full, upon presentation of each completed invoice. If payment is not made according to the terms above, Your Company Name will have the following rights and remedies. Your Company Name may charge a monthly service charge of one-and-one-half percent (1.5%), eighteen percent (18%) per year, from the first day default is made. Your Company Name may lien the property where the work has been done. Your Company Name may use all legal methods in the collection of monies owed to it. Your Company Name may seek compensation, at the rate of $_____ per hour, for attempts made to collect unpaid monies.

Page 1 of 2 initials _____

18-6 Proposal form.

Your Company Name may seek payment for legal fees and other costs of collection, to the full extent the law allows.

If the job is not ready for the service or materials requested, as scheduled, and the delay is not due to Your Company Name's actions, Your Company Name may charge the customer for lost time. This charge will be at a rate of $_____ per hour, per man, including travel time.

If you have any questions or don't understand this proposal, seek professional advice. Upon acceptance, this proposal becomes a binding contract between both parties.

Respectfully submitted,

Your name and title
Owner

Acceptance

We the undersigned do hereby agree to, and accept, all the terms and conditions of this proposal. We fully understand the terms and conditions, and hereby consent to enter into this contract.

Your Company Name   Customer

By _____   _____

Title _____   Date _____

Date _____

Proposal expires in 30 days, if not accepted by all parties.

18-6 Continued.

A written contract should be used for every job except small repairs. Even with repairs, it is best to use a written service order or some type of agreement. Contracts help protect you. They provide physical evidence of the understanding between you and the customer.

Addendums

Addendums (FIG. 18-8) are extensions of a contract. Addendums are used to add language to contracts after the contracts are written. In cases in which fill-in-the-blank contracts are used, addendums provide a means for making the contracts more explicit.

Change orders

Change orders are written agreements that are used when a change is made in a previous contract. For example, if you were contracted to paint a house white and the

Renaissance Remodeling
357 Paris Lane
Wilton, Ohio 55555
(102) 555-5555

REMODELING CONTRACT

This agreement, made this _____th day of _____, 19____, shall set forth the whole agreement, in its entirety, between Contractor and Customer.

Contractor: Renaissance Remodeling, referred to herein as Contractor.

Customer: _____, referred to herein as Customer.

Job name: _____

Job location: _____

The Customer and Contractor agree to the following:

Scope of Work

Contractor shall perform all work as described below and provide all material to complete the work described below. All work is to be completed by Contractor in accordance with the attached plans and specifications. All material is to be supplied by Contractor in accordance with attached plans and specifications. Said attached plans and specifications have been acknowledged and signed by Contractor and Customer.

A brief outline of the work is as follows, and all work referenced in the attached plans and specifications will be completed to the Customer's reasonable satisfaction. The following is only a basic outline of the overall work to be performed:

_____

_____

_____

_____

_____

_____

_____

_____

_____

_____

(Page 1 of 3 initials _____)

18-7 Remodeling contract.

Commencement and Completion Schedule

The work described above shall be started within three days of verbal notice from Customer; the projected start date is _____. The Contractor shall complete the above work in a professional and expedient manner, by no later than _____ days from the start date. Time is of the essence regarding this contract. No extension of time will be valid, without the Customer's written consent. If Contractor does not complete the work in the time allowed, and if the lack of completion is not caused by the Customer, the Contractor will be charged _____, per day, for every day work is not finished beyond the completion date. This charge will be deducted from any payments due to the Contractor for work performed.

Contract Sum

The Customer shall pay the Contractor for the performance of completed work, subject to additions and deductions, as authorized by this agreement or attached addendum. The contract sum is _____, ($_____).

Progress Payments

The Customer shall pay the Contractor installments as detailed below, once an acceptable insurance certificate has been filed by the Contractor, with the Customer:

Customer will pay Contractor a deposit of _____,
($_____), when work is started.
Customer will pay _____,
($_____), when all rough-in work is complete.
Customer will pay _____,
($_____), when work is _____ percent complete.
Customer will pay _____,
($_____), when all work is complete and accepted.

All payments are subject to a site inspection and approval of work by the Customer. Before final payment, the Contractor, if required, shall submit satisfactory evidence to the Customer, that all expenses related to this work have been paid and no lien risk exists on the subject property.

Working Conditions

Working hours will be ___ A.M. through ___ P.M., Monday through Friday. Contractor is required to clean work debris from the job site on a daily basis and to leave the site in a clean and neat condition. Contractor shall be responsible for removal and disposal of all debris related to their job description.

(Page 2 of 3 initials _____)

18-7 Continued.

Contract Assignment

Contractor shall not assign this contract or further subcontract the whole of this subcontract without the written consent of the Customer.

Laws, Permits, Fees, and Notices

Contractor is responsible for all required laws, permits, fees, or notices required to perform the work stated herein.

Work of Others

Contractor shall be responsible for any damage caused to existing conditions. This shall include work performed on the project by other contractors. If the Contractor damages existing conditions or work performed by other contractors, said Contractor shall be responsible for the repair of said damages. These repairs may be made by the Contractor responsible for the damages or another contractor, at the sole discretion of Customer.

The damaging Contractor shall have the opportunity to quote a price for the repairs. The Customer is under no obligation to engage the damaging Contractor to make the repairs. If a different contractor repairs the damage, the Contractor causing the damage may be back-charged for the cost of the repairs. These charges may be deducted from any monies owed to the damaging Contractor.

If no money is owed to the damaging Contractor, said Contractor shall pay the invoiced amount within _____ business days. If prompt payment is not made, the Customer may exercise all legal means to collect the requested monies. The damaging Contractor shall have no rights to lien the Customer's property for money retained to cover the repair of damages caused by the Contractor. The Customer may have the repairs made to his satisfaction.

Warranty

Contractor warrants to the Customer all work and materials, for one year from the final day of work performed.

Indemnification

To the fullest extent allowed by law, the Contractor shall indemnify and hold harmless the Customer and all of their agents and employees from and against all claims, damages, losses and expenses.

This Agreement entered into on _____, 19_____ shall constitute the whole agreement between Customer and Contractor.

_____   _____
Customer   Date   Contractor   Date

_____
Customer   Date

18-7 Continued.

Addendum

This addendum is an integral part of the contract dated _____, between the
Contractor, _____, and the Customer(s),
_____, for the work being done on real estate
commonly known as _____. The undersigned parties hereby agree to
the following:

_____

_____

_____

_____

_____

_____

_____

_____

_____

_____

_____

_____

_____

_____

The above constitutes the only additions to the above-mentioned contract. No verbal agreements
or other changes shall be valid unless made in writing and signed by all parties.

_____   _____
Contractor   Date   Customer   Date

                                                  _____
                                                  Customer   Date

18-8 Addendum.

customer asked you to change the color to beige, you should use a change order. If you paint the house beige without a change order, you could be found in breach of your contract, since the contract called for the house to be painted white. It might seem unlikely that the customer would sue you for doing what you were told, but it is possible. Without a written change order, you would be at the mercy of the court. If the court accepted the contract as proof that the house was to be painted white, and it probably would, you would be in trouble.

Service orders

Most companies that provide routine maintenance and repair services use service orders (FIG. 18-9). Service orders are the small tickets that customers are asked to sign, generally after the work is done, to acknowledge that the work is satisfactory. This is fine, except for the fact that most customers are not asked to sign the service order until the work is complete. This practice puts the business owner at risk. The customer should be asked to sign the service order before the work is started and again when the work is completed.

Liability waivers

Liability waivers are written releases of liability. These forms protect the contractor from being accused of an act that was nearly unavoidable. You should use liability waivers any time that you believe there may be a confrontation arising from your actions.

Written estimates

Written estimates and quotes (FIGS. 18-10 and 18-11) reduce the risk of confusion when you are giving prices to customers. They also present a more professional

Service Call

Date _____   Time _____

Customer name _____

Address _____

Type of service requested _____

Call taken by _____

Call assigned to _____

Service promised by _____

18-9 Service call ticket.

Green Tree Lawn Care
987 Willow Road
Wilson, Maine 55555
(101) 555-5555

WORK ESTIMATE

Date: _____

Customer name: _____

Address: _____

Phone number: _____

Description of Work

Green Tree Lawn Care will supply all labor and material for the following work:

_____

_____

_____

_____

_____

_____

_____

_____

Payment for Work as Follows

Estimated price: _____, payable as follows:

_____

_____

If you have any questions, please don't hesitate to call. Upon acceptance, a formal contract will be issued.

Respectfully submitted,

J. B. Williams
Owner

18-10 Work estimate.

Your Company Name
Your Company Address
Your Company Phone Number

Quote

This agreement, made this _____ day of _____ , 19_____, shall set forth the whole agreement, in its entirety, by and between Your Company Name, herein called Contractor and_____, herein called Owners.

Job name: _____

Job location: _____

The Contractor and Owners agree to the following:

Contractor shall perform all work as described below and provide all material to complete the work described below. Contractor shall supply all labor and material to complete the work according to the attached plans and specifications. The work shall include the following:

_____

_____

_____

Schedule

The work described above shall begin within three days of notice from Owner, with an estimated start date of _____. The Contractor shall complete the above work in a professional and expedient manner within ____ days from the start date.

Payment Schedule

Payments shall be made as follows:

_____

_____

This agreement, entered into on _____, shall constitute the whole between Contractor and Owner.

_____ ____	_____ ____
Contractor Date	Owner Date
	_____ ____
	Owner Date

18-11 Quote.

image than an oral estimate. Giving quotes verbally can lead to many problems. To eliminate these problems, make all of your estimates in writing, and keep copies of the estimates.

Specifications

Written specifications are another way to avoid confrontations with customers. With small jobs, the specifications can be included in the contract. When you are embarking on a large job, the specifications generally are too expansive to put in the contract. In these instances, make reference to the specifications in the contract and attach the specification sheets to the contract, making them a part of the contract.

Credit applications

Credit applications should be filled out by all customers to whom you plan to extend credit. These forms allow you to check into the individual's past credit history. Just because a person has good references now doesn't mean you are guaranteed of being paid, but your odds for collecting the money owed are better. There are many other possibilities for forms and paperwork to protect you. Let's take a look at some of the other ways you can use paper as an effective shield.

PROTECTING YOURSELF AND YOUR MONEY

Learn to protect yourself and your money with paperwork. Well-documented paperwork can be the business owner's best friend. To gain a higher appreciation for paperwork, you only have to need it once — when you don't have it. Let me explain further how paperwork can protect you and your assets.

Contracts

You already have a broad understanding of why contracts are needed. Now I am going to look more closely at how contracts can protect you. Contracts give a full description of the work that you are being engaged to perform. When written properly, contracts leave little room for misunderstandings. Working a job without running into confused confrontations with the customer is more pleasurable and more profitable.

Good contracts include the date work is scheduled to start and might even include the hours that may be worked. By inserting a starting date for the job in the contract, both you and the customer or your subcontractors know when to expect the work to begin. This clause keeps the customer from calling and hounding you about when the job will start. By having a written start date, you can organize your schedule and prepare for the job. Some contractors need this committed discipline to run their businesses effectively.

Contracts address how much you will be paid and in what increments you will receive your money. If you are to get a deposit before starting the job, the date the deposit is due and the amount of the deposit is stipulated in the agreement. Money is a major cause of job-related problems. The more documentation you have on the financial aspects of the job, the better off you will be.

Most contracts go on to cover a wide variety of other variables. These variable might include who is responsible for cleaning up after the job, what happens with the debris from the job, how long the job is guaranteed for, and so on. Generally speaking, a written contract is the foundation for your business.

Addendums

Addendums come in handy for significant changes in the job. Change orders can be used for minor revisions to an existing contract, but addendums should be used if the changes are substantial. By documenting all agreements between yourself, your subcontractors, and your customers, you eliminate confusion and many of the risks of confrontations.

Change orders

Most contractors have change-order forms (FIG. 18-12), but a large number of these business owners fail to use their change orders. Many changes requested by customers seem insignificant until a problem pops up; then you'll wish you had insisted on a written change order.

Change orders give you written documentation of the customers' decisions to alter original plans. Without using change orders, even for minor deviations, you are putting yourself at risk.

Change orders should include some basic information. All change orders should include the name and address of the customer. Additionally, the change order should identify the location of the job and the date and reference number of the original contract.

A description for the work being altered should be included in the body of the change order. This description should be specific. For example, don't write a change order that says the kitchen sink is changed from a single-bowl sink to a double-bowl sink. For this type of a change, include all pertinent data on the new sink. Information such as the make, model number, color, style, and whatever else is applicable should be included.

Always provide documentation on how the change order affects the cost of the job. If the change results in a credit to the customer, put the amount of the credit in the paperwork. If the change means you need to charge more money for the job, detail the extra charges in writing. Dictate how the credits or extra charges will be accounted for and when they will be dealt with.

Require all parties that signed the original contract to sign the change order. I have seen many contractors write up a change order and give it to the customer without ever receiving a signed copy for their files. I have also seen change orders signed by only one of the parties of the original contract. Both of these practices are bad business. Treat change orders with the same respect you would a contract, and keep signed copies on file.

Service orders

Service orders can serve many functions. They are most often used to document the acceptance of work by the customer. However, service orders can do much more.

Request for Substitutions

Customer name: _____

Customer address: _____

Customer city/state/zip: _____

Customer phone number: _____

Job location: _____

Plans & specifications dated: _____

Bid requested from: _____

Type of work: _____

The following items are being substituted for the items specified in the attached plans and specifications:

_____

_____

_____

_____

_____

_____

_____

_____

_____

_____

_____

Please indicate your acceptance of these substitutions by signing below.

_____   _____
Contractor   Date   Customer   Date

                                           _____
                                           Customer   Date

18-12 Form to acknowledge contract substitutions.

With the proper wording, service orders become mini-contracts. By having a customer sign a service order before any work is begun, you increase your protection.

The wording in your mini-contracts can be inclusive of payment terms, guarantees, liability restrictions, and much more. An attorney can help you to design a service order that best suits your needs.

Service orders can document the time your employees arrive on a job and the time they leave the job. This helps to keep your employees productive and gives you an edge in management.

Inventory control is another feature work orders can give you. By having service technicians list all materials used on the service ticket, you can maintain an accurate inventory of your rolling stock. When designed and used properly, service orders make your business run better.

Liability waivers

Liability waivers are becoming a fact of life. You can't take your car to the shop or your child to the hospital without signing some type of liability release. In today's lawsuit-happy world, you must protect yourself at all times. Liability waivers can help protect you.

You probably won't find many preprinted liability waivers. I suspect that in time, companies will make generic waivers, much like the fill-in-the-blank contract forms, but for now, you need to consult with an attorney.

When you talk with your attorney, explain all aspects of your business. The attorney cannot give you a comprehensive liability waiver unless you are specific about your work requirements. While you are talking with your lawyer, ask for some boilerplate language that you can use in on-the-spot cases. Even a customized waiver form won't fulfill all of the potential needs. There will come a time when special circumstances require you to draft a liability waiver on the job.

First, check to make sure you can draft your own waiver legally. Some states are very stringent on what individuals, other than attorneys, can do in preparing legal documents. If you can create your own liability waiver, obtain suitable language for the waiver from your attorney. If you are prohibited from drawing up your own document, let your attorney make a fill-in-the-blank waiver. Then you can add specific details to the waiver as required.

Written estimates

How can written estimates protect you? They can protect you in several ways. First, they make the price for your work known. Second, written estimates give the customer a written description of the work you propose to do. Written estimates also advise the customer of your terms and conditions. All of these factors help to protect you.

Specifications

Complete specifications are very beneficial during the course of a job. If you take the time to document all of the specifications for a job, you are less likely to lose time arguing with a dissatisfied customer. This not only saves time and money, it helps keep your customers happy.

When you are developing specifications for a job, be as detailed as possible. Include model numbers, makes, colors, sizes, brand names, and any other suitable description of the labor and materials you are providing.

Once you have a good spec list, have the customer review and sign it. Always have the customer sign the specifications list. Without a signature of acceptance from the customer, your specs are little more than a guide for you to work from. Unsigned specification sheets carry little weight in a legal battle.

Credit applications

I am surprised at how few contractors use credit applications. It is not unusual for a contractor to enter into a contract for a job worth thousands of dollars without running a credit check on the customer. Why do contractors take this risk? I don't know; banks run credit checks before loaning money, department stores check credit histories on their credit applicants, but most contractors don't.

As a business owner, you can subscribe to the services of a credit reporting bureau. Normally, for a small monthly fee and an inexpensive per-inquiry fee, you can pull a detailed credit history on your customers. You need, of course, the permission of your customers to check into their credit background. Credit applications provide documentation of this permission.

Even if you don't belong to a credit bureau, you can call references given by the customer on the credit application. This type of investigation is not as good as the reports you receive from credit agencies, but it is better than nothing. Credit applications should be used for all of your customers wishing to establish a credit account.

Cash receipts

Cash receipts for your cash purchases benefit you at tax time. By keeping records of all your cash purchases, you can take advantage of all your tax deductions. Receipts serve as documentation for your deductions. If the receipt does not state clearly what the purchase was, write the details on the receipt. You should include what the item was and what it was for. If the item was for a particular job, write the job name on the receipt. In a tax audit, your documented receipts might mean the difference between an easy audit and having to pay back taxes.

Inventory logs

Inventory logs (FIG. 18-13) can be used to maintain current information on your inventory needs and supply. If you take materials out of your inventory, write in the log what the items were and where they were used. At the end of the week, go over your log and adjust your inventory figures. If you need to replace the inventory, you know exactly what was used. If questions arise at tax time, you can identify where your inventory went. Something as simple as an inventory log can save you from lost money and time.

Inventory Log

Item	Quantity	ID Number	Checked By	Date

18-13 Inventory log.

WRITING GOOD CONTRACTS

Good contracts make for happy customers. Most customers start into a job happy, but by the time the job is done, many customers are disgruntled. This situation is bad for business. You need happy customers to get return business and referrals. How can you keep your customers happy? Easy-to-understand contracts go a long way in keeping consumers appeased. A majority of consumer complaints arise because of poor communication and confusion. Contracts that are worded clearly and contain all details of the job eliminate the likelihood of confusion. When you remove the risk of confusion, you improve the odds of ending the job with a happy customer.

When you prepare contracts, put as much information in them as possible. Once the contract is written, go over it with the customer. Allow the customer plenty of time to read and absorb the contents of the contract. Answer any questions the customer might have, and if necessary, reword the contract to eliminate confusion. Once the contract is agreeable to all parties, execute it, preferably in front of a witness.

After the contract is signed, don't deviate from its contents. If changes are to be made, use addendums or change orders, and make sure they are signed by all of the signatories on the contract. If you follow this procedure in all of your jobs, you should have more happy customers and fewer problems. A side benefit is making and keeping more money for your efforts.

19

Buying trucks, tools, equipment, and inventory

Equipment, vehicles, and inventory account for most of your start-up costs when you go into business. If you are already in business, these same items consume a large portion of your operating capital. There is no question that buying an $18,000 truck can put a dent in your bank account. Even with a 10-percent down payment and a five-year loan, at 10-percent interest, you are going to have to shell out $1800 in cash and about $344 a month for payments. Add to this the cost of registration, taxes, and insurance, and you have a pretty major expense.

Then, depending on the type of business you have, you might have to buy equipment. Hand tools are expensive and power tools are even worse. As for inventory, a home builder can easily invest $10,000, although some builders work with very little inventory. They buy what they need when they need it—a smart way to do business.

A large number of business are forced to close each year because of purchases for equipment, vehicles, and inventory. Some business owners turn to leasing vehicles. Leasing saves up-front money and the payments are usually less. But if the vehicles have excessive mileage or body damage when the lease is up, the costs might exceed those of purchasing the vehicle.

Not only is the cost of acquiring inventory steep, but keeping employees from stealing it can be a problem. Somewhere among the maze of options in these three big-ticket items is a happy medium. This chapter can help you find your way through the perilous maze to financial safety and success.

LEASING VS. PURCHASING

Most business owners contemplate the advantages of leasing versus purchasing. Both options offer advantages and disadvantages. It makes a lot of sense to rent tools that you only use a few times a year. Leasing vehicles can save you money and provide tax advantages. So, what should you do? Let's find out.

Renting tools

Renting expensive tools can be a very smart move for a new business. When you are first starting out, money can be especially tight, and you might not have a real handle on what tools you need. By renting what you need, when you need it, you derive several advantages. You can try it before you buy it. You can evaluate how much you need the tool, and you can determine how much extra money you can make with the tool.

Buying specialty tools

Buying specialty tools can be a mistake. If you buy a bunch of tools you don't use very often, you deplete your cash for a lost cause. For example, I used to do a high volume of basement bathrooms. Installing these basement baths required the use of a jackhammer. I started out renting the hammers and wound up buying one. For me, this was a good move. However, it might have been a superfluous expense for some remodelers.

If one of my employees had gone into business for himself, he might have thought he needed a jackhammer. This would make sense, since my company used its jackhammer frequently. But if the person going into business for himself didn't get the same kind of work I did, he might not have needed the jackhammer more than once or twice a year. Unless breaking up concrete floors was a routine part of his business, the cash outlay for a jackhammer would have been a mistake.

Before you buy specialty tools, make sure you need them. The best way to assess your need for specialty tools is to rent them when you need them, and keep track of how they affect your business. If your profits increase, consider purchasing the tool. If you find you only rent the tool a few times a year, continue to use rental tools.

Leasing vehicles

Leasing vehicles can be a good idea, but it can also be a bad decision. The good points of leasing are minimal out-of-pocket cash, normally lower payments, possible tax savings, and possible short-term commitments. The disadvantages are no equity gain, more concern for the condition of the vehicle, and possible cash penalties when the lease expires.

Most auto leases require only the first month's rent and an equal amount for the down payment. In other words, if the vehicle is going to cost you $250 a month, you need $500 for a down payment and the first month's rent. To purchase the same vehicle, you might need $1000 to $1500 for a down payment.

The monthly payments on leased vehicles are usually lower than the payments for a vehicle purchase. How much you save depends on how expensive the vehicle is, but the savings can add up. However, you don't own the vehicle, and, at the end of the lease, you have no equity in the truck or car.

Leases can usually be obtained for any term, ranging from one year to five, and sometimes more. Two-year leases are popular, and so are four-year leases. By leasing a vehicle for a short time, like two years, your company fleet can be renewed every two years. This practice keeps you in new vehicles and presents a good image. However, if the vehicle is not in good shape at the end of the lease, you pay a price for the abuse.

Most leases allow for a certain number of miles to be put on the vehicle during the term of the lease. If the mileage is higher than the allowance when the lease expires, you have to pay so much per mile for every mile over the limit, which can get expensive. Additionally, if the vehicle is beat up, you can be charged for the loss in the vehicle's value, which can also amount to a substantial sum of money.

Then there are the tax angles. Since I'm not a tax expert, I recommend you talk with someone who is. It is likely that leasing is more beneficial to your tax responsibilities than buying, but check it out.

Purchasing vehicles

Purchasing vehicles is the traditional way of getting them. Many companies lease cars and trucks, and a lot of companies buy their vehicles. When you purchase your vehicles, you are building equity in them. If you abuse the vehicle, you lose money when you sell or trade it, but you won't be penalized by a lease agreement. If you need trucks that can lead a hard life, purchasing them is probably a good idea.

NEEDS VS. DESIRES

Before you buy expensive equipment, vehicles, or inventory, separate your needs from your desires. Too many contractors get caught up in having the best and most expensive items available. This is not only unnecessary, it can kill your business. If you can do the job with a $15,000 van, don't buy a $20,000 truck. There is a big difference between a need and a desire. Let's examine how you can decide what to buy.

Let's look at needs and desires with inventory items. Most new business owners want to be well stocked with inventory. They often wind up being overstocked, which results in cash being tied up and sometimes in lost money. Before you buy your inventory items, you must figure out what you need, not what you want.

Let me give you an example of how bad inventory purchases can shackle your business. I consulted with a plumbing company a few years ago. This particular company had eight service plumbers on the road and a ton of unneeded inventory. The business owner couldn't understand why he was short on cash. When I assessed the business, I found several faults, but inventory was one of the major problems with the company cash flow.

Being a master plumber and a business owner for many years, I know what a plumber needs on a service truck. When I looked inside this company's trucks, I was amazed. Instead of having one box of copper ells, these trucks had as many as five boxes. One box contains 50 three-quarter copper ells. Granted, ells are a frequently used fitting, but you don't need 250 ells on a service truck.

My inspection turned up countless items in multiple quantities that were unlikely to be used in a year's time. When we streamlined the trucks to a viable rolling inventory, the company returned the excess inventory and received a credit from the supplier for more than $12,000, and that was just truck stock. When we went through the back room, we eliminated another several thousand dollars in inventory. Just by reducing the inventory to what was needed, not what was wanted, the company generated close to $20,000, even after paying restocking fees.

Tools and equipment are areas in which many business owners have little willpower. They seem to want every tool and piece of equipment they could ever hope to use. It's one thing to want it and quite another thing to buy it. It doesn't seem to matter if it is painters, plumbers, carpenters, or electricians, they all want the best tools and equipment available.

It is wise to have the tools and equipment you need, but it is senseless to buy expensive items that you will rarely use. Rent these occasional-use items. A carpenter who specializes in interior trim work does not need a set of pump jacks. A builder who subs out all grading work doesn't need a bulldozer or tractor. You must decide if you simply want something or if you need it.

The need-and-desire angle comes into play with vehicles too. For years I wanted my own dump truck. I thought it would be great to own such a truck. I came close to buying one, so close that I was sitting down with the salesman. But at the last minute, I bought a van instead. Why did I buy the van? The van would work for my needs, and it would get another crew on the road. In short, the van would pay for itself and the dump truck wouldn't. You need to give yourself time to make this type of consideration.

FINANCIAL JUSTIFICATION

Financial justification is the key to making a wise buying decision. Just like my dump-truck example, you have to see if what you are about to do is economically feasible. Some people are able to justify buying anything they want. I suppose if you work at it long enough, you can convince yourself of just about anything. However, justifying a purchase isn't enough. You must justify the purchase financially. There are different ways of doing this.

When you strive to justify a purchase, you can take the easy way out and simply talk yourself into buying something. A more realistic way, however, is to rent what you need long enough to establish a true need for it. Or you might use subcontractors for some part of your work until you see that it would be cost-effective to put your own people in place to take care of certain aspects of your job. For example, it might be feasible to hire your own plumber or electrician, but it probably won't pay to put a cleaning crew on your payroll. Test the waters before you jump in with both feet.

INVENTORY

How much inventory should you stock? The sarcastic answer would be to stock just as much as you need and not a bit more. But the real answer is harder to come by. Inventory requirements vary with different types of businesses. Where a remodeling contractor might need a rolling stock worth thousands of dollars, a builder has little need for an extensive inventory. You have to establish your inventory needs based on your customers' buying trends.

Planned jobs

Planned jobs, such as the construction of a new house, don't require much inventory. You can order a specific amount of material for the job when the job starts. It is usually best to have some inventory on hand to make up for any items you forgot in the big order, but on-truck inventory needs for this type of work are minimal. As an example, you might want a few extra boxes of nails on your truck, but you wouldn't be likely to haul around a bunch of wall studs.

Time savers

A good inventory of frequently used materials can be a real time saver. If you are on a job and need an extra roll of roof flashing, you save time and money by having it on the truck. It is a good idea to carry a rolling stock of your most frequently used items, but don't get carried away. Many builders don't carry excess material with them to job sites. This is okay. Plumbers and electricians who do service work need trucks with a lot of stock on them, but builders don't. Hauling around a lot of building supplies is not only heavy work, it requires a lot of money. You are better off if you figure your jobs accurately and order what you need on a schedule that allows you to stay in tune with the production schedule.

Shop stock

Shop stock is convenient, but it can be a drag on your cash flow. Limit shop stock to what you will use in a two-week period. As you use the material, order new stock. This practice keeps your inventory fresh and your money turning over.

Many builders have a tendency to create shop stock out of leftovers from various jobs. If you have a partial roll of insulation left over on a job, save it. However, if you ordered a range hood that is the wrong color, don't save it for another job. Return it and get credit for the item with your supplier. Money invested in inventory is money that is tied up. You should be able to find a better use for your cash.

INVENTORY THEFT

Controlling inventory theft can be a problem. This problem can hit any company that has employees. Whether you have one employee or 100 employees, you could be getting ripped off. Lost inventory is lost money. You must take steps to ensure that your employees are honest and that the material is going where it was meant to.

The best way to reduce inventory pilfering is to keep track of your inventory on a daily basis (FIG. 19-1). Have your workers fill out forms for all material used. Have the forms completed and turned in each day. Let employees know, in a nice way, that you check inventory disbursement every week. This tactic alone greatly reduces the likelihood of employees stealing from you.

Another way to control the loss of shop stock is to issue all stock yourself. If you don't allow employees access to your inventory, they can't steal it. Accounting for rolling stock is harder. However, if your employees know you keep daily records on your stock, they are less likely to empty your truck. It never hurts to do surprise inspections and inventories on the trucks. When you do it, make sure the employees see you inspecting the trucks. The fact that you go on the truck or on a job site and audit the inventory can reduce your losses.

Inventory Control for Trucks					
Item	Quantity	Job Name	Employee	Truck#	Date

19-1 Truck inventory form.

EFFICIENT TRUCK STOCKING

Stocking your trucks efficiently is crucial for success and maximum profits. Keep your rolling stock to a minimum. You never know when the truck might be broken into or stolen. It is more difficult to monitor employees who deal with mobile inventories. If the inventory on your trucks is collecting dust, get rid of it and don't reorder it.

A good way to establish your truck-stock needs is to keep track of the materials you use off the truck (FIG. 19-2). If you track your material usage, you can stock your truck with materials you are very likely to use. If you have some slow movers on the truck, don't replace them when they are sold. Conversely, if you have some hot items, keep them on the trucks.

Stocking your trucks efficiently takes a little time; start small and work your way up. If you have been in the trade for long, you have a good idea of what your inventory needs are. By tracking your material sales, you can perfect your rolling stock.

A remodeler has different inventory needs than a builder. Most builders have little need for inventory. Remodelers do often need odds and ends on their trucks. Your inventory needs are specific to your business. As your business and experience grows, so will your knowledge of what is and isn't required.

Truck Inventory

Truck number: _____

Driver: _____

Item	Size	Color	Brand	Quantity

Inventory taken on the _____th of _____, 19_____

19-2 Daily use truck inventory form.

20
Dealing with subcontractors, suppliers, and code officers

Subcontractors, suppliers, code officers, and materials are all part of most contracting businesses. When contractors don't know how to work with and control these aspects of business, their businesses suffer. All of these business elements can reflect on an individual's contracting company.

If subcontractors do substandard work, the quality is off and so is your company's reputation. If suppliers don't maintain delivery schedules, your crews can come to an abrupt halt. Code officers can also make your company look bad. When the work your company produces fails inspection, your business loses credibility. Inferior materials can make your best work look bad. If you want your business to be successful, you must learn how to handle subcontractors, suppliers, code officers, and materials.

ACTING AS THE CONTRACTOR

If you are a contractor, your subcontractors, suppliers, and materials have a strong reflection on your business. These three elements can make you look good or bad. How your business is perceived depends on your skills as a manager. Let's look first at how subcontractors can affect your business.

Hiring subcontractors

Subcontractors often represent the businesses of contractors. As a contractor, you will likely hire subcontractors to perform various forms of work. The quality of these subcontractors is very important; they represent your company.

Good subcontractors make you look good, and bad subs make your business look bad. Choosing subcontractors is much like hiring employees. When you put subcontractors in contact with your customers, you are trusting them to maintain the reputation you've worked hard to earn. However, many factors can make you wish you had never seen a subcontractor.

Good subcontractors can make you look great. If you have a deep stable of subs, you can respond to work quickly and efficiently. Customers love to get fast service, and subcontractors can give you this desirable dimension.

Subcontractors with good work habits and adequate people skills can build your business. While some subcontractors attempt to steal your customers, most are happy to maintain their relationship with your company. A lot of subs don't want all the responsibilities that go along with being a general contractor. If you take good care of your subs, most of them will take good care of your business.

Finding suppliers

You might not think that suppliers have much effect on the public opinion of your business, but they can. As a contractor, you are held responsible for everything that happens on the job. If your supplier's delivery truck damages the customer's lawn, you're going to catch the heat. When materials are not delivered on time, customers are not going to call the suppliers to complain; they are going to call you. As the general contractor, you are going to take all the abuse.

If you want your customers to remain happy, and what contractor doesn't, you must be in control of the job. This control extends from getting the permit to doing the punch-out work and everything in between.

When you have the right suppliers, they can improve your customer relations. If delivery drivers are courteous and professional, the customers appreciate it. When deliveries are made on time, customers are satisfied. Seeing to it that suppliers make and maintain good customer relations is up to you. You have to lay down the rules for your suppliers to follow. If the suppliers are unwilling to play by your rules, find new suppliers.

Buying materials

Materials can have a large effect on your customer's peace of mind. If shoddy material shows up on the job, your customer isn't going to be pleased. If the wrong materials are shipped, you lose time, and your customer loses patience. Don't overlook the important role that materials play in the way customers view your business.

CHOOSING YOUR PRODUCT LINES

Choosing your product lines carefully is important to the success of your company. If you pick the wrong products, your business can fizzle. When you choose the proper products, they sell themselves. As a business owner, you can use all the help you can get, so carry products that the public wants.

How do you know what products to carry? You can decide on what products to carry by doing some homework. Read magazines that appeal to the type of people you want as customers. For example, if you want to become known for the outstanding kitchen and bathroom designs used in your homes, read magazines that center around kitchens and baths. Look at advertisements in the magazines that you decide fit your business plan. By paying attention to these ads, you get a good idea of what your customers are interested in.

Take a walk through the local stores that carry products you sell or compete with. Take notes as you cruise the isles. Pay attention to what is on display and how much is being charged for the items. This type of research helps you to target your product lines.

Take a ride around town. Look at homes under construction. See what your competition is doing. By simply riding past a construction site you can probably tell what types of doors, windows, siding, shingles, and similar items are being used. This on-site investigating can put you in touch with what the public wants.

The most direct way to determine what customers want is to ask them. If you are thinking that you can't ask customers what they want until you have customers, you're wrong. As a matter of fact, you shouldn't wait until you have a strong customer base to establish your product lines.

How do you walk up to a stranger and ask them what type of windows they like best? You can go door to door and do a cold-call canvassing of a neighborhood. You might experience a lot of rejection, but you might also get some answers. If you don't like knocking on doors, you can use a telephone. You can even have a computer make the phone calls and ask the questions for you.

If you don't want to use face-to-face techniques or telephones, you can use direct mail. Direct mail is easy to target, and it's fast and effective. While mailing costs can get steep, the results can overcome the costs.

How would you use direct mail to establish your product lines? You could design a questionnaire to mail to potential customers. If it is done properly, your mailing will look like you have a sincere interest in what individuals want. It appears this way, because you do have a sincere interest.

The results of the answers to your questionnaire can tell you what products to carry. Why will people fill out and return your questionnaire? You can improve the odds of having the pieces returned by self-addressing the response card. You should also pick up the tab on the return postage. You can purchase a permit from the local post office and have it printed on your cards. If you don't like that idea, you can affix postage stamps to the cards, but this costs more. With the permit from the post office, you pay only for the cards that are returned, not counting the permit fee. If you use postage stamps, you are paying for postage that might never be used.

To convince people to fill out your questionnaire, you have to provide an incentive. One idea for an incentive is a discount from your normal fees. This idea might work, but it looks very commercial. A better idea might be to make the questionnaire look more like research. If you design the piece to look like a respectable research effort, more people might respond to your questions.

AVOIDING DELAYS IN MATERIAL DELIVERIES

Avoiding delays in material deliveries is crucial to the success of your business. If your materials are delayed, your jobs are delayed. If your jobs are delayed, your payments are delayed. If your payments are delayed, your cash flow and credit history can falter. If your cash flow dries up, your business is in trouble. Extended delays can result in the loss of your good credit rating or even your business. You can eliminate some of these potentially dangerous situations by avoiding delays in your material deliveries.

How can you avoid late material shipments? You can maintain a comfortable delivery schedule by being involved personally. This personal involvement can be time consuming, but it is worthwhile.

When you order your materials (FIG. 20-1), your job of maintaining the delivery schedule begins. You can apply some basic principles to keep your deliveries on schedule. Start by getting the name of the person taking your order. Use a phone log to document all of your calls (FIG. 20-2). Ask the order-taker to give written documentation for your delivery date. While you are at it, get the name of the store manager; you will probably need it.

Once you have the intended delivery date, stay on top of the delivery. If you have placed the order several days in advance, make follow-up phone calls to check the status of your material. Always get the names of the people you talk to; you never know when you might have to lodge a complaint. Keep clear records of your dealings in an order log (FIG. 20-3).

Your Company Name
Your Company Address

Dear Sir:

I am soliciting bids for the work listed below, and I would like to offer you the opportunity to participate in the bidding. If you are interested in giving quoted prices on <u>material</u> for this job, please let me hear from you, at the above address.

The job will be started in _____ weeks. Financing has been arranged and the job will be started on schedule. Your quote, if you choose to enter one, must be received no later than
_____.

The proposed work is as follows:

_____

_____

_____

Plans and specifications for the work are available upon request.

Thank you for your time and consideration in this request.

Sincerely,

Your name and title

20-1 Form letter for soliciting material quotes.

Phone Log

Date/Time	Company Name	Contact Person	Remarks

20-2 Phone log.

Material Order Log

Supplier: _____

Date order was placed: _____

Time order was placed: _____

Name of person taking order: _____

Promised delivery date: _____

Order number: _____

Quoted price: _____

Date of follow-up call: _____

Manager's name: _____

Time of call to manager: _____

Manager confirmed delivery date: _____

Manager confirmed price: _____

Notes and Comments

_____

_____

_____

_____

_____

_____

_____

_____

_____

_____

_____

_____

_____

20-3 Material order log.

By maintaining a presence on the phone or in person, the employees handling your deliveries won't forget you. They will assume that if you are this attentive now, you will be horrible to deal with if they mess up the order. This intimidation works in your favor.

With a lot of effort and a little luck, your deliveries will be made on schedule. If the shipment does go astray, contact the store's manager. Advise the manager of the problem and the ripple effect it is creating for your business. Produce your documentation on the order. By showing the manager your written delivery date, employee names, and supporting documentation, such as material specifications (FIG. 20-4), you will make a strong impression. This tactic sets you apart from the customers who complain incoherently. You will come across as a serious professional.

With this type of approach, the manager will take you seriously and work to help you. If you don't get satisfactory results, move up the ladder to higher management. If you have created a strong paper trail, you will get results.

CHOOSING SUBCONTRACTORS

Choosing subcontractors can require extensive time and effort. Knowing the importance of subcontractors, you cannot tackle the task with a carefree attitude. You must be serious, and you must give this part of your job the attention it deserves.

When it comes to selecting subcontractors, there are some rules you should follow. If you take the time to screen your subcontractors, your business will fare better.

Making initial contact

You can learn a lot about your subcontractors from the initial contact. Most people act, at least partially, on gut instinct. When you meet subcontractors for the first time, you will develop an opinion. Your first impression might not be accurate, but you are sure to formulate one.

Not only do you form a first impression, but so does the sub. The conditions under which this first contact is made can influence your future. With the potential importance of this meeting, you want to control the circumstances.

Material Specifications

Phase	Item	Brand	Model	Color	Size
Plumbing	Lavatory	WXYA	497·	White	19"×17"
Plumbing	Toilet	ABC12	21	White	12" rough
Plumbing	Shower	KYTCY	41	White	36"×36"
Electrical	Ceiling fan	SPARK	2345	Gold	30"
Electrical	Light kit	JFOR2	380	White	Standard
Flooring	Carpet	MISTY	32	Grey	14 yards

20-4 Sample of material specifications.

As a contractor, you probably need subcontractors. If you need subs, you can't afford to blow the initial contact. Just as subcontractors might alienate you, you might scare off the subs. This, obviously, is not what you want.

You want your first contact with subcontractors to be productive. If it is not, you are wasting time—yours and the subcontractors'. Wasted time is potentially wasted money. With so much at stake, you must choreograph your first contact carefully.

Just as you want to engage professional subcontractors, subcontractors want to work with successful contractors. Subcontractors have a legitimate fear that they might not be paid. If, as a general contractor, you come across as an unorganized, shaky business, subs will not be thrilled at the possibility of working with you.

Subcontractors and general contractors are meant to go together like peanut butter and jelly. If you don't establish a comfort level between the two parties, the work resulting from the business marriage will not be the best.

What can you do to attract quality subcontractors? If you project a professional image, subcontractors will seek you out. Once subs find you, you must maintain the business image. Whether you are talking on the phone or in person, send the right messages. Let subcontractors know you are a professional and you will accept nothing less from them.

Requiring application forms

Application forms (FIG. 20-5) can come in handy when searching for new subcontractors. While subs are not going to be traditional employees, it is not unreasonable to ask them to complete an employment application. The applications you use might not resemble those used for employees, but you want to know as much about your subcontractors as possible.

The application might contain questions pertaining to the types of work the subcontractor is equipped to do. Asking for credit and work references is a reasonable request. Having the subcontractors list their insurance coverage is beneficial. You can customize your applications to suit your needs. It might be wise to discuss the form and content of your subcontractor applications with an attorney. You don't want to be guilty of asking questions you are not supposed to ask.

Conducting basic interviews

You want to ask many questions in the basic interviews with subcontractors. When you conduct your interviews, you want to derive as much insight into the qualities of the subcontractors as possible. These interviews are the basis for your decision to use or eliminate subcontractors (FIG. 20-6).

Where should you meet with prospective subcontractors? If you have a professional office, your office is a fine place to meet subcontractors. If your office conditions don't reflect the image you want to give, meet the subcontractors on neutral ground. You could meet the subcontractors at lounges, restaurants, or almost any other place. Pick a meeting place that allows you to project your best image.

Subcontractor Questionnaire

Company name _____

Physical company address _____

Company mailing address _____

Company phone number _____

After-hours phone number _____

Company president/owner _____

President/owner address _____

President/owner phone number _____

How long has company been in business? _____

Name of insurance company _____

Insurance company phone number _____

Does company have liability insurance? _____

Amount of liability insurance coverage _____

Does company have worker's comp. insurance? _____

Type of work company is licensed to do _____

_____

List business or other license numbers _____

Where are licenses held? _____

If applicable, are all workers licensed? _____

Are there any lawsuits pending against the company? _____

Has the company ever been sued? _____

Does the company use subcontractors? _____

Is the company bonded? _____

Who is the company bonded with? _____

Has the company had complaints filed against it? _____

Are there any judgments against the company? _____

20-5 Subcontractor questionnaire.

20-5 Continued.

During the interview, control the conversation. Let the subcontractor talk, but don't let the sub run the show. You should set the pace for the interview. If subcontractors run over you in the interviews, they will run over you in the normal course of business.

Checking references

Checking references should be standard procedure when selecting subcontractors. If a subcontractor has been in business long, references should be available. Ask for these references, and follow up on them. If you don't confirm the qualities of subcontractors by checking references, you might not get the service you are paying for.

Checking credit

Another part of screening subcontractors is checking credit. Why should you check the credit histories of subcontractors? By checking the credit ratings of subcontractors, you can determine much about the individuals and their businesses.

If subcontractors have bad credit, it doesn't mean they are bad subcontractors or poor workers. However, if their companies are in trouble, you probably don't want to entrust your business to them. Credit reports can tell you a lot about the people you might be doing business with.

Reading between the lines

In all of your business endeavors, you must learn to read between the lines. Credit reports are a good example where the facts might not tell the whole story. Let's say

Contractor Rating Sheet

Category	Contractor 1	Contractor 2	Contractor 3
Contractor name			
Returns calls			
Licensed			
Insured			
Bonded			
References			
Price			
Experience			
Years in business			
Work quality			
Availability			
Deposit required			
Detailed quote			
Personality			
Punctual			
Gut reaction			

Notes

_____

_____

_____

_____

_____

20-6 Contractor rating sheet.

you are reviewing a credit report and see that a subcontractor has filed for bankruptcy; would you subcontract work to this individual? If you wouldn't, you might be missing out on a good worker.

The fact that someone has filed for bankruptcy is not enough to rule out doing business with the individual. Individuals can get into financial trouble without fault of their own. You must be willing and able to decipher what you are seeing. When you learn to read between the lines, you become a more effective businessperson.

Setting guidelines

If you plan to use the services of the subcontractors you are interviewing, set the guidelines for doing business with your firm. If you require all of your subcontractors to carry pagers, let this fact be known in the interview. If your rules require subcontractors to return your phone calls within an hour, make the point clearly. Remember, you are in control, but you can't expect people to read your mind. You have to let your desires be known.

Coming to terms

Coming to terms is a key issue in selecting subcontractors. What you want and what the subcontractors want might not be the same. If you are going to do business with subcontractors, you should work out the terms of your working arrangements in advance.

Discussing contracts

Discussing contracts is a vital topic in meeting with subcontractors. You should go over your subcontract agreements (FIG. 20-7) with the subcontractors. If either party has any questions or hesitations, resolve them in the meeting. You don't want to get into the middle of a job and find out that your subs won't play by the rules.

The more detail you go into in the early stages of your relationships, the more likely you are to develop good working conditions with your subcontractors. Just like your contracts with homeowners, you want your subcontractor agreements to be free of confusion. Insert whatever clauses are appropriate to make sure subs understand what you expect (FIGS. 20-8 and 20-9). Take as much time as necessary, but remove any doubts about the meaning of your contracts.

Maintaining relationships

Once you find new subcontractors, you must concentrate on maintaining the relationships. When you find good subs, you have probably invested a significant amount of time in your acquisitions. To avoid having this time investment wasted, you must work to keep the relationship on the friendly side of the scale. This doesn't mean you have to become buddies with your subs, but you do have to fulfill your commitments.

If you tell a subcontractor that you pay bills within five days of receiving them, you had better be prepared to pay the bills. When you agree to terms in

Subcontract Agreement

This agreement, made this ____th day of _____, 19___, shall set forth the whole agreement, in its entirety, between Contractor and Subcontractor.

Contractor: _____, referred to herein as Contractor.

Job location: _____

Subcontractor: _____, referred to herein as Subcontractor.

The Contractor and Subcontractor agree to the following:

Scope of Work

Subcontractor shall perform all work as described below and provide all material to complete the work described below.

Subcontractor shall supply all labor and material to complete the work according to the attached plans and specifications. These attached plans and specifications have been initialed and signed by all parties. The work shall include, but is not limited to, the following:

_____

_____

_____

_____

_____

_____

_____

_____

Commencement and Completion Schedule

The work described above shall be started within three days of verbal notice from Contractor; the projected start date is _____. The Subcontractor shall complete the above work in a professional and expedient manner by no later than _____ days from the start date. Time is of the essence in this contract. No extension of time will be valid without the Contractor's written consent. If Subcontractor does not complete the work in the time allowed, and if the lack of completion is not caused by the Contractor, the Subcontractor will be charged <u>fifty dollars ($50.00)</u> per day, for every day work extends beyond the completion date. This charge will be deducted from any payments due to the Subcontractor for work performed.

Page 1 of 3 initials___

20-7 Subcontract agreement.

Contract Sum

The Contractor shall pay the Subcontractor for the performance of completed work subject to additions and deductions as authorized by this agreement or attached addendum. The contract sum is _____($_____).

Progress Payments

The Contractor shall pay the Subcontractor installments as detailed below, once an acceptable insurance certificate has been filed by the Subcontractor with the Contractor.
Contractor shall pay the Subcontractor as described:

_____

_____

_____

_____

_____

_____

_____

_____

_____

_____

_____

_____

_____

_____

_____

_____

_____

All payments are subject to a site inspection and approval of work by the Contractor. Before final payment, the Subcontractor shall submit satisfactory evidence to the Contractor that no lien risk exists on the subject property.

Page 2 of 3 initials___

20-7 Continued.

Working Conditions

Working hours will be 8:00 A.M. through 4:30 P.M., Monday through Friday. Subcontractor is required to clean his work debris from the job site on a daily basis and leave the site in a clean and neat condition. Subcontractor shall be responsible for removal and disposal of all debris related to his job description.

Contract Assignment

Subcontractor shall not assign this contract or further subcontract the whole of this subcontract, without the written consent of the Contractor.

Laws, Permits, Fees, and Notices

Subcontractor shall be responsible for all required laws, permits, fees, or notices required to perform the work stated herein.

Work of Others

Subcontractor shall be responsible for any damage caused to existing conditions or other contractor's work. This damage will be repaired, and the Subcontractor charged for the expense and supervision of this work. The Subcontractor shall have the opportunity to quote a price for said repairs, but the Contractor is under no obligation to engage the Subcontractor to make said repairs. If a different subcontractor repairs the damage, the Subcontractor may be back-charged for the cost of the repairs. Any repair costs will be deducted from any payments due to the Subcontractor. If no payments are due the Subcontractor, the Subcontractor shall pay the invoiced amount within 10 days.

Warranty

Subcontractor warrants to the Contractor all work and materials for one year from the final day of work performed.

Indemnification

To the fullest extent allowed by law, the Subcontractor shall indemnify and hold harmless the Owner, the Contractor, and all of their agents and employees from and against all claims, damages, losses, and expenses.

This agreement, entered into on _____, 19_____, shall constitute the whole agreement between Contractor and Subcontractor.

_____   _____
Contractor   Date   Subcontractor   Date

20-7 Continued.

Commencement and Completion Schedule

The work described above shall be started within _3_ days of verbal notice from the customer, the projected start date is _____. The Subcontractor shall complete the above work in a professional and expedient manner by no later than <u>twenty (20)</u> days from the start date.

Time is of the essence in this subcontract. No extension of time will be valid without the General Contractor's written consent. If Subcontractor does not complete the work in the time allowed and if the lack of completion is not caused by the General Contractor, the Subcontractor will be charged <u>one hundred dollars ($100.00)</u> for every day work is not finished after the completion date. This charge will be deducted from any payments due to the Subcontractor for work performed.

20-8 Commencement and completion clause.

Subcontractor Liability for Damages

Subcontractor shall be responsible for any damage caused to existing conditions. This shall include new work performed on the project by other Contractors. If the Subcontractor damages existing conditions or work performed by other Contractors, said Subcontractor shall be responsible for the repair of said damages. These repairs may be made by the Subcontractor responsible for the damages or another Contractor, at the discretion of the General Contractor.

If a different Contractor repairs the damage, the Subcontractor causing the damage may be back-charged for the cost of the repairs. These charges may be deducted from any monies owed to the damaging Subcontractor, by the General Contractor. The choice for a Contractor to repair the damages shall be at the sole discretion of the General Contractor.

If no money is owed to the damaging Subcontractor, said Contractor shall pay the invoiced amount, to the General Contractor, within <u>seven</u> business days. If prompt payment is not made, the General Contractor may exercise all legal means to collect the requested monies.

The damaging Subcontractor shall have no rights to lien the property where work is done for money retained to cover the repair of damages caused by the Subcontractor. The General Contractor may have the repairs made to his satisfaction.

The damaging Subcontractor shall have the opportunity to quote a price for the repairs. The General Contractor is under no obligation to engage the damaging Subcontractor to make the repairs.

20-9 Subcontractor liability clause.

your subcontractor agreements, stick to them. If you breach your agreements with subcontractors, you will have a great deal of difficulty in getting and keeping good help.

RATING SUBCONTRACTORS

Rating subcontractors might take a little extra effort, but it's worth it. Why should you rate subcontractors? You should rate subs to have a better chance of finding subcontractors that suit your needs. This rating procedure starts in the interview, but it goes deeper. What do you have to look for when rating subcontractors? You should look for a number of qualities. Let's take a look at some of the factors you should consider.

Evaluating work history

One of the first qualities you should evaluate is the work history of the subcontractor. When it comes to hiring subcontractors, experience counts. It might not be important if the sub has just started in business, but it is meaningful for the individual to have work experience. For example, a person with 15 years of experience that has just gone into business might be a better sub than the one who has been in business for two years but only has five years of experience. However, there are advantages to choosing a subcontractor that has an established business.

If the subcontractor has been in business for a while, there is a better chance that the business will last. New businesses often fail, but businesses that have been around for three to five years have a better chance of survival. Business owners that survive these early years have business experience and dedication; these are admirable traits in a subcontractor.

Checking business procedures

The business procedures employed by subcontractors can affect their desirability. This factor can be hard to assess in a single meeting. However, if you are willing to do some research, you can ascertain much about how the sub does business.

One of the most crucial aspects of subcontractors is how easy they are to reach by phone. If you can't communicate with your subs, you'll have problems. Most subs tell you how attentive they are, but you should verify their claims. After the contractors leave your office, call them. You know they have just left the meeting and are not in their offices. You can find out how their phones are answered and how quickly they return your call.

Before you commit to using a subcontractor, conduct a test. Call three contractors and schedule a bid meeting. It doesn't matter if the job is real or a dummy. What matters is that the subcontractors believe they are inspecting and bidding a real job.

Why should you play this hoax on innocent contractors? You should do it to see how the contractors respond. Are they punctual? How long does it take to get the quotes you want? How do the subcontractors behave around the stand-in customer? All of these are conditions that you should investigate before allowing the subs on a real job. If they fail your test, you might have saved a job. As you know, undependable subcontractors can cost you a job. By testing the subs with a mock estimate, you can determine how they will perform on a real estimate.

Asking about tools and equipment

Tools and equipment are another consideration when you are judging subcontractors. If your subcontractors don't have the necessary tools and equipment, they won't be able to give you the service you desire. Don't hesitate to inquire about the tools and equipment the subcontractors possess.

Verifying insurance coverage

Insurance coverage is a big deal. With the way society is obsessed with lawsuits, you cannot afford to use subcontractors who are not properly insured. It is easy to lose your business to a court decision. For your own protection, you must make sure your subcontractors carry all the insurance they should.

What type of insurance should you be concerned with? Liability insurance should be a mandatory requirement. If the sub has employees other than close family members, workman's compensation insurance is needed. Even if the subcontractor is not required to carry worker's comp, you should have a waiver signed by the business owner. The waiver, which should be prepared by your lawyer, can protect you from claims and insurance audits.

If you use the services of subcontractors who are not properly insured, you might have to pay up at the end of the year. When your insurance company audits you, as they usually do, you are responsible for paying penalties if you used improperly insured contractors. These penalties can amount to a substantial sum of money. To avoid losing money, make sure your subcontractors are currently insured for all necessary purposes.

Considering specialists

Many subcontractors have specialties. When you are dealing with specialists, you might pay extra, but the end result could be a bargain. How can you pay more and come out of the job with more profits? While specialists might charge higher fees, they are often worth the extra cost. Why are they worth more? Specialists are just that—specialists. As specialists, these subcontractors can often do a better job and do it faster. Remember, time is money, and when you save time, you have a chance to make more money. With this in mind, ask potential subcontractors what they specialize in. You might find it cost-effective to use different subs for different jobs.

Checking licenses

Licenses are another issue you should investigate when rating subcontractors. If subcontractors are not licensed legally, you can get into deep trouble. It is imperative for you to engage only subcontractors who meet standard licensing requirements. If you use unlicensed subcontractors, you are flirting with disaster.

Checking on the work force

A subcontractor's work force is another consideration in rating the desirability of the sub. How much work can the contractor handle? You don't want to take on a sub that cannot handle your workload. For this reason, you must know what the capabilities of the subcontractors are.

If you give a small contractor too much work, you can find yourself in a bind. The small contractor might frantically add employees to keep your business. In doing this, the subcontractor is likely to pick up some undesirable help. This undesirable help can make your company look bad. It is better to have multiple subcontractors than it is to have one contractor who can't give the service you need.

While it is more convenient to work with a single subcontractor, it might not be feasible. At times you need more than one sub in each trade. As a safety precaution, you should have at least three subcontractors in each trade. This depth of subcontractors gives you more control.

Controlling subcontractors

Controlling subcontractors is much easier when you follow some simple rules. What are the rules? The most important rule is to document your dealings in writing. Other rules include:

- Create and use a subcontractor policy.
- Be professional and expect professionalism from the subs.
- Use written contracts with all of your subcontractors.
- Use change orders for all deviations in your agreement.
- Dictate start and finish dates in your agreement.
- Penalize subcontractors for being late in finishing jobs.
- Always have subs sign lien waivers when they are paid.
- Keep certificates of insurance on file for each sub.
- Don't allow extras, unless they are agreed to in writing.
- Don't give advance contract deposits.
- Don't pay for work that hasn't been inspected.
- Use written instruments for all your business dealings.

Subcontractors can take advantage of you, if you let them. However, if you establish and implement a strong subcontractor policy, you should be able to handle your subs. It is imperative that you remain in control. If subcontractors have the lead role, your company will be run by the subs.

DEALING WITH SUPPLIERS

Dealing with suppliers is not as simple as placing an order and waiting. Your business depends on the performance of suppliers, and it is up to you to set the pace for all of your business dealings. This is not to say that you should be a radical dictator, but you should call the shots.

Establish a routine with your suppliers. If you are going to use purchase orders, use them with every order. When you want job names written on your receipts, insist that they are always included. Are you going to allow employees to make purchases on your credit account? If so, set limits on how much can be purchased, and make sure everyone at the supply house knows which employees are authorized to charge on your account.

Get to know the manager of the supply house. Without a doubt, at some time you and the manager will have a problem to solve. At these times it helps to know each other.

When you begin to use a new supplier, make sure you understand the house rules. What is the return policy? Will you be charged a restocking fee? Will you get a discount if you pay your bill early? What is your discount percentage? Will the discount remain the same regardless of the volume you purchase? These are just some of the questions you need answered.

If all goes well, you will be doing a lot of business with your suppliers. Since each of you depends on the other to make money, you should develop the best relationship possible.

Cutting your best deal

How do you know when you are cutting your best deal? Is price the only consideration in the purchase of materials or the selection of subcontractors? No; price is not the only consideration; service and quality are two factors that must be pondered.

Getting the lowest price doesn't always mean you are getting the best deal. If you don't get quality and service to go along with a fair price, you are probably asking for trouble. Let me give you a few examples.

Assume you have requested bids (FIG. 20-10) from five painters. You accept the lowest bid based on price alone. When the painters are scheduled to start the job, they don't show up. After calling and insisting that they be on the job by the next morning, you get some satisfaction. The painters show up and start to work. You go back to your office, and at noon, the homeowner calls, wanting to know where the painters are. You find out the painters left for a morning break and never came back.

How does this make you and your company look to the customer? Not very good, I'm afraid. What does the slow-down do to your cash flow? It crimps it, of course. You got the cheapest painter you could find, but your great deal doesn't look so good now. This type of problem is common, and you have to do a better job of finding suitable subcontractors in the future.

For the next example, assume you have ordered roof trusses from your supplier. After shopping prices, you decided to go with the lowest price, even though

Bid Request

Customer name: _____

Customer address: _____

Customer city/state/zip: _____

Customer phone number: _____

Job location: _____

Plans and specifications dated: _____

Bid requested from: _____

Type of work: _____

Description of material to be quoted: _____

All quotes to be based on attached plans and specifications. No substitutions allowed without written consent of customer.

Please provide quoted prices for the following: _____

_____

All labor, materials, permits, and related fees to complete plumbing as per attached plans and specifications.

All bids must be submitted by: _____

20-10 Bid request.

you had never dealt with the supplier before. The trusses are ordered and you are given a delivery date. All of your work is scheduled around the delivery of the trusses.

The trusses are to be used to replace a rotted roof structure. You can't tear off the old roof until you know the trusses are available. On the day of delivery, you call the supplier and inquire about the status of the trusses. You're told the trusses are on a delivery truck and will be on your job by mid-morning.

Your crew finishes framing work and is waiting to set trusses. It's nearly noon, and the trusses have not yet been delivered. Your crew is at a standstill. A phone call to the supplier reveals that the delivery truck broke down on the way to the job. You're told the trusses won't arrive until the next morning. Now what are you going to do? You should have gotten the trusses on the job earlier, but you didn't. Needless to say, you've got a problem.

When you ask the supplier to transfer the trusses to a different delivery truck, so you can get them immediately, you're told that the supplier doesn't have another truck capable of transporting the trusses. As it turns out, you have to abandon the job, losing time and money. You also lose credibility with your customers.

Would this have happened if you had used your regular supplier? Probably not, because your regular supplier has enough trucks to make a switch if necessary. Your great deal on inexpensive trusses has turned into a major flop. So, you see, price isn't everything.

Expediting materials

Learning the secrets of expediting materials keeps your business running on the fast track. For most businesses, work cannot get done unless materials are available. If work doesn't get done, money isn't earned. Since business owners are in business to make money, they need to keep materials available.

Material handling accounts for much of the time that contractors lose on a daily basis. If a worker has to leave the job to go pick up materials at a supply house, time and money are lost. Inaccurate take-offs and deliveries that are poorly managed can cost contractors thousands of dollars. Can you do anything to reduce these losses? Yes, by expediting materials, you can save time and make more money.

Large companies have people that do little more than expedite materials. For these companies, expediting materials is a full-time job. However, most contractors don't have an employee with the sole responsibility of getting materials on the job. These contractors must do their own material acquisitions and handling. When a person has to tend to multiple tasks, it is easy for some part of the job to be neglected. In contracting, the expediting of materials is often pushed aside to make room for more pressing duties.

All too many contractors call in a material order and forget about it. They don't make follow-up calls to check the status of the material. It is not until the material doesn't show up that these contractors take action. By then, time and money is being lost.

A large number of contractors never inventory materials when they are delivered. If 100 sheets of plywood were ordered, they assume they received 100 sheets of plywood. Unfortunately, mistakes are frequently made with material deliveries. Quantities are not what they are supposed to be. Errors are made in the types of materials shipped. All of these problems add up to more lost time and money.

If you want to make your jobs run smoother, take some time to perfect your control over materials. When you place an order, have the order-taker read the order back to you. Listen closely for mistakes. Call in advance to confirm delivery dates. If a supplier has forgotten to put you on the schedule, your phone call can correct the error before it becomes a problem.

When materials arrive, check the delivery for accuracy. Ideally, this should be done while the delivery driver is present. If you discover a problem with your order, call the supplier immediately. By catching blunders early, you can reduce your losses.

Keeping a log of material orders and delivery dates is one way of staying on top of your materials. One glance at the log lets you know the status of your orders.

When you talk to various salespeople, record their names in your log. If there is a problem, it always helps to know who you talked to last. Get a handle on your materials, and you can enjoy a more prosperous business.

AVOIDING COMMON SUPPLIER AND SUBCONTRACTOR PROBLEMS

By avoiding common supplier and subcontractor problems, you can spend more time making money. The two biggest reasons for problems between contractors and suppliers or subcontractors are poor communication and money. Money is usually the largest cause of disputes, and communication breakdowns cause the most confusion. If you can conquer these barriers, your business will be more enjoyable and more profitable.

There are few excuses for problems in communication if you always use written agreements. When you give a subcontractor a spec sheet that calls for a specific make, model, color, and whatever, you eliminate confusion. If the subcontractor doesn't follow the written guidelines, an argument might ensue, but you will be the victor.

As for money, written documents can solve most of the problems caused by cash. When you have a written agreement that details a payment schedule, there is little room for disagreement. By using written agreements, you can eliminate most of the causes of aggravation and arguments. It's a good idea to create a boilerplate bid form to use in conjunction with plans and specifications. Create separate forms that list each item that must be completed at each phase of work. You'll have separate forms for insulation, flooring, site work, framing, septic work, and so on. When these forms are geared to specific trades, you can eliminate confusion and mistakes during the bidding process (FIGS. 20-11 through 20-25).

Job Name: _____

Phase: Insulation

Contractor: _____

All Work To Be Done According To Attached Specifications

Bid Item

    Supply and install all insulation

20-11 Insulation bid sheet.

Job Name: _____

Phase: Framing

Contractor: _____

All Work To Be Done According To Attached Specifications

Bid Item

    Supply labor to frame house to a dried-in condition
    If a crane is needed, it will be at the framing contractor's expense
    Install all exterior windows and doors
    Subfloors are to be glued and nailed
    Provide access for bathtubs and showers
    Install ceiling strapping
    Install all steel beams and plates as might be required
    Install all support columns
    Build stairs during initial framing

20-12 Framing bid sheet.

Job Name: _____

Phase: Flooring

Contractor: _____

All Work To Be Done According To Attached Specifications

Bid Item

    Provide price for supplying and installing underlayment
    Provide price for labor and material to prepare all floor surfaces
    Provide price to supply and install flooring as specified

20-13 Flooring bid sheet.

Job Name: _____

Phase: Well

Contractor: _____

All Work To Be Done According To Attached Specifications

Bid Item

Supply labor and material to install drilled well with steel casing and cap
Supply labor and material to install submersible pump and related equipment
Bid job on a per-foot basis and on a flat-fee basis

20-14 Well bid sheet.

Job Name: _____

Phase: Siding

Contractor: _____

All Work To Be Done According To Attached Specifications

Bid Item

Install siding materials provided by general contractor

20-15 Siding bid sheet.

```
Job Name: _____

Phase: Tree Clearing

Contractor: _____

All Work To Be Done According To Attached Specifications

Bid Item

    Cut all trees marked with blue ribbons
    Remove all wood, branches, brush, and debris from cutting procedure.
```

20-16 Tree clearing bid sheet.

```
Job Name: _____

Phase: Plumbing

Contractor: _____

All Work To Be Done According To Attached Specifications

Bid Item

    Supply and install all rough plumbing and plumbing fixtures, including bathing
    units
    Bid a separate price for well pump and related equipment
```

20-17 Plumbing bid sheet.

Job Name: _____

Phase: Drywall

Contractor: _____

All Work To Be Done According To Attached Specifications

Bid Item

Supply and install all materials needed to drywall all interior walls and ceiling to code requirements
Provide separate labor only price for hanging, taping, and finishing drywall
If heat is needed, drywall contractor shall supply it
Provide separate price for texturing ceilings

20-18 Drywall bid sheet.

Job Name: _____

Phase: Heating

Contractor: _____

All Work To Be Done According To Attached Specifications

Bid Item

Supply and install all rough heating materials and finished heating equipment including boiler and baseboard units

20-19 Heating bid sheet.

Job Name: _____

Phase: Foundation

Contractor: _____

All Work To Be Done According To Attached Specifications

Bid Item

Supply labor and material for footings
Supply labor and material for foundation walls and piers
Supply and install foundation windows/vents
Supply labor and material to create bulkhead opening, ready for door installation
Supply and install foundation bolts
Remove all foundation clips
Waterproof foundation to finished grade level
Supply labor and material to install concrete basement floor

20-20 Foundation bid sheet.

Job Name: _____

Phase: Paint

Contractor: _____

All Work To Be Done According To Attached Specifications

Bid Item

Provide price for labor and material to paint, stain, and/or seal all surfaces specified
Price should include all preparation work required (i.e., filling nail holes)

20-21 Painting bid sheet.

Job Name: _____

Phase: Electrical Work

Contractor: _____

All Work To Be Done According To Attached Specifications

Bid Item

    Supply and install temporary power pole
    Supply and install all rough wiring
    Install light fixtures supplied by general contractor
    Electrical contractor to provide GFI devices and smoke detectors
    Supply all needed permits and inspections

20-22 Electrical bid sheet.

Job Name: _____

Phase: Site Work

Contractor: _____

All Work To Be Done According To Attached Specifications

Bid Item

    Remove all tree stumps and debris from any excavation
    Supply and install metal culvert pipe for driveway
    Install driveway—site contractor to furnish all materials
    Dig foundation hole
    Provide rough grading
    Backfill foundation
    Perform final grading
    Seed and straw lawn
    Install septic system
    Supply and install foundation drainage
    Supply and install crushed stone for foundation
    Dig trenches for water service and sewer
    Backfill trenches for water service and sewer

20-23 Site work bid sheet.

Job Name: _____

Phase: Trim

Contractor: _____

All Work To Be Done According To Attached Specifications

Bid Item

Supply labor to install trim materials supplied by general contractor
Provide separate price for installing counters and cabinets in all areas
Trim price should include hanging all interior doors, installing window and
door hardware, and bath accessories

20-24 Trim bid sheet.

Job Name: _____

Phase: Roofing

Contractor: _____

All Work To Be Done According To Attached Specifications

Bid Item

Install roofing materials provided by general contractor

20-25 Roofing bid sheet.

BUILDING GOOD RELATIONS WITH CODE OFFICERS

Building good relations with code officers is an important part of most contracting businesses. If your business depends on the approval of code officers, you'll do well to get to know the inspectors.

Code officers are often scorned. Contractors who have problems with inspectors cuss them and buck against the system. If these contractors would direct the same amount of energy in a more productive direction, they could solve their problems.

Like it or not, code officers are a fact of life for most contractors. The relationship between code officers and contractors can go one of two ways—good or bad. As a contractor, you can influence which way the pendulum swings.

If you want to make your life easier, get to know your code officers. I'm not saying you have to become best buddies, but at least be civil. Your attitude has a great deal of influence on the posture assumed by the code officer. Don't be afraid to smile and talk with your inspectors. If you get to know each other, problems can be easier to resolve.

AVOIDING REJECTED CODE-ENFORCEMENT INSPECTIONS

One of your goals must be avoiding rejected code-enforcement inspections. This goal is not difficult to achieve. Work gets rejected because it is not in compliance with local code requirements. If you know and understand the code requirements, you shouldn't get many rejection slips. If you don't understand a portion of the code, consult with a code officer. It is part of an inspector's job to explain the code to you. When the work of your subcontractors is rejected, notify them, in writing (FIG. 20-26), of their code violation.

Again, attitude can have a bearing on the number of rejections you get. If you walk around with a chip on your shoulder, inspectors might look a little more closely for minute code infractions. If you play by the rules, you won't have much trouble with the officials. But don't ever try to put one over on a code officer. If you get caught, your life on the job can be miserable for a long time to come. Inspectors can be a close-knit group. When you con one, others get the word, and your work will be put under a microscope.

Learning to work well with subcontractors, suppliers, and code officers is essential to the success of a building business. Work hard to develop and keep good relationships. You are dependent on a lot of people when you build homes for a living.

Code Violation Notification

Contractor: _____

Contractor's address: _____

City/state/zip: _____

Phone number: _____

Job location: _____

Date: _____

Type of work: _____

Subcontractor: _____

Address: _____

Official Notification of Code Violations

On March 22, 1993, I was notified by the local code enforcement officer of code violations in the work performed by your company. The violations must be corrected within two business days, as per our contract dated March 1, 1993. Please contact the codes officer for a detailed explanation of the violations and required corrections. If the violations are not corrected within the allotted time, you may be penalized, as per our contract, for your actions in delaying the completion of this project. Thank you for your prompt attention to this matter.

_____

General Contractor   Date

20-26 Code violation notification.

21
Bidding: Methods that really work

Building clientele is one of your most important jobs as a business owner. Learning how to win bids is one of the most effective ways to build your business clientele. How many times have you bid a job and never heard back from the potential customer? Many contractors never figure out how to win bids. This chapter shows you how to win bids and build up your business.

GETTING WORD-OF-MOUTH REFERRALS

Word-of-mouth referrals are the best way to get new business. Of course, you need some business before you can benefit from word-of-mouth referrals. But every time you get a job, you need to work that customer for referrals that can lead to more work.

Getting referrals from existing customers is not only the most effective way to generate new business, it's the least expensive. Advertising is expensive. For every job you get from advertising, you are losing a percentage of your profit to the cost of advertising. If you can turn up new work from talking with existing customers, you eliminate the cost of advertising.

If you do good work and take care of your customers, referrals are easy to get, but you might have to ask for them. People sometimes might give your name and number to friends, and they occasionally write nice letters. However, to make the most of word-of-mouth referrals, you have to learn to ask for what you want. Let's see what it takes to get the most mileage out of your existing customers.

The groundwork

Laying the groundwork is an important step in getting a strong portfolio of customer referrals. If you don't make your customers happy, they are going to talk to their friends, but they won't be saying what you want people to hear. People quickly reveal their bad experiences, but they are not so quick to spread the good word. To get out the message you want, you have to work hard at pleasing your customers.

To lay the groundwork, you must start with the first contact you have with customers and maintain your efforts through to the end. Many contractors start off on the right foot, only to stumble before the job is done. I have seen many jobs go sour in the final days of completion. One of the largest mistakes I have seen contractors make is not responding promptly to warranty calls. If you are in business for the long haul, you don't want to alienate customers even after the job is done. Old customers often become repeat customers. If you don't respond to call-backs, you won't be called when new, paying work needs to be done.

On the job

During the job you must cater to the customer. Most contractors don't have a problem with this aspect of customer satisfaction, but making customers happy takes more than doing good work. You have to fulfill your promises, be punctual, be respectful, and be professional.

At the end of the job

At the end of the job you have to ask for referrals. Don't expect customers to run up to you and hand you a letter of reference. Often asking for a letter of reference isn't enough. It helps if you provide a form for the consumer to fill out. People never seem to know what to say in a reference letter. They are much more comfortable filling out forms.

If you design a simple form, almost all satisfied customers will complete and sign it. I'm sure you have seen these quality-control forms in restaurants and with mail-order shipments. You can structure the form in any fashion you like.

Once you have designed and printed your forms, use them. When you are completing a job, ask your customer to fill out and sign your reference form. Do it on the spot. Once you are out of the house, getting the form completed and signed is more difficult.

As you begin building a good collection of reference forms, don't hesitate to show them to prospective customers. Use an attractive three-ring binder and clear protective pages to store and display your hard-earned references. When you get enough reference letters, you have strong ammunition to close future deals.

ENSURING CUSTOMER SATISFACTION

Customer satisfaction is a key to success. Business builds on itself when customers are satisfied. Oh sure, you might never be able to satisfy some people. Every business owner seems to have these hard-to-please customers. If you haven't run into a habitual complainer, you will if you're in business long enough. Leaving this small segment of the population out of the picture, let's concentrate on how to please the majority of your customers.

Customers like to feel comfortable with their contractors. To make customers comfortable, you have to deal with them on their level. You have to learn how to play the give-and-take game. Communication skills are essential to a good relationship. If you and the customers can't communicate, you won't get far in your business dealings. Occasionally, you have to babysit customers. You might not like having to soothe customers, but at times you have to smooth the feathers of ruffled clients.

BUILDING A NEW CUSTOMER BASE

You can start reaching out for a new customer base through bid sheets. Bid sheets are open to all reputable contractors. When a job is placed on a bid sheet, somebody is going to get the job. The jobs put out to formal bids are almost always done. Unlike common residential estimates, where potential customers can change their minds, formal bid sheets are rarely changed. This type of work is very competitive and the percentage of profit is usually low, but bid work can pay the bills.

Where do you get bid sheets?

Where do you get bid sheets? Bid sheets can be obtained by responding to public notices in newspapers and by subscribing to services that provide bid information. If you watch the classified section of major newspapers, you'll see advertisements for jobs going out for bids. You can receive bid packages by responding to these advertisements. Normally, you get a set of plans, specifications, bid documents, bid instructions, and other needed information. These bid packages can be simple or complicated.

What is a bid sheet?

What is a bid sheet? A bid sheet is a formal request for price quotes. There is a difference between a bid sheet and a bid package. The bid sheet gives a brief description of the work available. A bid package gives complete details of what is expected from bidders. Most contractors start with a bid sheet and if they find a job of interest, order a bid package. Bid sheets are usually provided free of charge. Bid packages often require either a deposit or a nonrefundable fee.

What are bidder agencies?

Bidder agencies are businesses that provide listings of bid opportunities. These listings are normally published in a newsletter form. The bid reports are generally delivered to contractors on a weekly basis. Each bid report might contain five jobs or 50 jobs. These publications are an excellent way to get leads on all types of jobs.

What type of jobs are on bid sheets?

What type of jobs are on bid sheets? All types of jobs appear on bid sheets. They can range from small residential jobs to large commercial jobs. The majority of the jobs are commercial. The sizes of the jobs range from a few thousand dollars on up into the millions of dollars.

What about government bid sheets?

Government bid sheets are another opportunity for finding an abundance of work. Like other bid sheets, government bid sheets give a synopsis of the job description and provide information for obtaining more details. Government jobs can range from replacing a dozen lavatory faucets to building a commissary. Building new base housing units can provide many months of work for a home builder.

Government jobs are a safe bet for getting your money. The money might be slow in coming, but it is coming. The paperwork involved with government jobs can be excessive. If you are not willing to deal with mountains of paperwork, stay away from government bids.

CHECKING YOUR BONDABILITY

Are you bondable for large jobs? Many of the jobs found on bid sheets require contractors to be bonded. Bonds are obtained from bonding companies and insurance companies, but not all contractors are bondable. Before you try bidding jobs that require bonding, check to see if you are bondable. The requirements for being bonded vary. Check in your local phone book for an agency that does bonding and call to inquire about the requirements.

Performance and security bid bonds are a necessity with many major jobs. If you order a bid sheet or package, you'll almost certainly see that a bond is required. Some listings on bid sheets might not require a bond. It is common for bid requirements to be tied to the anticipated cost of the job. The bigger the job, the more likely it is a bond is required.

Bonds are required to ensure the success of the job. The people offering the work want to be sure that the job will be done right and that it will be completed. When the person or firm issuing the work requires a bond, they establish a degree of safety.

It is very difficult for some new businesses to obtain a bond. If the new company doesn't have strong assets or a good track record, getting a bond can be tough. Unfortunately, a bond can be a hurdle you can't get over until you don't care whether you have it or not.

There are three basic types of bonds to be considered. These bonds are bid bonds, performance bonds, and payment bonds. Each type of bond serves a different purpose. A bid bond is put up to assure the person receiving bids that the bidding contractor will honor the bid if a contract is offered.

Performance bonds prevent contractors from abandoning a job and leaving the customer in dire financial straits. If a contractor reneges on completing the job, the customer can hold the performance bond for financial damages.

Payment bonds are used to guarantee payment to all subcontractors and suppliers used by a contractor. These bonds eliminate the risk of mechanic and materialman liens being filed against the property where work is being done.

When you put up a bond, the value of the bond is at risk. If you default on your contract, you lose your bond to the person who contracted you for the job. Since many people use the equity in their home for collateral to get a bond, they can lose their houses if they default on bonds. Bonds are serious business. If you can get a bond, you have an advantage in the business world. Talk with local companies that issue bonds to see if you can qualify for bonding.

TAKING ON THE RISKS OF BIG JOBS

Are big jobs surrounded by big risks? You bet they are. All jobs carry risk, but big jobs carry big risks. Should you shy away from big jobs? Maybe, but if you go into the deal with the right knowledge and paperwork, you should survive and possibly prosper.

Cash-flow problems

Cash flow is frequently a problem for contractors doing big jobs. Unlike small residential jobs, big jobs don't generally allow contractors to receive cash deposits. If you tackle these jobs, you have to work with your own money and credit. For a new business, the money needed to get to a draw disbursement in big jobs can be the undoing of the company. It's wonderful to think of signing a million-dollar job in your first year of business, but that job could put your business into bankruptcy court.

Before you dive into deep water, make sure you can get to the other side. Some lenders allow you to use your contract as security for a loan, but don't bet your business on it. If you want to take on a big job, get your finances in order first.

Slow pay

Slow pay can be another problem with big jobs. Large jobs are notorious for slow pay. It's not that you won't get paid, but you might not get paid in time to keep your business going. New businesses are especially vulnerable to slow pay. When you move into the big leagues, be prepared to hold your financial breath for a while. The check you thought would come last month might not show up for another 90 days.

No pay

Slow pay is bad, but no pay is worse. Just as new contractors can get in trouble with large jobs, developers and general contractors can get into financial difficulty with big jobs. Most of the people spearheading big jobs don't intend to stick their subcontractors, but sometimes they do. When the top dogs get in over their heads, they can't pay their bills.

If the subcontractors don't get paid, suppliers don't get paid. The ripple effect continues. Anyone involved with the project is going to lose. Some lose more than others. Generally, when these big jobs go bad, the banks or lenders financing the whole job foreclose on the property. These lenders normally hold a first mortgage on the property.

If you're working as a subcontractor for a large outfit, filing mechanic liens is the best course of action when your customer refuses to pay you. If a contractor hasn't been paid for labor or materials, a mechanic lien can usually be levied against the property where the labor or materials were invested. If you have to file a lien, make sure you do it right. There are rules you must follow in filing and perfecting a lien. You can file your own liens, but I recommend working with an attorney on all legal matters.

Even after you file and perfect your lien, you might not get your money. If you get any money, it will likely be a settlement for a reduced amount. You can never quite get the taste of sour jobs out of your mouth.

Completion dates

Completion dates can also wreak havoc with the inexperienced contractor. Big jobs often include a time-is-of-the-essence clause. Along with this clause is usually a penalty fee that must be paid if the job is not finished on time. The penalty is normally based on a daily fee. For example, you might have to pay $200 per day for every day the job runs past the deadline.

Penalty fees and the possible loss of your bond can ruin your business. Contractors with limited experience in big jobs are often unprepared to project solid completion dates. Don't sign a contract with a completion date you are not sure you can meet.

PARTICIPATING IN THE BID PROCESS

When you learn how to eliminate your competition in the bid process, you are on your way to a successful business. There is no shortage of competition in most fields of contracting. There are, however, often shortages of work. With the combination of limited work and unlimited competition, a new business owner, or any business owner for that matter, can get discouraged. But don't—there are ways to thin out the competition.

Beating the competition with bid sheets

Beating the competition with bid sheets is hard. Unless you have a track record and are well known, money talks. Low prices are what most decision-makers are looking for in bids that are the result of bid sheets. Being the low bidder can get you the job, but you might wish you had never seen the job. Don't bid a job too low. It doesn't do you any good to have work if you're not making money.

How can you improve your odds in mass bidding? If you can get bonded, you have an edge. A lot of bidders can't get bonded. This fact alone can be enough to cull the competition. When you prepare your bid package for submission, be meticulous. All you have going for you is your bid package. If you want the job, spend adequate time to prepare a professional bid packet.

Making in-person bids

If you are dealing with in-person bids, follow the guidelines found throughout this book. The basic keys include dress appropriately, drive the right vehicle, be professional, be friendly, get the customer's confidence, produce photos of your work, show off your letters of reference, give your bid presentation in person, and follow up on all your bids.

MAKING ACCURATE TAKE-OFFS

To make sound bids, you must be adept at preparing accurate take-offs (FIG. 21-1). It doesn't matter if you use a computerized estimating program or a pen and paper; you must get your facts straight. If you miss items on the take-off and get the job, you can lose money. If you overestimate the take-off, your price will be too high. An accurate take-off is instrumental in winning a job.

A take-off is a list of items needed to do a job. Take-offs are the result of reading blueprints or visiting the job site and making a list of everything you need to do the job. Some estimators are wizards with take-offs, and others have a hard time trying to project all of their needs. If you can't discipline yourself to learn how to do an accurate take-off, your venture into contracting is going to be a rough road to travel.

Using take-off forms

You can reduce your risk or errors by using a take-off form (FIG. 21-2). If you use a computerized estimating program, the computer files probably already contain forms. You might want to customize the standard computer forms. Whether you are using stock computer forms or making your own forms, you must be sure they are comprehensive.

Materials Take-off List.

Item name or use of piece	No. of pieces	Unit	Length in place	Size	Length	No. per length	Quantity
1. Footers	45	Pc	1'5"	2×6	10'	7	7
2. Spreaders	30	Pc	1'4"	2×6	8'	6	5
3. Foundation post	15	Pc	3'0"	6×6	12'	4	4
4. Scabs	20	Pc	1'0"	1×6	8'	8	3
5. Girders	36	Pc	10'0"	2×6	10'	1	36
6. Joists	46	Pc	10'0"	2×6	10'	1	46
7. Joist splices	21	Pc	2'0"	1×6	8'	4	6
8. Block bridging	40	Pc	1'10⅜"	2×6	8'	4	10
9. Closers	12	Pc	10'0"	1×8	10'	1	12
10. Flooring	800	BF	RL	1×6	RL	—	—

Department of the Army

21-1 Take-off list.

Take-off forms should have every item you might use in various types of jobs listed. It's best if you create forms that list every expense that you might incur on a job (FIGS. 21-3 through 21-10) to improve your odds of reducing ommissions when figuring bid prices. In addition, the form should have places that allow you to fill in blank spaces with specialty items.

Job Take-Off Form

Job Name: _____

Job Address: _____

Item	Quantity	Description
2" pipe	100'	PVC
4" pipe	40'	PVC
4" clean-out w/plug	1	PVC
2" quarter-bend	4	PVC
2" coupling	3	PVC
4" eighth-bend	2	PVC
Glue	1 quart	PVC
Cleaner	1 quart	PVC
Primer	1 quart	PVC

21-2 Take-off form.

Phase	Vendor	L/M	Price	Notes
Credit check	_____	_____	_____	_____
Plans/specs	_____	_____	_____	_____
Loan application fees	_____	_____	_____	_____
Appraisal & credit	_____	_____	_____	_____
Points/closing, 1st loan	_____	_____	_____	_____
Construction interest	_____	_____	_____	_____
Loan fees for 2nd loan	_____	_____	_____	_____
Points/closing 2nd loan	_____	_____	_____	_____
Survey costs	_____	_____	_____	_____
Homeowner's insurance	_____	_____	_____	_____
Land cost	_____	_____	_____	_____
Builder's risk insurance	_____	_____	_____	_____
Permits	_____	_____	_____	_____
Zoning approval	_____	_____	_____	_____
DEP approval	_____	_____	_____	_____
Title work	_____	_____	_____	_____
Driveway permit	_____	_____	_____	_____
Access granted	_____	_____	_____	_____
Roads posted	_____	_____	_____	_____
Soils test	_____	_____	_____	_____
Ledge to deal with	_____	_____	_____	_____
Utilities available	_____	_____	_____	_____
Sewer fees	_____	_____	_____	_____
Water fees	_____	_____	_____	_____
Association fees	_____	_____	_____	_____
Trees marked for cutting	_____	_____	_____	_____

21-3 Estimating form for prepurchase phase.

Phase	Vendor	L/M	Price	Notes
Site & septic work 1				
Site & septic work 2				
Site & septic work 3				
Tempory power				
Electrical 1				
Electrical 2				
Electrical 3				
Well 1				
Well 2				
Well 3				
Foundation 1				
Foundation 2				
Foundation 3				
Framing lumber 1				
Framing lumber 2				
Framing lumber 3				
Framing crew 1				
Framing crew 2				
Framing crew 3				
Windows & doors 1				
Windows & doors 2				
Windows & doors 3				
Siding 1				
Siding 2				
Siding 3				
Notes				

21-4 Estimating form for site work.

Phase	Vendor	L/M	Price	Notes
Drywall 1	_____	_____	_____	_____
Drywall 2	_____	_____	_____	_____
Drywall 3	_____	_____	_____	_____
Paint 1	_____	_____	_____	_____
Paint 2	_____	_____	_____	_____
Paint 3	_____	_____	_____	_____
Trim material 1	_____	_____	_____	_____
Trim material 2	_____	_____	_____	_____
Trim material 3	_____	_____	_____	_____
Trim crew 1	_____	_____	_____	_____
Trim crew 2	_____	_____	_____	_____
Trim crew 3	_____	_____	_____	_____
Cabinets 1	_____	_____	_____	_____
Cabinets 2	_____	_____	_____	_____
Cabinets 3	_____	_____	_____	_____
Flooring 1	_____	_____	_____	_____
Flooring 2	_____	_____	_____	_____
Flooring 3	_____	_____	_____	_____
Bath accessories 1	_____	_____	_____	_____
Bath accessories 2	_____	_____	_____	_____
Bath accessories 3	_____	_____	_____	_____
Light fixtures 1	_____	_____	_____	_____
Light fixtures 2	_____	_____	_____	_____
Light fixtures 3	_____	_____	_____	_____
Notes	_____	_____	_____	_____
Notes	_____	_____	_____	_____

21-5 Estimating form for drywall and painting phase.

Phase	Vendor	L/M	Price	Notes
Siding crew 1				
Siding crew 2				
Siding crew 3				
Roof material 1				
Roof material 2				
Roof material 3				
Roofer 1				
Roofer 2				
Roofer 3				
Steel 1				
Steel 2				
Steel 3				
Plumbing 1				
Plumbing 2				
Plumbing 3				
Heating 1				
Heating 2				
Heating 3				
Insulation 1				
Insulation 2				
Insulation 3				
Flue/fireplace 1				
Flue/fireplace 2				
Flue/fireplace 3				
Notes				
Notes				

21-6 Estimating form for insulation and exterior work.

Phase	Vendor	L/M	Price	Notes
Appliances 1	_____	_____	_____	_____
Appliances 2	_____	_____	_____	_____
Appliances 3	_____	_____	_____	_____
Appliances 4	_____	_____	_____	_____
Appliances 5	_____	_____	_____	_____
Appliances 6	_____	_____	_____	_____
Appliances 7	_____	_____	_____	_____
Appliances 8	_____	_____	_____	_____
Appliances 9	_____	_____	_____	_____
Appliances 10	_____	_____	_____	_____
Appliances 11	_____	_____	_____	_____
Appliances 12	_____	_____	_____	_____
Appliances 13	_____	_____	_____	_____
Appliances 14	_____	_____	_____	_____
Appliances 15	_____	_____	_____	_____
Final survey	_____	_____	_____	_____
Builder contributions	_____	_____	_____	_____
Notes	_____	_____	_____	_____
Notes	_____	_____	_____	_____
Notes	_____	_____	_____	_____

21-7 Estimating form for appliances.

Option	Vendor	L/M	Price	Notes
Gutters				
Rain diverters				
Shutters				
Window screens				
Garage				
Deck				
Fireplace				
Power venter				
Flue				
Domestic coil				
Hot-water tank				
Cleaning				
Trash removal				
Landscaping				
Walkways				
Porches				
Vented range hood				
Tub/shower doors				
Sump pump & piping				
Flood lights				
Ceiling lights				
Ceiling fans				
Overhead/underground electrical service				
Wall wrap				
Rigid foam insulation				
Pull-down attic stairs				

21-8 Estimating form for miscellaneous exterior and interior work.

Option	Vendor	L/M	Price	Notes
Solid trim	_____	____	_____	_____
Stained trim	_____	____	_____	_____
Solid doors	_____	____	_____	_____
Stained doors	_____	____	_____	_____
Valance over kitchen cabinets	_____	____	_____	_____
Dishwasher	_____	____	_____	_____
Garbage disposer	_____	____	_____	_____
Automatic garage door	_____	____	_____	_____
Opener	_____	____	_____	_____
Tile work	_____	____	_____	_____
Wood floors	_____	____	_____	_____
Fancy handrails	_____	____	_____	_____
Wood stairs	_____	____	_____	_____
Security system	_____	____	_____	_____
Cable television pre-wire	_____	____	_____	_____
Telephone pre-wire	_____	____	_____	_____
Intercom	_____	____	_____	_____
Additional oil tank	_____	____	_____	_____
Additional attic/crawl lighting	_____	____	_____	_____
Ground cover in crawl	_____	____	_____	_____
Mirrors	_____	____	_____	_____
Plywood instead of wafer board	_____	____	_____	_____
2 layers of subfloor or ¾ t & G	_____	____	_____	_____
Insulated wall sheathing	_____	____	_____	_____
Ridge vent	_____	____	_____	_____
Soffit vents	_____	____	_____	_____
Gable vents	_____	____	_____	_____

21-9 Estimating form for options.

Option	Vendor	L/M	Price	Notes
Bulkhead door	_____	____	_____	_____
Doorbell	_____	____	_____	_____
Outdoor lighting	_____	____	_____	_____
Range	_____	____	_____	_____
Refrigerator	_____	____	_____	_____
Wallpaper	_____	____	_____	_____
HOW warranty	_____	____	_____	_____
Deadbolt locks	_____	____	_____	_____

21-10 Estimating form for additional optional items.

The advantage of using take-off forms is that you are prompted on items you might otherwise forget. However, don't get into such a routine that you only look for items on your form. It is very possible a job might require something that you haven't yet put on the form. Forms help, but they are no substitute for thoroughness.

Keeping track of what you've already counted

Keeping track of what you've already counted is a problem for some contractors. If you are doing a take-off on a large set of plans, say a shopping mall, it can be tedious work. The last thing you need to have happen is to lose your place or forget what you've already counted. To avoid this problem, mark each item on the plans as you count it.

Building in a margin of error

You should build in a margin of error on your take-off. If you think you are going to need 100 sheets of plywood, add a little to your count. How much you add depends on the size of the job you are figuring. A lot of estimators build in a float figure of between 3 and 5 percent. Some contractors add 10 percent to their figures. Unless the job is small, I think a 10-percent add-on can cause you to lose the bid.

Of course, much of your cushion for mistakes depends on your ability to make an accurate take-off. If you are good with take-offs, a small percentage for oversights should be sufficient. If you always seem to get on the job and run short of materials, you'll need a bigger slush pile.

Keeping records

Keep records of your material needs (FIG. 21-11) on each job. Don't throw away your take-off. When the job is done, compare the material actually used (FIG. 21-12) with what you estimated in your take-off. This habit not only helps you to see where your money is going, it makes you a better estimator. By tracking your jobs and comparing final counts with original estimates, you can refine your bidding techniques and win more jobs.

Cost Projections

Item/Phase	Labor	Material	Total
Plans			
Specifications			
Permits			
Trash container deposit			
Trash container delivery			
Demolition			
Dump fees			
Rough plumbing			
Rough electrical			
Rough heating/ac			
Subfloor			
Insulation			
Drywall			
Ceramic tile			
Linen closet			
Baseboard trim			
Window trim			
Door trim			
Paint/wallpaper			
Underlayment			
Finish floor covering			
Linen closet shelves			
Closet door & hardware			
Main door hardware			
Wall cabinets			
Base cabinets			
Countertops			
Plumbing fixtures			
Trim plumbing material			
Final plumbing			
Shower enclosure			
Subtotal			

21-11 Cost projections.

Item/Phase	Labor	Material	Total
Light fixtures			
Trim electrical material			
Final electrical			
Trim heating/ac material			
Final heating/ac			
Bathroom accessories			
Clean up			
Trash container removal			
Window treatments			
Personal touches			
Financing expenses			
Miscellaneous expenses			
Unexpected expenses			
Margin of error			
Subtotal from first page			
Total estimated expense			

21-11 Continued.

PRICING YOUR SERVICES AND MATERIALS

Pricing your services and materials is an essential part of running a profitable business. If your prices are too low, you might be very busy, but your profits will suffer. If your prices are too high, you will be sitting around, staring at the ceiling, and hoping for the phone to ring. Somewhere between too low and too high is the optimum price for your products and services. The trick is finding out what those prices are.

You must learn how to make your prices attractive without giving away the store. How do you know what price is the right price? Well, you can't pull your prices out of thin air. You must establish your pricing structure with research—lots of research.

What kind of research is necessary to pick the proper pricing? Talking to real estate brokers and appraisers is one of the most effective ways to establish the market value of homes you plan to build. While real estate brokers can be very helpful, I would spend most of my time consulting with appraisers.

When I come up with a new house plan that I'm interested in building, I meet with a licensed appraiser to get an opinion of value. Since I am a licensed broker, I don't have to talk with real estate agents. I have access to multiple listing services that keep me apprised of what's on the market and what's been sold in recent months. By reviewing comparable sales and getting information from appraisers, I can target what the most realistic price should be for a home that I plan to build.

Some newcomers to the contracting business make a serious mistake. When these people learn what their competitors are charging, the rookies price their services far below the crowd, hoping to grab all the business. However, setting extremely low prices can be the same as setting a trap for yourself. If you set your prices

Job Cost Log

Job name _____

Mechanic _____

Truck _____

Date _____

Item	Quantity	Size

21-12 Job cost log.

far below your competitors, you alienate yourself from the competition and might lose respect from potential customers. If your prices are too low, customers might be afraid to use your services. As for your competition, if your prices are extremely low, they see you not only as competition but as a target. As a new company with low overhead, it's fine to work for less that the well-established companies, but don't price yourself into a deep hole.

When it comes to picking attractive prices, you must look below the surface. Many factors can control what you are able to earn. Let's take a look at what is considered a profitable markup.

Profitable markup on materials

What is a profitable markup on materials? This can be a hard question to answer. It is not difficult to project what a reasonable markup is, but defining a profitable markup is not so easy.

Some contractors feel a 10-percent markup is adequate. Others try to tack 35 percent onto the price of their materials. Which group is right? Well, you can't make that decision with the limited information I have given you. The contractors who charge a 10-percent markup might be doing fine, especially if they deal in big jobs and large amounts of materials. The members of the 35-percent group might be justified in their markup, especially if they are selling small quantities of lower-priced materials.

Markup is a relative concept. Ten percent of $100,000 is much more than 35 percent of $100. For this reason, you cannot blindly pick a percentage of markup to be your firm figure. The figure must to be adjusted to meet the changes in market conditions and individual job requirements. You can, however, pick percentage numbers for most of your average sales.

If you were in a repair business and were typically selling materials that cost around $20, a 35-percent markup would be fine, if the market would bear it. If you are building and selling houses, a 10-percent markup on materials should be sufficient. To some extent, you have to test the market conditions to determine what price consumers are willing to pay for your materials.

If you are selling common items that anyone can go to the local hardware or building supply and price out, you must be careful not to inflate your prices too much. Customers expect you to mark up your materials, but they don't want to be gouged. If you installed light bulbs with my new light fixture and charged me twice as much for the bulbs as what I could have bought them for in the store, I'm not going to be happy. Even though the amount of money involved in the light-bulb transaction is puny, the principle of being charged double for a common item annoys customers.

If you typically install specialty items, you can increase your markup. People are not as irritated to pay a well-marked-up but reasonable price for an unusual product. A markup of 20 percent is almost always acceptable on small residential jobs. When you decide to go above the 20 percent, do so slowly and while testing the response of your customers. Realistically, as a builder, you'll probably find that a 10-percent markup works best. You might be able to stretch it to 15 percent, but that's usually about the limit for big-ticket items.

Competitors' low prices

How can your competitors do work for such low prices? This question is one almost every contractor considers. There always seems to be a company that has a knack for winning bids and beating out the competition. If you know the bids are being won with low prices, you can't help but wonder how the winner of the bids can do it.

Low prices can keep companies busy, but that doesn't mean the low-priced companies are making a profit. Gross sales are important, but net profits are what business is all about. If a company is not making a profit, there is not much sense in operating the business.

Companies that work with low prices fall into several categories. Some companies work on a volume principle. By doing a high volume, the company can make less money on each job and still make a profit by doing so many jobs. This type of company is hard to beat.

When I was building in Virginia, I operated on a volume basis. My profit from a house was only about $7000, but I was building as many as 60 of them a year. If you do the math, you'll see that I wasn't working for peanuts. The volume principle worked well for me in Virginia, but it won't work for me in Maine. I probably built more houses in a year while working than Virginia than are built by all of the builders in my area of Maine in any given year. There simply isn't enough demand for housing in Maine to allow the volume approach to work very well.

Some companies sell at low prices out of ignorance. Many small business owners are not aware of the overhead expenses involved in business. For example, a carpenter who is making $16 an hour at a job might think that going into business independently and charging $25 an hour would be great. The carpenter might even start into business charging only $20 an hour. From the carpenter's perspective, he is making at least $4 an hour more than he was at his job. But is he really making $4 an hour more? Yes and no. He is being paid an extra $4 an hour, but he is not going to get to keep much of it.

When this low-priced carpenter comes into the business world, he might take a lot of work away from established contractors. Experienced contractors know they can't make ends meet by charging such low fees. But the carpenter, an inexperienced business owner, doesn't know all of the expenses he might encounter. Once the overhead expenses start eating away at what he thought was a great profit, the deal might not look so good.

This new businessman soon learns that overhead expenses are a force to be reckoned with. Insurance, advertising, call-backs, self-employment taxes, and a mass of other hidden expenses can erode any profits the plumber thought he was making. This type of inexperienced businessperson either goes out of business quickly or adjusts the prices of services and materials. For established contractors competing against newcomers, being able to endure the momentary drop in sales is enough to weather the storm. In a few months the new business will either be gone or up into a reasonably competitive range.

PRICING YOUR SERVICES FOR SUCCESS AND LONGEVITY

Pricing your services and materials is related directly to your success and longevity. It can be very difficult to decide on what the right price for your time and material is. Books give formulas and theories about how to set your prices, but these guides are not always right. Every town and every business dictate different factors in the prices the public is willing to pay. You can use many methods to find the best fees for your business to charge; let's look at some of them.

Pricing guides

Pricing guides can be a big help to the business owner with little knowledge of how to establish the value of labor and materials. However, these guides can cause you frustration and lost business. Most estimating guides provide a formula for adjusting the recommended prices for various regions. For example, a two-car garage that is worth $7500 in Maine might be worth $10,000 in Virginia. The formulas used to make this type of adjustment usually provide a number to multiply against the recommended price. By using the multiplication factor, you can derive a price for services and materials in any major city.

The idea behind these estimating guides is a good one, but the system has flaws. I have read and used many of these pricing books. From my personal experience, the books have not been accurate for the type of work I was involved in. I don't say this to mean the books are no good. At times the books are accurate, but I have never been comfortable depending solely on a mass-produced pricing guide.

I have found estimating books to be very helpful as a piece of the pricing puzzle. While I don't use pricing figures from these books as my only means of setting a price, I do use them to compare my figures and to ensure I haven't forgotten items or phases of work. Most bookstores carry some estimating guides in their inventory.

Research

Research is one of the most effective ways to determine your pricing. When you look back at historical data, you can find many answers to your questions. You can see how the economy swings up and down. You can see how prices have fluctuated over the years, and you can start to project the curve of the future. Historical data can be found by reading old newspaper ads, by researching tax evaluations on homes, and by talking with real estate appraisers.

Real estate appraisers are an excellent source for pricing information. Most appraisers are willing to consult with clients on an hourly basis. By spending a little money to talk with an appraiser, you can save thousands of dollars in lost income. If your business involves providing goods and services for homes and businesses, appraisers are one of your best sources of pricing information.

Let's say that you are a general contractor and that you build houses, garages, additions, decks, and related home improvements. One way for you to establish the value of these services is to consult with a licensed real estate appraiser. The appraiser can tell you what value a service might have on an official appraisal of the property. This doesn't guarantee that the values given by the appraiser are the best prices for you to use, but they are an excellent reference.

Assume the appraiser gives you a value of $1000 for a 10-foot-square deck. You could ask the appraiser to provide a written statement of value. Of course, the value will be generic and might not apply to all conditions. Many factors affect the value of real estate and improvements, but the $1000 figure would be a solid average. You can use this written statement as a sales tool when you sit down with customers. If you are willing and able to sell the deck for $800, you can show the customer how you are giving them a 20-percent discount off the average retail value of the deck.

You can also use the appraiser's figures as a piece of your puzzle rather than the last word. By combining the appraised value with numbers given in pricing guides and your own estimate, you can come up with solid numbers to work with. In the case of new houses, the appraised value is typically the most you can expect to get for a house. Unless your buyer is willing to put up unborrowed money for any overage in the price, you are limited to the appraised value.

Combined methods

Combined methods are best when setting your pricing principles. Use as much research and as many resources as possible to set your fees. Once you feel comfortable with your rates, test them. Ask customers to tell you their feelings. There is no feedback better than that from the people you serve.

MAKING PROPER PRESENTATIONS

Proper presentation is crucial for business success. Even if your price is higher than the competition's, you can still win the job with an effective presentation. On many occasions the low bidder does not get the job. As a contractor, you can set yourself apart from the crowd by using certain presentation methods. What are these methods? There are numerous ways to achieve an edge over the competition. The ways to win the bid battle can include the way you dress, what you drive, your organizational skills, and much more.

Why you are worth more money

When you learn to show a customer why you are worth more money, you are more likely to make the most of your time and effort. People are often willing to pay a higher price to get what they want. You have to convince the customer to want you and your business, not the competition. How can you sway customers your way? Let's study some methods that have proven effective over the years.

Mail mistakes

You might be surprised how many contractors mail estimates to customers and wait to be called to do the job. Many of these contractors never hear from the customers again. Mailing bids to potential customers is usually not the best way to get the job. When consumers spread estimates out and go over them, it is difficult to see much difference other than price. You want to influence customers with the extras you can bring to the table, so you need a better method of presentation for your quote.

If you must mail your proposal, make sure you prepare a professional package. Use printed forms and stationery, not regular paper with your company name rubber-stamped onto it. Use a heavyweight paper and professional colors. Type your estimates and avoid using obvious correction methods to hide your typing errors. If you are mailing your proposal to the customer, you must make it neat, well-organized, attractive, professional, and convincing.

Phone facts

One of the worst ways to present a formal proposal is by phone. Telephones are great tools for prospecting and following up on estimates, but they are a poor vehicle to use when delivering initial estimates. When you call in your price, people can't see what you are giving them. They can't linger over the estimate and evaluate it like they can with a written one. Chances are good that the customer might write down your price and then lose the piece of paper. Phone estimates also tend to make contractors look lazy, since they don't even take the time to present a professional, written proposal. In general, use the phone to get leads, set appointments, and follow up on estimates, but don't use it to give prices and proposals to possible customers.

Dress code

The dress code for contractors can span a wide range. In whatever you wear, wear it well. Be neat and clean. Dress in a manner that you can be comfortable with. If you are miserable in a three-piece suit, you won't project as well to the customer. If you normally wear uniforms, you can wear your uniform when presenting your proposal. Jeans are acceptable and so are boots, but both must be clean and neat. Avoid wearing tattered and stained clothing. You don't want the customer to be afraid to ask you sit on the furniture.

When you are deciding on what to wear, consider the type of customer you are meeting. If the customer is likely to be dressed casually, then you should dress casually. If you suspect a suit might fit in with your customer's attire, consider wearing a suit. Choose a wardrobe that blends in with the customer's. If you dress too well, you might intimidate the customer. If your clothes and jewelry are too expensive, the customer might think you make too much money.

What you drive

What you drive says a lot about you. If you pull up in front of the average house in a high-priced sports car, you are sending signals to the customer. When the customers look out the window and sees the contractor getting out of a fancy car, they are going to believe that the rumors about the outrageous prices contractors charge are true. If you crawl out of a beat-up, ancient truck, they might assume that you are not very successful.

Choosing the best vehicle for your sales calls is a lot like choosing the proper clothing. You want a vehicle that makes the right statement. A clean van or pick-up should be fine for most any occasion. Cars are okay for sales calls, but stay away from fancy ones, unless you are dealing with a clientele that expects you to spend $40,000 for your work car.

Confidence, the key to success

Confidence is the key to success. You must be confident in yourself, and you must create confidence in the mind of the consumer. If you can get the customer's confidence, you can almost always get the sale. You gain your own confidence through experience, but you must learn how to build confidence with your customers.

You can often gain the confidence of your customer by talking. If you are able to sit down with customers and talk for an hour, your odds of getting the job increase greatly. By showing customers examples of your work, you can build confidence. Letters of reference from past customers help in establishing trust. But if you have the right personality and sales skills, you can create confidence by simply talking.

As a business owner you are also a sales professional, or at least you had better be. Unless you hire outside sales staff, you are the one customers deal with. If you learn basic sales skills you'll have much more work than the average contractor, even if your prices are higher.

KNOWING THE COMPETITION

You must know your competition. How you price your services and materials is affected by the competition. Your prices should be in the same ballpark as your reputable competitors. If your prices are too high, you won't get much work. If your prices are too low, you might be flooded with work and low profits, not to mention angry competitors.

USING EFFECTIVE ESTIMATING TECHNIQUES

Effective estimating techniques come in many forms. What works for one person doesn't work for another. By learning from your mistakes and successes, you can mold your own profitable estimating methods. Once you have ways that work, you can use them over and over again. You might have to alter your techniques to match them with specific customers, but the same basic principles that work for you in one sale should work in another.

How do you develop effective estimating techniques? You can learn technical aspects of estimating by reading. Your estimating skills can be further perfected by referring to estimating handbooks and pricing guides. Much of what you learn will come from experience. Learning from your mistakes can be expensive, but you don't soon forget your costly lessons. One of the most important factors in effective estimating is organization. When you are well organized, you are more likely to complete a thorough estimate.

22

Running smooth
and profitable jobs

To move your business along the road to success, you must learn to work with schedules, budgets, and job costing. These three areas of your business can have a strong influence on how much money you make and how long you stay in business.

Schedules should be used for many facets of your business. You should have a production schedule, a daily schedule, and a delivery schedule, to name a few.

Budgets should also be used for multiple aspects of your business. You should have an overall business budget, an advertising budget, and an inventory budget. Additionally, you need budgets for individual jobs and the projected growth of your business.

Without the skills to make and follow schedules and budgets, a business owner is likely to make mistakes and lose money. One way to tell if a company is losing money is by using job costing. As jobs are in progress and as they are completed, you need to perform and monitor job costing. This practice allows the business owner to determine future pricing, forecast cash-flow needs, and derive an overall view of the company's on the-job performance.

Job costing, budgets, and schedules all play a vital role in the development of a contracting business. Without these elements, the business is running in the dark. You can't afford to blindly wander through the business arena; if you do, your competitors will knock you out of the game. Let's see how you can benefit from working intelligently.

ESTABLISHING PRODUCTION SCHEDULES

By establishing production schedules, you can make your business more organized and more profitable. Production schedules (FIGS. 22-1 and 22-2) allow you to plan and track your work. This advantage helps you keep track of your workload. Without production schedules, you'll have trouble completing your jobs on time. When the jobs run past their completion dates, you'll have angry customers on your hands.

How should you go about setting up a production schedule? Working out a viable schedule is not hard and it doesn't take long to do. You might, however, need the help of your subcontractors.

You can rough up your production schedules on a computer or on paper. You need a separate schedule for each job you have, but start with just one job.

Subcontractor Schedule

Type of service	Vendor name	Phone number	Date scheduled
Site work			
Footings			
Concrete			
Foundation			
Waterproofing			
Masonry			
Framing			
Roofing			
Siding			
Exterior trim			
Gutters			
Pest control			
Plumbing/R-I			
HVAC/R-I			
Electrical/R-I			
Insulation			
Drywall			
Painter			
Wallpaper			
Tile			
Cabinets			
Countertops			
Interior trim			
Floor covering			
Plumbing/final			
HVAC/final			
Electrical/final			
Cleaning			
Paving			
Landscaping			

Notes/Changes

22-1 Subcontractor schedule.

Once you have a schedule for the first job, scheduling the remainder of the jobs is easier. If your jobs overlap, you might have to schedule multiple jobs at the same time.

Start by putting the job name and address on the schedule. Lay out the schedule with headings for each phase of work for which you are responsible. Create spaces

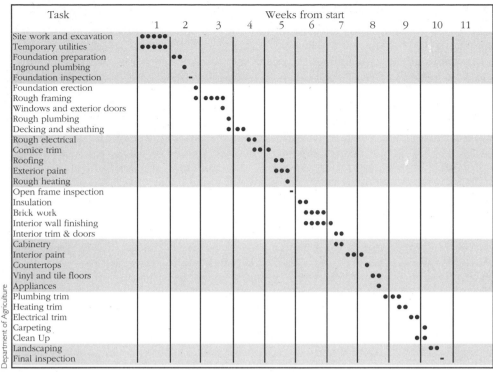

22-2 Typical work schedule.

to fill in dates for the various work phases. You want space for many dates. The first space should contain your anticipated start date. Another space should contain the date work is actually started. You should have an entry for the estimated time needed to complete the task. Then you should have a spot to enter the date of actual completion. In addition to your start and finish dates, you should allow space to write in dates for material deliveries.

The production schedule should have provisions for listing the names and phone numbers of subcontractors and suppliers you are depending on. Write these names and numbers next to each work phase or delivery. Having the names and numbers on the schedule make it easier to make follow-up confirmation calls.

Now all you have to do is fill in the appropriate dates. Choosing the dates for your schedule might not be simple. Attempting to work out schedules that don't conflict can be an arduous task, but it is not impossible. The best way to coordinate your jobs is to talk openly with all the people to be involved in the jobs.

Call and talk to your subs and suppliers. Go over the first draft of your schedule with them. Ask if they are available on the dates you want to have them on the job. If you don't know how long it should take the plumber to rough-in the job, ask. When you are concerned about how long it might take to have roof trusses manufactured and delivered, get advice from your supplier. By following these techniques, the numbers you pencil into the production schedule will be as accurate as you can make them. However, don't expect the dates to remain static. Plans change and so do scheduled dates.

STAYING ON SCHEDULE

Staying on schedule is going to take effort. You cannot maintain a schedule without working at it. What do you have to do? You have to make confirmation calls, review the progression of your jobs, stay on top of subs and suppliers, and much more. Let's take a look at some of your scheduling responsibilities.

Making confirmation calls

Making confirmation calls is essential to staying on schedule. If you don't make follow-up calls, subs and suppliers may forget their responsibilities to you. As a general contractor, you cannot allow your subcontractors and suppliers to overlook their obligations. It is your job to assemble and manage all the elements of a successful job.

Calling suppliers to confirm orders and deliveries can be done during normal business hours, but this isn't always true with subcontractors. Many subcontractors work in the field. These people don't spend their days sitting next to a phone. To get through to busy subs, you might have to spend time at night making your calls. This can get old fast.

The nature of your business determines how and how much you have to work, but general contractors often work from dawn into the late evening. You can reduce the need to work so many hours if you have your subcontractors check in with you at specific times. By asking your subs to call you when they take breaks or lunch, you can eliminate some of the night calls.

Even with the best of efforts, making follow-up calls can become a pain. It might be tempting to let some of your calls slide, but don't do it. If you don't make confirmations on your schedule, the schedule is useless.

Tracking production

Tracking production is one way to keep your schedule accurate. Schedules require adjustments. To make these adjustments, you must keep track of your production. If the drywall contractor is running behind, you have to reschedule the painter. If a special-order item, like a custom window, is past due, you have to adjust the schedule. To make these adjustments, you must keep your finger on the pulse of the job. This generally means checking each job's status on a daily basis. After checking your jobs, you can move the schedule up or down to allow for the field production.

Keeping subcontractors and suppliers in line

Keeping subcontractors and suppliers in line is another responsibility associated with staying on schedule. If subs or suppliers don't fulfill their obligations in a timely fashion, your schedule can be blown out of the water. You must control your subs and suppliers to stay on schedule.

If you follow my earlier advice and include start and finish dates (FIG. 22-3) in your subcontractor agreements, you are on the right track. If you include a penalty clause that allows you to charge a daily penalty for every day the job is not finished, you have an even better chance of staying on schedule. Couple these two contract clauses with persistent prodding, when needed, and you should do fine. If all else fails, exercise your ejection clause, and bring in a new sub or supplier to fulfill your needs.

Certificate of Subcontractor Completion Acceptance

Contractor: _____

Subcontractor: _____

Job name: _____

Job location: _____

Job description: _____

_____

_____

_____

_____

_____

Date of completion: _____

Date of final inspection by contractor: _____

Date of code compliance inspection & approval: _____

Defects found in material or workmanship: _____

_____

_____

_____

_____

Acknowledgment

Contractor acknowledges the completion of all contracted work and accepts all workmanship and materials as being satisfactory. Upon signing this certificate, the contractor releases the subcontractor from any responsibility for additional work, except warranty work. Warranty work will be performed for a period of one year from the date of completion. Warranty work will include the repair of any material or workmanship defects occurring between now and the end of the warranty period. All existing workmanship and materials are acceptable to the contractor and payment will be made, in full, according to the payment schedule in the contract, between the two parties.

_____   _____
Contractor   Date   Subcontractor   Date

22-3 Certificate of subcontractor completion acceptance.

Juggling your schedule

At times juggling your schedule is necessary. While job juggling might be a necessity, it can be very detrimental; be careful. It is one thing to juggle to accommodate unforeseen production changes, but it is quite another thing to juggle for cash flow or greed. Before we go on, let's establish the difference between the two types of job-juggling.

Job-juggling is necessary and acceptable when materials are late or subcontractors don't perform as expected. This type of juggling is usually associated with extending the estimated completion date, but it might mean moving the completion date up. For example, if you thought it was going to take 10 days to complete the drywall work, and it only took seven days, you can move your schedule up by three days. This, of course, is your choice—you could let the existing schedule remain intact.

Juggling your work schedule to generate cash flow or to take on more work is a much more dangerous form of job-juggling. If you move your crews from job to job to start more jobs or to collect more cash draws, you'll probably be out of business within a year. This type of job-juggling gives your business a bad reputation and results in more trouble than it is worth.

ADJUSTING FOR UNFORESEEN SCHEDULING OBSTACLES

Adjusting for unforeseen scheduling obstacles gets easier with experience. After you've been in the business for awhile, you start to anticipate the unexpected. Seasoned contractors are so good at projecting problems that they rarely have them. Unfortunately, contractors who are new to the business suffer through some tough times and pay a price for their lack of experience.

Many people assume that because they know their trade, they know their business. Knowing how to do your job as a tradesperson is not the same as knowing how to run a business. While you might be very competent at estimating the time you need for the work you are doing yourself, if you're not an experienced contractor, you might not know what allowances to make for subcontractors and suppliers. These two variables can destroy your scheduling plans.

During my many years in business I have never ceased to be amazed at how some people conduct business. I am a master plumber, licensed remodeler, licensed builder, and licensed real estate broker. I have been involved with property management, remodeling, plumbing, and building. During these experiences I have seen a number of different trades function. I have a good feel for how long it should take to rough-in the electrical wiring for a home or to install a new roof.

After building up to 60 houses a year, I believe I am well qualified to project the time needed for various trades to do their jobs. You would think I could estimate these work requirements with extreme accuracy, but it doesn't always work out that way.

I know that I can rough-in the plumbing for an average two-and-a-half bath house in two days if I have a helper. It takes about half a day to test the rough-in and have it inspected. Then, I need one day and a helper to set all of the plumbing fixtures on the trim-out. Now this is what I can and have done. So, shouldn't most experienced plumbers be able to perform the work in a similar amount of time? It seems to me they should, but it is surprising how long some plumbers can take to do the same job.

I have hired employees and subcontractors to perform plumbing work. I have had some very good plumbers. During my prime, I rarely found anyone who was substantially faster than I was, but I constantly find plumbers who are extremely slow. Their slowness is caused by being lazy, inexperienced, unfocused, and a number of other reasons.

For the same work I can accomplish with a helper in two and a half days, some crews take a full five days. As a contractor, you cannot assume that all subcontractors can complete their tasks in the same periods of time. It is very possible that one sub might need twice as long to get the job done as another. As a contractor, you have to find the secret to work your way through the maze of subcontractors and suppliers.

As you gain experience and get to know your subcontractors and suppliers, you'll become more adept at adjusting your schedule. You can almost anticipate how your schedule might need to change. This is not to say that you can avoid all of your scheduling problems, but you can learn to take control of them.

PROJECTING YOUR JOB BUDGET

Projecting your job budget must be one of your priorities. Job budgets are the basics for setting your profit goals. If you miscalculate your job budgets, you won't make your expected profits. Occasionally, you might make more money than you expected, but in most cases, you'll leave the job with less money than you projected.

For some people, the term *budget* is a dirty word. This mindset extends from personal budgets right into business budgets. While it might not be fun to follow the guidelines of a financial budget, budgets are often the only way to reach your monetary goals.

If you don't project a budget for each of your jobs, you are sure to be disappointed with some of your income results. Budgets might not be enough to keep you from losing money on a job, but they can help you spot your problems and avoid making the same mistake twice.

If you sell a job too cheaply, setting a budget is not going to solve your problem. The best budget going cannot compensate for a poorly estimated job. However, by budgeting your jobs, you can evaluate your results and price future jobs to keep your profit margin where you want it, within reason.

How should you structure a job budget? To make a job budget, you must project and list all anticipated costs of the job. For some contractors, thinking of all the expenses to be incurred in a job is a chore they cannot fathom. These people don't have the natural ability to put all steps of a job into chronological order. This doesn't mean that contractors can't learn to forecast a budget, they just are not naturally talented in this aspect of running a business. Building a job budget is not difficult if you proceed in a logical way. Let's see how you can structure a stable job budget.

First label your form with the job name and address like you did when you built the production schedule. Continue by listing all aspects of the work phases. For example, if you are going to forecast your plumbing expenses, include spaces for sewer work, water-service work, groundwork, rough-in work, and trim work.

Most contractors don't have problems listing the large phases of work, like carpentry, plumbing, heating, electrical, painting, and so on. However, many neglect to list the smaller phases of work required on a job. These minor phases can add up to quite a lot of money.

Consider all of your anticipated costs. Let's look at some of the costs that might be incurred that often go undetected in a budget. These costs could include:

- Clearing trees
- Trash-disposal fees
- Clean-up costs

- Drafting or architectural fees
- Code-enforcement permits
- Administration costs
- Office overhead expenses
- Charges for electrical power
- Inspection fees
- Field supervisors
- Advertising costs incurred to obtain the job
- Time spent estimating and selling the job
- Potential warranty work

This list could go on for pages. There are many costs associated with doing a job that do not show up on most contractors' budgets. If these soft costs are left out, the profits from the job shrink. You must be thorough in itemizing your job expenses; if you are not, the budget is bogus.

PROJECTING YOUR BUSINESS BUDGET

Projecting your business budget is similar to forecasting a job budget. In both instances, you must account for all anticipated expenses. Unlike a job budget, a business budget must include projections for a much longer period of time. This extended time can make the creation of a good business budget more difficult to master.

Developing a business budget that works requires extensive thought. You must consider your immediate needs and your projected future needs. Your desires come into play with a business budget. For example, if you want to retire in 20 years, you must make an allowance for this desire in your budget. Let's take a look at some of the factors you must incorporate into the budget for your business.

Salary

Many business owners don't pay themselves a set salary. They make as much money as they can and use the money they need. This works for some people, but it is not the way to make a business budget.

When casting your business budget, you need to establish a set amount for your salary. You might not always be able to take as much money for your salary as the budget reflects, but you need a number plugged in for your income.

When you set your salary, don't forget your responsibility for income taxes. You have to look at your net income needs as well as your gross income requirements. There is a substantial difference between your net needs and your gross needs.

Office expenses

Office expenses can cause your budget to balloon. Think about all the various expenses that you incur to keep your office running. These expenses might include rent, utilities, phone, cleaning, equipment rentals, office supplies, furniture, and much more. The costs of your office expenses can account for a high percentage of your annual expenses.

Many contractors have excessive office expenses. When you build your business budget, check over your office expenses; you will likely find ways to save money.

Field expenses

For a number of contractors, field expenses are a large part of the annual budget. Field expenses can be broken down into more specific categories and usually are. These costs include vehicles, fuel, field supervisors, signs, and other related expenses.

Vehicle expenses

Almost every contracting business is affected by vehicle expenses. The cost of trucks, cars, fuel, tires, and similar expenses can amount to thousands of dollars. Of course, your vehicle expenses are related directly to the size and structure of you business. If you rely on subcontractors for all of your field work, your vehicle expenses should be minimal. However, if you have a fleet of trucks and an army of employees, vehicle expenses can be astronomical.

Tool and equipment expenses

There are two types of tool and equipment expenses: the cost of initial acquisitions and the cost of replacement. Even if you have all the tools and equipment you need, you have to budget for replacement costs. Your hardware can get broken, stolen, or just worn out, but sooner or later, you'll need to replace it. If you are not prepared financially for these replacements, you can find yourself unable to continue doing business. Take a look at your tools and equipment. Estimate the life expectancy of your hardware, and put a figure in your budget for its replacement.

Employee expenses

Employee expenses can be devastating. When you look at employee expenses, you have to look much deeper than the hourly rates earned by your employees. You must look at the taxes involved with having employees. Other considerations are employee benefits. If you provide insurance benefits for your employees, you are spending some serious money. Paid vacations are another major expense. Sick leave, paid holidays, and other similar employee benefits can amount to thousands of dollars quickly.

If your business is top-heavy with employee expenses, you might do well to consider engaging independent contractors. When employees are involved, don't neglect to include the cost of employee benefits in your budget.

Insurance expenses

Insurance expenses can run some businesses into the ground. Insurance needs vary, depending on the nature of the businesses. Liability insurance is needed for all businesses. When you have employees, worker's compensation insurance can become an expensive factor in the business budget. Insurance to protect against theft and fire is another common business need. When a company is properly insured, the expense of the insurance can account for much of the total business budget.

Advertising expenses

Advertising expenses are one of the few expenses from which businesses can see a direct gain. When used efficiently, advertising can pay for itself and then some. However, a business owner can never be sure how effective advertising will be.

How much will your advertising needs cost? While you cannot be sure what your advertising results might be, you can project how much you are willing to spend on generating new business. Many companies dedicate a percentage of their gross sales to advertising. Most small companies pick a dollar amount rather than a percentage. Whichever method you use, make sure to include your anticipated advertising costs in your business budget.

Loan expenses

Loan expenses can be overlooked when building a budget. This is especially true if you use short-term, interest-only loans. The cost of these loans might not register as an expense, since the interest is usually paid in a lump sum, not on a monthly basis. Don't forget to include the fees you'll incur with your business financing.

Taxes

Taxes are another expense that is often left out of a business budget, but it shouldn't be. The impact of taxes can cause severe stress on a business. Unless regular tax deposits are made, the burden of coming up with enough money to satisfy the tax authorities can be overwhelming.

To avoid being caught in a cash bind, include your estimated tax liabilities in your budget projections. If you are unable to forecast your taxes, consult a tax professional; you cannot afford to be left high and dry on tax day.

Growth expenses

To see business growth, you must plan for growth expenses. Few businesses grow without a plan. Part of any growth plan involves expansion capital. Consider your intent for company growth when building your budget, and include provisions for the cost of expansion.

Retirement goals

Retirement goals frequently get pushed to the back of the list. When money is tight, it is hard to invest in retirement plans. The impulse is to fight the fires of today and worry about retirement later. Unfortunately, age and retirement creeps up on all of us, often sooner than we planned for.

New business owners often see the business they are building as their retirement. In some cases their beliefs are well-founded, but more often, the business is not enough for retirement. In fact, many new businesses won't be in operation at the time the owners decide to retire. For these reasons, investments in your business are not enough for safe retirement planning.

To build a suitable retirement portfolio, you need a plan. The plan should involve diversified investments. Examples could include stocks, bonds, real estate, coins, art, and a number of other possibilities. It is a wise idea for most people to talk with professionals to set up a viable retirement plan. Once you have a solid path to retirement, include the needs for funding your retirement in the business budget.

As you build your business budget, you might discover other items you need to account for, but this description of routine expenses should get you started in the right direction. Remember, you wouldn't take a trip to a strange destination without directions, and your budget is to your business what a map is to your travel plans.

MAINTAINING YOUR FINANCIAL BUDGET REQUIREMENTS

Having a budget does you little good unless you have the discipline and skills to maintain your financial budget requirements. It helps to have a natural ability to maintain budget constraints, but if you don't, you can learn the skills and discipline.

Self-discipline is a factor throughout our lives. It is this quality that allows us to live among the laws of society. Some people have strong discipline towards certain aspects of life and no discipline in others. Compulsive gamblers might discipline themselves to going to work on time and maintaining an average life until it comes to gambling. People with an obsession for food might be perfectly normal until food is set in front of them. Most of us have some weak areas in our self-discipline. Your job as a business owner is to find your weak areas and reinforce them.

If you find the deals on tools in mail-order catalogs irresistible, you might find that you are spending far too much on tools. When you are infatuated with having the most high-tech office equipment available, you might be investing too much money on the wrong items. The list of potential weaknesses could go on, but you get the idea.

All of these potential weaknesses must be evaluated. If you are quick to give in to impulse buying, you must set rules for yourself. Make yourself wait for a reasonable period of time before buying that new tool or office equipment. Wait and see if you need the item or if you only want the item.

Look into every aspect of your business habits. Are you going to have lunch at expensive restaurants on a regular basis? If you are, you might be spending enough to pay for a more justified business expense.

Document all of your spending for the next two weeks; include every item you spend a dime on. At the end of the two weeks, go over your list of expenditures. Compare these expenses with your budget. How many of your recent expenditures are not reflected in your budget? My guess is that you'll find several areas where you are spending money that wasn't budgeted. This is a danger signal; you can't allow yourself to run rampant with your company's financial resources. If you do, the business will fail.

After tagging all the unbudgeted expenses, decide if you will continue to make similar purchases. If you come to the conclusion that your expenses are justified, include them in the budget. If the spending habits are not necessary, eliminate them. Repeat the two-week test periodically. By monitoring and refining your budget, you can stay within its framework.

COSTING JOBS TO DETERMINE FUTURE JOB PRICING

Job costing is your way of assessing future job pricing. To make the most of your business, you must obtain the best price possible for your services. Sometimes this means increasing your prices, and sometimes it means lowering your fees. How do you know which way to move your price pointer? You can develop a strong sense of direction by reviewing and assessing your profit performance on past jobs. Job costing gives you the information you need to make sound decisions on your pricing structure.

Job costing is the act of adding up all the costs incurred to complete a job. Depending on your methods, job costing might only account for labor and materials used on the job. In more sophisticated reports, the numbers include the cost of overhead and operation expenses.

Your routine business expenses must be factored into the cost of jobs, but you don't have to include them in individual job-cost reports. You can, instead, use a percentage of your on-job profit to defer your other costs of doing business. Let's see how each of these methods can be used.

Simple job costing

Simple job costing includes all expenses incurred for a specific job except overhead expenses. Overhead expenses might include office rent and utilities, general insurance, advertising, and so forth.

To do a simple job cost, you need nothing more than pencil and paper and information. First, list all materials used on the job. Then, price these materials based on what you paid for them. Next, calculate all the labor that went into the job. Once you have all the labor accounted for, give it a value. The value should be your cost for the labor. Soft costs are your next category. Soft costs could include permit fees, the cost of blueprints, and so on.

Once you have all the costs for the job listed, check over the list for omissions. When you are sure you haven't forgotten anything, add up the total of the costs. Subtract the total of your costs from the money you received for doing the job, and you've got your gross profit.

In real life, it is not so simple to account for overhead expenses. It is not uncommon to have several jobs running at once and many jobs overlapping each other. This complicates the division of overhead between jobs. Some businesses project a percentage of their overhead and apply it to every job. Other businesses use complex methods to detail the exact cost of their overhead to each job. How you do your job costing is up to you, but you must allow for the off-job expenses to get a true impression of your profit margin. Now, let's see how a job-costing report might be done using a blanket percentage for overhead expenses.

Percentage job costing

Percentage job costing is another common method for determining profits and losses. To perform this task, duplicate the instructions given for simple job costing. Once this is done, subtract a percentage of the contract price for the job. The percentage you subtract represents an estimate of all your overhead expenses.

What percentage should you use? The percentage applied for overhead depends on your business structure. Some businesses carry heavy overhead expenses and others are streamlined to maximum efficiency. You have to experiment with your personal circumstances to determine what percentage covers your off-job expenses.

Adjusting your prices is much easier when you are working with the results of accurate job-costing reports. If after reviewing the job costs on three similar jobs you see a pattern, you can determine if you are charging enough, too much, or not enough.

TRACKING YOUR PROFITABILITY FROM JOB COSTING

Tracking your profitability is possible from job costing, but much more can be accomplished. Job costing allows you to monitor all of your jobs for maximum profit potential. It is easy to say you want to make a gross profit of 20 percent on every job, but it is not so easy to accomplish. Some jobs yield a higher profit potential than others. This is where job costing can lead you in the right direction.

To see how different jobs produce various results, let's consider the circumstances of a general contractor. This contractor deals only in residential jobs. The jobs range from minor remodeling to major renovations. New construction, such as houses, garages, and additions, also makes up a part of this business.

The general contractor has allocated $15,000 for advertising over the next year. Knowing how much he has to spend on advertising, the contractor must decide what to spend it on. One consideration is the type of advertising media to use, but beyond that is the question of what types of jobs produce the most profits.

By going over past job-cost reports, the contractor can determine what types of jobs offer the most profit potential. In reviewing past performances, the contractor might find that attic conversions are the most profitable jobs to undertake. He might discover that building garages is a quick and easy source of substantial profits. Perhaps the construction of additions provides the most net income. Building houses yields the highest overall income, but it can take a long time for all of the money to be earned.

There are some industry standards for which jobs should produce the most money, but all business owners must identify what jobs offer the most benefits to their companies. While building decks might make the most economic sense for Larry and Ann, Joe might find his highest profits come from kitchen and bathroom remodeling. Your big money maker might be contemporary homes. I've found that I make less money on one-level, ranch-style homes than I do on other styles. Unfinished Cape Cods provide me with a good profit percentage, and bigger houses pay better profits than small ones do.

The degree of profits from various jobs depends on the companies performing them. Some companies are more efficient at some jobs than others. For your company, you can look over the results of your completed projects to forecast your future.

Job costing does much more than just tell you how much you made or lost. It allows you to outline a plan for your business. Advertising is more effective when it is the result of carefully studying job-cost reports. Budget distributions are easier to establish when you have detailed job costs to pull from. There is almost no limit to how accurate job-cost reports can help your business.

COSTING JOBS ACCURATELY FOR LONG-TERM SUCCESS

Accurate job costing is essential to long-term success. To endure the test of time, businesses must be financially sound and flexible.

One reason so many new businesses fail is the inexperience of their owners and operators. These people might have solid work skills but lack business skills. It is not uncommon to find first-time contractors working hard and believing they are making money, only to find at the end of the year that they would have made more as an employee than as an owner.

These inexperienced business owners are not aware of the time, effort, and skills needed to guide a business to success. Many of them think that making $25 per hour is fantastic. Even when these hourly earnings are double what the individual was making as an employee, the end result can be lower net pay.

Because so many expenses are never considered by rookie contractors, they go about their business with stars in their eyes. Then, one day, the clouds move in and cover the stars. This is the day when a new contractor realizes how much money is needed to keep the business operating. All of the sudden, that $25 per hour is nowhere near enough to cover the business and personal needs.

You can avoid this trap by building a budget of all the businesses expenses you will incur and maintaining that budget. Job costing can show you where you stand financially. You can discover categories that were left out of your budget. Job-cost reports can bring the reality of overhead expenses to light. While you might have never considered charging a portion of your rent to a job, you will see that you must.

The information you gain from accurate job costs makes you a better estimator. If you lost money or made a minimal profit on your last job, the next job should go better. Since you can pinpoint your problems with job costing, you can adjust your next estimate to protect yourself.

One area that causes a lot of problems for contractors is travel time. Many business owners underestimate the amount of time that they spend on the road. Job-cost reports can open your eyes to mistakes in this category, too.

Since the labor on your job costs is broken down into phases, you can see how time was spent. One problem area to look for is time wasted on trips to the supply house. This is a common robber of profits for many contractors.

ELIMINATING LOST TIME RUNNING TO THE SUPPLY HOUSE

Eliminate lost time running to the supply house, and you make more money. Wasted time on the road is one of the most prevalent causes of lost income to contractors. It might not be one of the larger financial losses a company experiences, but it is often the most regular loss of income.

Contractors who are not able to make accurate take-offs and schedule deliveries properly lose money on the road. When they or their employees leave a job for missing materials, profits are eroding.

During my time as a contractor and a consultant, I have witnessed countless situations when runs to the supply house pulled the job profits down. Most of these trips were avoidable. Even circumstances that are not reasonably avoidable can be made better with the use of logic. As a business owner, you cannot afford this type of wasted time. Except in rare cases, there is no suitable excuse for it. If you spend time preparing for a job, you should not waste time running for materials.

23
Keeping your customers happy

Customers and public relations are two facts of business that every business owner must consider. Without customers, a business is worthless. Without good public relations, customers can be lost. These two pieces of the business puzzle go together; you can't have one without the other.

Most contractors recognize the need for customers, but many underestimate the importance of public relations. By being ignorant of or blind to the importance of public relations, many contractors lose business. The business they lose goes past without leaving tracks. These unfortunate contractors don't know why they lost business, only that their business volume is down.

If you don't want to fall into the category of contractors dazed by loss of business, hone your public relations skills. When you improve these skills, you'll see an improvement in your business. The improvement might not be obvious, but if you look closely, you will see it.

MEETING YOUR CUSTOMERS ON THEIR LEVEL

Meeting your customers on their level is important. When you are working with people, you want those people to be as comfortable as possible. For this reason, you have to be able to change clothes and personalities.

The most successful salespeople are chameleons. These gifted people can move up and down the social ladder to serve any prospect that comes along. This flexibility provides an advantage the average salesperson doesn't possess. Being able to assume the personality of a customer makes it easier to close a deal.

Fitting in, with what you wear and drive, is another key to sales success. I talked about this issue earlier, but it is worth repeating. When you are dealing with customers, attempt to fit into what the customers are comfortable with. This might mean wearing a suit in the morning and jeans in the afternoon. You might find it advantageous to switch from your family car to your pick-up truck. The more effort you put into blending in with the customers, the more sales you can make.

Without customers, your business isn't worth much. Customers are what give your business value. How you treat existing customers can influence the long-range success of your business. If you alienate customers, they won't give you return business or referrals. It is much less expensive to keep good customers than it is to find new ones. Once you have established a customer base, work hard to maintain it.

QUALIFYING YOUR CUSTOMERS

When qualifying your customers, don't be afraid to ask questions. While it is true that the customer is hiring you, you have the right to know a little about who you are working for.

What do I mean by qualifying your customers? Qualifying customers can include numerous angles. One of the first questions you might be concerned with is the customer's ability to pay for the work being requested. This might seem like a silly curiosity, but it is not.

It is not uncommon to have customers not pay their bills. The reasons for non-payment are extensive, but the end result is that you don't get paid. No business owner can afford to work for free. Let's look at some of the reasons you might not get paid by your customers.

Loan denial

Loan denial can cause a well-meaning customer to be unable to pay your fees. If you specialize in large jobs, like building homes, your customers will usually be depending on borrowed funds to pay you. If the loan isn't approved, the customer won't have enough money to settle your bill.

This might not seem like a high risk, but many contractors put themselves into positions to take a direct hit from refused loan requests. Contractors, especially new contractors, are often anxious to work. When a contract is signed for a big job, some contractors start the job immediately before a loan has been arranged. This is bad business.

To avoid getting stuck on a big job, ask your customers to show evidence of the money required to do the job. This could amount to seeing a loan agreement or a bank statement for the customer's account. You might feel awkward asking to see proof of available funds, but you'll feel worse if you don't get paid.

Deadbeats

There are always deadbeats looking to get the best of someone. These people are good at beating honest people out of their hard-earned money. Avoiding deadbeats can be difficult; they are not always easy to identify. To protect yourself from this undesirable group of customers, get permission to run a credit check on the customer and do it. This is good business for all of your customers. You never know when the sweetest, most trusting person is going to turn out to be a bad debt.

Dissatisfied customers

Sometimes nonpayment can be your fault. If you have not made the customer happy, you could have trouble collecting your cash. When you qualify your customers, try to read them for trouble signs. If you feel friction in the meeting, perhaps you should pass on the job and look for another customer.

Death

Death is always a good excuse for not paying bills. Of course, death is no laughing matter, but neither is not getting paid. While you can't avoid a customer's demise with qualifying, you can make arrangements in your contract to cover the death contingency. Ask your attorney for help in drafting a clause that holds the heirs and estate responsible for your fees if the client passes away.

Bankruptcy

If a customer owes you money and files for bankruptcy protection, your chances of being paid are all but nonexistent. During the qualifying stage you can screen the customer's credit rating and financial strength. If the customer is financially healthy when you start the job, you run limited risk of losing your money in the bankruptcy courts.

SATISFYING YOUR CUSTOMERS

Do you know how to satisfy your customers? The answer to this question is complex. Not all customers are alike, and what satisfies one customer can infuriate another. When you are learning how to keep your customers satisfied, you must look at each customer individually. There are, of course, similarities between customers, but each person will have at least a slightly different opinion of what is required of you.

Here are some basic principles to follow when working with customers:
- Keep your promises.
- Return phone calls promptly.
- Maintain an open and honest relationship.
- Be punctual.
- Listen to their requests.
- Do good work.
- Don't overcharge.
- Stand behind your work.
- Be professional at all times.
- Don't take your customers for granted.
- Put priority on warranty work.
- When feasible, give customers what they want.
- If possible, give customers more than they expect.

If you follow the basic rules of customer satisfaction, you should have a high ratio of happy patrons. There are always, of course, some customers you can't satisfy. Some of these people won't know themselves what it would take to make them happy. When you run into this type of client, grin and bear it.

Even if customers are being irrational, go to extremes to please them. One angry customer spreads more word-of-mouth advertising than 10 satisfied customers, and you don't want the type of publicity a disgruntled customer gives you.

If you reach a point where you can't deal with an unreasonable person, end the connection as quickly and professionally as you can. Avoid name-calling and arguments. If the customer is wrong, defend your position, but if the circumstances are questionable, cut your losses and get out of the game. Avoiding conflicts and striving for customer satisfaction do you much more good than standing on a soapbox.

LEARNING WHEN TO GIVE AND WHEN TO TAKE

Learn when to give and when to take. A business relationship is similar to a marriage. In any relationship, there is a certain amount of give and take required. If you are hardheaded and bullish, you might win the battle only to lose the war. For success in business, you msut learn the fine art of flexibility.

If you make a habit of being too considerate and giving, some people might take advantage of you. If you are cold and take a stout stand against giving in, you can lose business. To have a harmonious business, you must learn to blend give and take into a masterful mix.

It's funny, in a way; people go into business to be their own boss, but as a business owner, you have more bosses than ever before. Every customer you serve is your boss. Being in business is not the easy ride some people think it is. Owning a business is hard work, and sometimes you have to do things you don't want to.

Learning to compromise, especially on money, can be a sobering experience. But if you are going to stay in business, you are going to have to make compromises. The key is learning the difference between compromise and giving in.

When you are in disagreement with a customer, talking is your best course of action. Actually, listening might be the best move to make. If you listen carefully to what your customer is complaining about, you can often find a simple solution to the problem. If, on the other hand, you do all the talking, you probably will only worsen the situation.

Becoming a good listener is one of the best ways to keep your customers happy. Once you have heard the customer out, propose a reasonable plan to resolve the dispute. You might not come to terms on the first attempt, but if you follow this procedure, you have a good chance of finding common ground.

LEARNING PUBLIC RELATIONS SKILLS

Public relation skills are essential in service businesses. To make your business successful, you need the support of the public. How do you gain this needed support? You win over the public by demonstrating strong public relations skills.

How can you learn the skills needed to build public favor? You can start by reading books. If you want to build on your knowledge, you can take college courses and seminars. If you don't want to take the time to go to classes or read, you can buy cassette tapes to train yourself. How you learn the skills is not the issue; the fact that you learn the skills is what's important.

Public relations is a career field in itself; people specialize in this area of work. If you think you are going to master these skills overnight, you're wrong. But that's okay, you don't have to learn the skills all at once. In fact, most people refine their skills over years of work experience.

Every time you deal with the public, you can build on your public relations skills. This earn-while-you-learn process might not be the fastest way to develop good people skills, but it is one of the best. Reading and going to classes can give you the basics, but you have to deal with people to put those basics into motion. Don't be intimidated by having to learn to become a public relations officer. Study the principles used in public relations and refine them as you go.

ESTABLISHING CLEAR COMMUNICATION CHANNELS

Establishing clear communication channels with customers is essential to making your business better. If you and your customers are on the same wavelength, you'll have far fewer problems.

How should you go about opening the communication lines? Your first contact with customers often comes over the phone lines. In this first contact, you cannot read body language and facial expressions. What is said and the way it is said are all either of you have to judge each other. For this reason, it is important to speak clearly and project a cheerful attitude.

Once you get past the phone conversation, a face-to-face meeting will likely occur. During this meeting, pay attention to the words and actions of your prospective customers. Also, be selective in your conversation. If you use four-letter words on the job site, keep them out of the meeting. If you smoke, refrain from doing so in the meeting. If you watch and listen to the people you are meeting with, you can learn how to handle the potential clients.

When you get to the estimate and contract stage, keep your thoughts concise and in writing. Once the written documents are drawn up, go over them with your customers. If the customers want to have the documents reviewed by an attorney, by all means, let them.

You want to establish a comfort level for the customer. By keeping your contracts clear and easy to read, you keep your customers happy. Clear communication is essential to good business.

DECIDING WHO SHOULD DEAL WITH THE CUSTOMERS

Who in your organization should deal with the customer? Are you the best person for meeting with clients? Do you have the skills necessary to sell jobs? Obviously, no one can be the best at every aspect of running a business. Smart business owners know their strengths and their limitations.

You have to evaluate your personal skills. Being a fantastic tradesperson does not make you great at clerical tasks. Just because you can hammer a nail or hang a door doesn't mean you can sit down at a computer and be an instant wizard on the keyboard. You are bound to be better at some chores than others. To maximize your business, you must itemize your strong points and weaknesses.

Once you know where you could use some help, you can work towards getting that help. This doesn't necessarily mean you have to hire employees. You can farm much of your work out to independent contractors. For example, you can engage commissioned salespeople, independent bookkeepers, and other specialists.

By finding and using skilled independent professionals, you can make your business more efficient. While your employees or independent contractors are doing their jobs, you can devote more of your time to doing what you do best.

DEFUSING TENSE SITUATIONS

Learning how to defuse tense situations is well worth the time it takes. If you choose not to work on ways to handle difficult situations, you'll have a tough time in business.

All business deals have the potential to turn into tense situations. It doesn't take much to make a good deal turn sour. Many times jobs run smoothly until the last few days of the project. For some reason, the end of the job seems to be the most difficult hurdle to get over.

When you have a job nearly complete, you have expended a lot of time, money, and energy to get and keep a happy customer. If you lose control at the end of the job, all of your previous efforts are wasted. Normally, talking through a bad time can solve the problem.

Sometimes just being a good listener is enough to resolve disputes. Some people get anger out of their systems by yelling and screaming. Some people deal with anger by beating their head. When a human being is consumed with anger, results are unpredictable.

Most on-the-job confrontations don't elevate to extreme limits. Generally, work-related disputes can be resolved with open communication and compromise. Every angry customer reacts a little differently. It might be necessary to step away from the argument and look at the real reasons why the blow-up has occurred. The best thing you can do is remain calm and reasonable.

CALMING A DISGRUNTLED CUSTOMER

Calming a disgruntled customer is similar to defusing a tense situation, but it might not always be the same. A customer can be dissatisfied without being physically upset. In fact, the calm customer whom is displeased can be more difficult to work with than the customer who is shouting in rage.

When you have an unhappy customer on your hands, you have to find a remedy that appeases the client. If you are able to carry on a normal conversation with the customer, resolving the problem shouldn't be too difficult.

Ask your customer why he or she is dissatisfied. Before making a rebuttal, consider the other person's position. Does he or she have a legitimate gripe? If so, take action to rectify the situation. If you disagree with the customer's opinion, discuss the problem in more depth.

Before starting a debate, put the customer at ease. Assure him or her that you are willing to be reasonable, but that you need more facts to understand his or her position fully. If you start this way, the customer is likely to remain calm and businesslike. On the other hand, if you open your defense aggressively, the situation could escalate to a tense and unpleasant shouting match.

It is usually best to attempt a settlement of disputes in a relaxed atmosphere. If your crews are banging hammers and buzzing saws, ask the customer if the two of you can find a more suitable place to discuss your differences. This accomplishes two goals—you get the customer in a congenial setting, and your workers do not witness the disagreement.

After relocating, ask the client to repeat the grievance. Pay close attention and see if the story remains the same. If the customer comes up with additional complaints or a great variance from the initial comments, you might have some additional trouble. This behavior indicates a person who is going to be hard to please. However, if the complaint is essentially the same as it was when you first heard it, you have a good change of ending your discussion in concurrence.

Think before you speak. If you are good at thinking on your feet, you'll do better than people who must meditate before coming to a conclusion. Your customer is going to expect answers now, not next week. Once you know what you want to say, say it sincerely and with conviction. Let the customer know you believe strongly in your position, but that you are willing to compromise.

You might find that you and your customer exchange several opinions and offers before reaching an amicable decision. Once the two of you agree on a plan, put the plan in writing. By writing a change order for the compromise, you seal the deal and reduce the risk of having to negotiate it further at a later date.

BUILDING A REFERENCE LIST

As you know, building a reference list from existing customers is an excellent way to produce new business. By asking your customers to give a letter of reference or to complete a performance-rating card, you are compiling a valuable stack of ready references.

Asking customers for the names and phone numbers of friends or relatives is another way to get new business. You ask your customers for the names and phone numbers of anyone who might be interested in your professional services. That evening, you call the people and introduce yourself as the contractor working on Mrs. Smith's house. By knowing the neighbor's name, you have a good chance of starting a conversation without being hung up on. By giving Mrs. Smith's name as a reference, you have a good chance of keeping the conversation going. You might be surprised at how many of the people your present customer refers you to might be interested in your services.

MAKING ON-THE-JOB DECISIONS

Making on-the-job decisions can be dangerous. When you make snap decisions, you are likely to make some mistakes. These mistakes could run the gamut from upsetting the homeowner to causing extra work for yourself. When possible, give yourself some time to consider your decisions before they are made. Of course, at times you have to shoot from the hip, but avoid quick decisions when you can.

LIEN RIGHTS AND WAIVERS

Lien rights and waivers can have a significant impact on your business. These factors can work for you or against you. What is a lien right? A lien right is the right to place a lien against property for which you have supplied labor or material and have not been paid. What is a lien waiver? A lien waiver is a legal document that, when signed, relinquishes your lien rights.

Two types of liens generally come into play with contractors. A mechanic's lien can be placed by tradespeople not being paid for labor provided on a job. A materialman's lien can be levied by suppliers who have not been paid for materials supplied to a job.

Lien-right laws vary from jurisdiction to jurisdiction, but they exist to protect workers and suppliers. To gain a full insight into the lien rights available to you, consult a local attorney.

As a contractor, you'll probably be asked to sign lien waivers and you'll most likely ask subcontractors to sign lien waivers. When a job is being financed, the lender often require lien waivers to be signed for every cash disbursement. When the person being paid signs a lien waiver, that person gives up rights to lien the property for the labor or materials being paid for. This protects the property owner and the lender.

Some contractors are approved for short-form lien waivers (FIG. 23-1) and others must use long-form lien waivers (FIG. 23-2). A short-form lien waiver is a form that only the general contractor signs. In signing a short-form waiver, the contractor swears that all subcontractors have been paid for work done to a certain point.

The contractors who are approved for short-form lien waivers have usually been in business for a while and normally have strong company assets. Property owners and lenders take a bigger risk in allowing contractors to sign short-form waivers. If the general contractor signs the waiver but has not paid the subcontractors or

---

Short-Form Lien Waiver

Customer name: _____

Customer address: _____

Customer city/state/zip: _____

Customer phone number: _____

Job location: _____

Date: _____

Type of work: _____

Contractor: _____

Contractor address: _____

Subcontractor: _____

Subcontractor address: _____

Description of work completed to date: _____

_____

_____

_____

Payments received to date: _____

Payment received on this date: _____

Total amount paid, including this payment: _____

The contractor/subcontractor signing below acknowledges receipt of all payments stated above. These payments are in compliance with the written contract between the parties above. The contractor/subcontractor signing below hereby states payment for all work done to this date has been paid in full.

The contractor/subcontractor signing below releases and relinquishes any and all rights available to place a mechanic or materialman lien against the subject property for the above described work. All parties agree that all work performed to date has been paid for in full and in compliance with their written contract.

The undersigned contractor/subcontractor releases the general contractor/customer from any liability for nonpayment of material or services extended through this date. The undersigned contractor/subcontractor has read this entire agreement and understands the agreement.

_____   _____

Contractor/Subcontractor   Date

23-1 Short-form lien waiver.

Long-Form Lien Waiver

Customer name: _____

Customer address: _____

Customer city/state/zip: _____

Customer phone number: _____

Job location: _____

Date: _____

Type of work: _____

The vendor acknowledges receipt of all payments stated below. These payments are in compliance with the written contract between the vendor and the customer. The vendor hereby states that payment for all work done to this date has been paid in full.

The vendor releases and relinquishes any and all rights available to said vendor to place a mechanic or materialman lien against the subject property for the described work. Both parties agree that all work performed to date has been paid for, in full and in compliance with their written contract.

The undersigned vendor releases the customer and the customer's property from any liability for nonpayment of material or services extended through this date. The undersigned contractor has read this entire agreement and understands the agreement.

Vendor Name	Signature of Co. Rep.	Signature Date	Service Performed	Date Paid	Amount Paid

_____
Plumber (Rough-in)

_____
Plumber (Final)

_____
Electrician (Rough-in)

_____
Electrician (Final)

_____
Supplier (Framing lumber)

*This list should include all contractors and suppliers. All vendors are listed on the same lien waiver, and sign next to their trade name for each service rendered, at the time of payment.

23-2 Long-form lien waiver.

suppliers, the unpaid parties can still lien the property. The general contractor is responsible for having the lien removed, but there is some risk that the general contractor might not have the funds to settle the issue and have the lien removed.

Long-form lien waivers are more time-consuming for contractors, but the property owners and lenders are in a much safer position. With a long-form lien waiver, anyone providing labor or materials for a job must sign the lien waiver at the time of payment. This way, no one can say he or she wasn't paid, because their signature is on the lien waiver.

As a general contractor, it is good business to have subcontractors and suppliers sign lien waivers even if a lender or property owner does not require the paperwork. By having your vendors sign off on the waivers when they are paid, you ensure that liens won't be placed against the properties you are working on. This is just one more paperwork step that can help you avoid conflicts and trouble.

SOLIDIFYING PLANS AND SPECIFICATIONS

Solidifying agreed-upon plans and specifications is a step that should be taken before starting a job. Once you and your customer agree on a set of plans and specifications, you should memorialize the agreement.

When you have a finished and accepted set of plans and specifications, have your customers sign them. Date the documents and make notations that the documents are the final and working plans and specifications. Further note that no changes can be made to the documents unless all parties agree to the changes in a written change order. When your customers sign below these notes, you have a solid set of plans and specifications. This procedure eliminates the possibility of the customers coming to you later and saying that the job is not in compliance with the plans and specs. If this happens, all you would have to do is produce the signed documents and show that your work is in accordance with the agreed-upon plans and specifications.

It is hard to put a price on the value of signed agreements. Having clear contracts and supporting documents keeps you out of ambiguous arguments. Your signed documentation proves who is in the right. Keeping customers happy is not too difficult if you have a good system by which you work. Most contractors can enjoy friendly relationships with their customers throughout the course of long jobs. To succeed in keeping customers happy, you might have to go out of your way and bite your tongue from time to time, but the end result is worthwhile.

24

Creating and promoting an attractive business image

How much is your public image worth? Well, the type of business image you present can mean the difference between success and failure. Does it really matter if you don't have a logo? People quickly learn to associate a logo with its owner. Logos can do a lot for you in all your display advertising. How important is the name you choose for your company? A company name should be given careful consideration. Some names are easier to remember than others. A name can conjure up a mental image, an image you want for your business. If you have any intent of selling your business in the future, the company name should not be too personalized; the new owner might not like owning a business with your name as part of the company name.

A company image can affect the type of customers the business attracts and the rates that can be charged for the business services. Your business advantage can come from a high profile in community organizations. How you shape your business image might set you apart from the crowd, helping to eliminate normally heavy competition.

The value of a public image is hard to put a price on. While it might be almost impossible to determine a monetary value for your company image, it is easy to see how a bad image can hurt your business. Your image is made with many facets. Your tools, trucks, signs, advertising and uniforms all have a bearing on your corporate image. This chapter details how these and other factors work to make or break your business.

INFLUENCING PUBLIC PERCEPTION OF YOUR BUSINESS

How the public perceives your business is half the battle. If you give your customers the impression that you have a successful business, you'll probably be successful. On the other hand, if you don't take an active interest in building a strong public image, your business might sink into obscurity.

How does the public judge your image? Many factors can contribute to how the public perceives your business. Take your truck as an example. If you had a contractor come to your house to give an estimate, how would you feel if he or she arrived in an old, battered pick-up with bald tires and the license plate hanging from baling wire? Would you rather do business with this individual or a person pulling up in a late-model, clean van with the company name professionally lettered on the side? Which truck points to the most company success and stability? Most people would prefer to do business with a company that gives the appearance of being financially sound. You don't have to have flashy new trucks, but they should be modern and clean.

Is it important to have your company name on the business vehicles? You bet it is. The more people see your trucks around town, the more they remember your name and develop a sense of confidence. It is acceptable to use magnetic signs or professional lettering, but don't letter the truck with stick-on letters in a haphazard way. Remember, you are putting your company name out there for all the world to see. You want to attract attention, but not as the laughingstock of the town.

Designing your ad for the phone directory is another major step. As people flip through the pages of the directory, a handsome ad might stop them in their tracks. An eye-appealing ad can get you business that would otherwise be lost to competitors.

While we are talking about phone directories, let's not forget about phone manners. Telephones often provide the first personal link between your business and potential customers. If you lose customers at the inquiry stage, your business will sag and falter. How do you lose these customers? You could lose customers by allowing small children to answer your business phone. A rude answering service is a sure way to lose potential business. Answering machines cost you some business, too, but they are an acceptable form of doing business. However, don't try to get too cute on your machine's outgoing message. Customers are calling and expecting a professional response. If you put a tape in your answering machine that is ridiculous or offending, would-be customers are sure to hang up.

Any professional salesperson can tell you that to be successful, you must always be in a selling mode. It doesn't matter where you are or what you're doing, you must be ready to cultivate sales. Compare this to your public image. Your image is being made and presented every day and in every way. You can't afford to let your business image slip.

If your company image is strong enough, customers will come to you. They will see your trucks, job signs, and ads and call you. When customers call a contractor, they are usually serious about having work done. Whether your company does the work or one of your competitors gets the job depends on many factors. Your company image is one of those factors. By building and presenting the proper image, you are halfway to making the sale.

PICKING A COMPANY NAME AND LOGO

Picking a company name and logo (FIG. 24-1) is a major step in building your business image. The name and logo you choose will be with the company for many years, you hope. Before you decide on a name or a logo, you should do some research and some thinking. Ask yourself some questions. For example, do you plan to sell the business in later years? If you do, pick a name that anyone could use comfortably.

A name like Pioneer Plumbing can be used by anyone, but a name like Ron's Remodeling Services is a little more difficult for a new owner to adapt to.

Company names

Company names can say a lot about the business they represent. For example, High-Tech Heating Contractors might be a good name for a company specializing in new heating technologies and systems. Solar Systems Unlimited could be a good name for a company that deals with solar heating systems. Authentic Custom Capes would make a good name for a builder that specialized in building period-model Cape Cods. How would a name like Jim's Custom Homes do? It might be all right, but it doesn't say much. A better choice might be Jim's Affordable First-Time Homes. Now the name tells customers that Jim is there for them with affordable first homes. See how a name can influence the perception of your business?

What qualities should you look for in a company name? Your company name should be one that you like, but it should also work for you. If you can imply something about your business in the name, you have an automatic advantage. It has been proven that people remember things that they are shown in repetition. When you run an ad in the paper every week, people will absorb your ad. Even though readers might not realize they are committing your company name or logo to their subconscious, they are. Then, when these same people scan the pages of a phone directory for a contractor and run across you name or logo, you have an edge.

Since these people have been given a steady dose of your advertising, your name or logo sticks out from the crowd. Without thinking about it, people who have had regular exposure to your advertising remember something about the ads. While

24-1 Logo.

the customer might not know why her or she is going to call your company, the odds are good that you'll get the call.

Since advertising is expensive, it only makes sense to get as much for your advertising dollar as possible. If you were scanning through the newspaper and noticed a company name like Tanglewood Enterprises, what would you associate the name with? This name could apply to any business. The name is nondescriptive and could be used for any number of different types of businesses.

On the other hand, if you saw a name like Deck Masters, Inc., you would probably associate the name with decks. If the name was White Lighting Electrical Services, you would think of an electrical company. A name like Homestead Homes gives a clear impression of a company that offers warm, comfortable housing.

The more you can tie the name of your business to the type of business you are in, the better off you are. If you can add in descriptive words, your potential customers will know more about your business just from the name.

Also, when choosing a name for your company, find words that flow together smoothly. How does Pioneer Plumbing sound to you? Both words start with a "P," and the words work well together. A name like Ron's Remodeling sounds good and so does a name like Mike's Masonry. In contrast, a name like Englewood Heating And Air Conditioning is not bad, but also is not as good. A name like Septic Suckers flows well and might be fitting for a company offering service to pump out septic tanks, but the name is not a good one for generating business. (Although, the more I think about it, the more I like the name. It might just generate enough talk around town to keep your phone ringing for months.) But you get the drift—your company name says a lot about your business. Maybe your company name should be something like Best Builders or Built To Perfection. Find a name that identifies what you do best.

Logos

Logos can be as important as your company name. Logos, the symbols that companies adopt to represent themselves, play an important role in marketing and advertising. Although people might not remember a specific ad or even a company name, they are likely to remember distinctive logos. If you put your mind to it, I'll bet you can come up with at least ten logos that stick in your mind.

Major corporations know the marketing value of logos and invest considerable time and money in coming up with just the right symbol to represent their corporation. While you might not go to the expense of hiring an ad agency or graphic artist to design a custom logo for your business, it might still pay off in the future to use a logo in your advertising. Like slogans and jingles, logos are often much easier to remember than company names.

Do you remember what gas company puts a tiger in your tank? In the cola wars, who has the right thing? If you were going shopping for tires, which ad would you remember? My guess is that you might remember a baby riding around in a certain brand of tire. What company lets you reach out and touch someone, by phone, of course? You see, it is easy to remember segments of ads, slogans, and jingles. It is equally easy to remember logos.

Your logo doesn't have to be complex. In fact, it might be nothing more than the initials of your company name. Then again, you may have a very complex logo, one that incorporates an image of what your business does. One of my favorite logos was used by a real estate company. The logo was a depiction of Noah's ark, complete with animals. In the ad featuring the ark were the words, "Looking For Land?" The business was selling land, and the ark logo was humorous and fitting.

If you're not at your best when it comes to creative images and marketing, it might serve you well to consult a specialist in the field. Choosing the proper name and logo is important enough to warrant investing some time and money. Of course, you know your financial limitations, but if you can swing it, get some professional advice in designing the image of your company.

AFFECTING YOUR CLIENTELE AND FEE SCHEDULE

How does your image affect your clientele and fee schedule? Image might not be everything in business, but it is a big part of your success. People have become afraid of contractors. The public has read all the horror stories of ripoffs and contractor con artists. Unfortunately, much of what has been printed about unscrupulous contractors is true, and the public does have a right to be concerned. With the growing awareness of consumers, image is more important than ever before. Let's take a quick look at three examples of a contractor going out to do an estimate.

The visual image

The visual image of you and your business can influence the profits of your company. In our first example, a contractor goes on an estimate in well-worn work clothes driving a truck that has seen better days. While this image won't offend or alienate some customers, it will surely turn many customers to other contractors.

For the second example, the contractor drives up in an expensive luxury car and goes into the home wearing a suit that cost more than the first contractor's truck. A select group of homeowners might relate and respond well to the image, but most consumers will resent the contractor's financial flaunting. These consumers will assume the contractor charges far too much or else couldn't afford such expensive clothing and transportation. Further, the contractor might intimidate or embarrass the customer with his or her assumed financial standing. For yet another group of consumers, the fancy suit might make the statement that its owner doesn't know the first thing about hands-on work and is only a sales professional who wants their hard-earned money to pay for more suits and a more expensive car.

If a contractor dresses neatly in casual clothing and drives a respectable, but not lavish, vehicle, his or her odds of appealing to the masses improve. By wearing clothes that make the contractor believable as a skilled tradesperson, he or she gives the impression of someone who knows the contracting business. The vehicle looks professional and successful without reeking of money. For me, this combination has always worked best.

The same basic principles apply to your office. If your office is little more than a hole in the wall with an answering machine, an old desk, and two chairs, people might be concerned about the financial stability of your business. However, if your office is staffed with several people, decorated in expensive art and furnishings, and in the high-dollar district, customers might assume your prices are too high. It generally works best to hit a happy medium with your office arrangements.

Fee factors

What does image have to do with the fee factor—what you charge for your services? To a large extent, people feel that they get what they pay for. While this might be a misconception, it is a popular belief. The image you present has a direct effect on your fees.

If you convince potential customers you are a professional, they'll be willing to pay professional fees. Extend your image by making the customers feel safe in doing business with you, and you have leverage for even higher fees. When you become a specialist, you can demand higher fees. Think about it, who gets a higher hourly rate, your family doctor or a heart specialist? When you convince the customer that nobody builds a better house than you do, you are building a case for higher fees.

Is there any truth in the idea that you are worth more than the next guy to come along for the job? There certainly could be. For years I specialized in kitchen and bath remodeling. My crews did nothing but kitchen and bathroom remodeling. When you do the same type of work day in and day out, you get good at it.

With my experience in this specialized field of remodeling, I could anticipate problems and find solutions that most of my competitors could not. This specialized experience made me more valuable to consumers. I could snake a 2-inch vent pipe up the wall from their kitchen to their attic without cutting the wall open. I could predict with great accuracy how long it would take to break up and patch the concrete floor for a basement bath. In general, I became known as a competent professional in a specialized field, and I could name my own price, within reason, for my services. You can do the same thing with home building.

If you have a special skill and can show consumers why you are more valuable than your competitors, they are very likely to pay a little extra for your expertise. Building a solid image as a professional who specializes in a certain field has its advantages.

CHANGING AN IMAGE

Once you cast an image, it is difficult to change. If you are already running an established business, don't be discouraged by this comment. While it is harder to change an established image than it is to create a new one, it is not impossible. If during the building of your company image you have found flaws, work to change them. With enough time, effort, and money, you can make a difference in your company image.

Let's say you started your business without much prior thought. You picked a name out of thin air and you never got around to designing a logo. Now you realize that you have hurt the prospects of getting your business off the ground. What should you do? You must make changes to correct your identified mistakes.

You can create a logo, and you can change the direction of your company, but changing the name can get tricky. If you change the name abruptly, you might lose existing customers. How can you accomplish your goal of changing the name? The procedure is not as difficult as you might think.

When you want to change the name of your existing company, do a direct-mail campaign on your existing customer base. Send out letters to all of your customers advising them of your new company name. Explain that due to growth and expansion you are changing the name of the company to reflect your growth. Impress upon the existing customers that the company has not been sold and is not under new management. Of course, if you have a bad image to overcome, the new-management announcement might be a good idea.

Start running your new advertisements with the new company name and logo. Build new business under your new name and convince past customers to follow you in your expansion efforts. By taking this approach, you get a new public image without losing the bulk of your past customers. While this approach works, it is best to create a good image when you begin the business. It is always easier to do the job right the first time than it is to go back and correct mistakes.

SETTING YOURSELF APART FROM THE CROWD

To make your business better than average, you must set yourself apart from the crowd. How do you do that? You do it with a logo, company colors, slogans, and any other ways that are appropriate for your business. I've already talked about logos, so let's look at some other ways to give your company a unique identity.

Company colors

Company colors are one way to attract attention and become known all over town. How important are colors when it comes to company recognition? Well, if you don't think color makes a difference, ask the cab drivers who ride around in yellow cars. Need another example? How about the colors red, white, and blue—what do they mean to you? See, colors can have a strong impact on what we think of and the context we think of it in.

There are consultants who specialize in colors. These professionals work with companies to design colors to influence consumers. Different colors affect how people think, the mood they are in, and what they do. We've all heard how bulls charge a red flag. Would they charge a green flag? Well, I don't think the flag has to be red to attract the bull's attention, but if I were forced to enter the pasture wearing either red pants or green pants, I'd choose green. Since we are taught that red makes the bull mad, we tend to believe it, but I'll bet there are some bulls that would be just as happy charging a blue flag.

Look at how our culture has classified the personalities of people based on hair color. People with red hair are said to have short fuses and high tempers. Blondes are supposed to have more fun, even if they are sarcastically considered dumb. Obviously, the color of a person's hair doesn't automatically make him or her dumb, fun, or hot-tempered.

Choosing the right colors for your company is important. How seriously would you take a builder who pulled up in a pink van with flowers painted on it? The color and decoration of the van has no bearing on the technical ability of the builder, but it does cast an immediate impression. You should choose your company colors with care.

The color of your trucks might be dictated by the color of the truck you presently own. It is more impressive to see a fleet of trucks that are uniform in color and design than it is to see a parade of trucks that include various makes and colors. A unified fleet gives a better impression.

Colors are also important for your business stationery. It would generally be considered inappropriate to use fluorescent orange for your letters and lime green for your envelopes. Certainly these colors would attract attention and be remembered, but the impression would not likely be the one you wanted to create. For most businesses, tan, ivory, light blue, or off-white are acceptable stationery colors.

Color is also important in your truck lettering and job-site signs. If your truck is dark blue, white letters show up better than black letters. If the truck is white, black or blue letters would be fine. For job signs, it is important to pick a background color and a letter color that contrast well. You want the sign to be easy to read. When you talk with your sign painter or dealer, you can review samples of how different colors work together. Now, let's examine the value of company slogans.

Slogans

Slogans are often remembered when company names are not. If you'll be advertising on radio or television, slogans are especially important. Since radio and television provide audible advertising, a catchy slogan can make its mark and be remembered. When advertising in newspapers or other print ads, slogans give readers key words to associate with your company.

I want you to try a simple test. When I ask you the question in the next sentence (no fair peeking), think of one company as fast as you can. What pizza company delivers? If you thought of Domino's, my guess was right. Domino's has been a great success in a highly competitive business. The logo on its box is a domino. They have the Noid character, and one of their slogans says that Domino's delivers. Now a lot of pizza places deliver, but if you live in an area where Domino's is available, you can't think of pizza delivery without thinking of Domino's. This fact is no accident. I'm sure the brains behind Domino's spent huge sums of money to develop this image.

The golden arches is another example of master marketing. Who has the slogan that lets you have it your way? All of these major food franchises have established logos, slogans, colors, and more. Their marketing and advertising is expensive, but it works. Advertising only costs a lot of money when it doesn't work. If expensive advertising works, it makes you money.

Thinking of a slogan for your business might take a while, but it's worth the effort. If you need inspiration, look around at other successful companies. Examine their slogans to gain ideas for yours, but never use someone else's slogan.

BUILDING DEMAND FOR YOUR SERVICES

You can build demand for your services with a strong image. People like to deal with winners. If your company has the reputation of being fair, professional, competent, and dependable, customers will seek you out.

Building business demand with a strong image can be done in several ways. One way is to build your business a little each time you serve a customer. This builds a word-of-mouth referral system. Word-of-mouth referrals are the best business you can get. But, if you don't want to wait for the results of customer recommendations, you can use advertising to put your company in demand.

Advertising is a very powerful business tool. In skilled hands, advertising can produce illusions and fantastic sales results. Consider this situation: You are about to move to a new city. What real estate brokerage in the new city will you call for relocation help? I would guess you might call a company whose brokers and agents all wear gold coats. There we go again with company colors, but the gold-coat brokers get a lot of visibility on TV, radio, and in print ads. Once you hear a hundred times how they are the best real estate team around, you might start to believe it.

You might not know anything about a particular brokerage, but advertising plants the seed that the gold coats mean success. If you buy into the advertising, you are likely to call these brokers. If the broker doesn't make a good personal impression, you might choose another company, but at least you called the gold team. This same strategy can work for you.

When you are running a contracting business, advertising alone does not get the job done. You or your company representatives have to keep the ball rolling once you are in touch with potential customers, but advertising can get you in a selling posture. Talk to some professionals in the field of marketing, and I think you might be surprised at the results you can achieve.

JOINING CLUBS AND ORGANIZATIONS TO GENERATE SALES LEADS

Joining clubs and organizations is an excellent way to generate sales leads. Whether you like it or not, as a business owner, you must also be a salesperson. When you join local clubs and community organizations, you meet people. These people are all potential customers.

By becoming visible in your community, your business has a better chance of survival. If you support local functions, children's sports teams, and the like, you become known. You can use the local opportunities to build your business image. When citizens see your company name on the uniforms of the local kids' baseball team, they remember you. Further, they respect you for supporting the children of the community. You can take this type of approach to almost any level. After you have established a public awareness of your business, you should get busy. But you can't afford to let down on your marketing and advertising needs.

Marketing and advertising might well be the most important lessons for new business owners to learn. While it is true that marketing and advertising alone do not make a business a success, they are crucial elements in building a thriving business. If you don't do a good job with your marketing and advertising, you won't have much of a business.

FINDING OPPORTUNITIES TO MAKE SALES

You have no business without sales. To get the opportunity to make sales, most businesses must advertise. Since public exposure is paramount to the success of your business, so is a strong marketing plan and effective advertising.

Too many contractors fail to see the importance of marketing and advertising. For some reason, many contractors think the public will seek them out. Let me repeat myself, if the public doesn't know you exist, they can't very well seek you out. Regardless of how good you are at what you do, you won't get much work without making people aware of your services.

To get busy and stay busy, you need regular sales. Marketing and advertising can provide you with sales leads. It is up to you or your salespeople to convert the leads into closed sales, but you must start by getting prospects to want what you have to offer. Advertising is the most effective way to generate leads quickly.

MARKETING

Marketing is pivotal to any business. If you have the ability to perform a good market study, you should be able to generate a vast amount of business. Marketing is not just advertising. Marketing is reading the business climate. When you track your advertising results, design your ads, develop sales strategies, and define your target market, you are marketing.

Marketing is much more complicated than advertising. Advertising your business requires little more than the money needed to pay for your ads. Marketing demands an extension of the normal senses. You must be able to read between the lines and determine what the buying public wants. Many books on marketing are available. Professional seminars teach marketing techniques. Many community colleges offer courses in marketing. With enough effort and study, you can become very efficient with your marketing ploys. If you want to have a business with a long life, you need to develop marketing skills.

COMMISSIONED SALESPEOPLE

Should you enlist commissioned salespeople? This is a good question, and the answer lies within your business goals. Commissioned salespeople can make a dramatic difference in your business. On the plus side of the deal, commissioned salespeople can generate a high volume of gross sales. Since you are paying the sales staff only for what they sell, an army of sales associates can be mighty enticing. However, a high volume of sales can create numerous problems. You might not have enough help to get the jobs done on time. You might have to buy new trucks and equipment. The increased business might tie you to the office and cause your field supervision to suffer. You need to consider many angles before bringing a high-powered sales staff online.

Benefits of a sales staff

The benefits of a sales staff are many. If you find the right people to represent your company, you can enjoy increased sales. By having commissioned salespeople, you don't have the normal overhead of employees, and you only pay for what you get. Good salespeople can hustle up deals that would otherwise never come your way. A strong closer can make deals happen on the spot, so you have quick sales and no downtime. Sales professionals can take a simple estimate and turn it into a major job for your tradespeople. With the right training and experience, sales professionals can get more money for a job than the average contractor would. It is clear that for some businesses a sales staff is a powerful advantage.

Drawbacks to commissioned salespeople

The drawbacks to commissioned sales people might outweigh the advantages. Some salespeople tell customers anything they want to hear to get a signature on the contract.

As the business owner, you have to deal with this form of sales embellishment at some point during the job. A customer might tell you that the salesperson assured her she would get screens with her replacement windows when you had not figured screens into the cost of the job. The salesperson might have promised that the job could be done in two weeks, when, in reality, the job will take four weeks. This type of sales hype can cause some serious problems for you and your workers.

Most sales associates are not tradespeople. They don't know all the ins and outs of a job; they only know how to sell the job, not how to do it. A salesperson might tell a prospect that putting a bathroom in the basement is no problem, when in fact, such an installation requires a sewer pump that adds nearly $800 to the cost of an average basement bath. Many times outside salespeople undersell jobs. Sometimes they sell the job cheap to get a sale. At other times the wrong price is quoted out of ignorance. In either case, you, as the business owner, have to answer to the customer.

Getting too many sales too quickly can be as devastating as not having enough sales. If the salesperson you put in the field is good, you might be swamped with work. This can lead to problems in scheduling work, the quality of the work you turn out, field supervision, cash flow, and a host of other potential business killers.

Sending the wrong person out to represent your company can have a detrimental effect on your company image. If the salesperson is dishonest or gives the customer a hard time, your business reputation suffers. Deciding if and when to use commissioned salespeople is your decision. But, let me tell you, don't make the decision lightly. There is no question that the right salespeople can make your business more profitable; however, there is also little doubt that the wrong sales staff can drive your business into the ground.

If you decide to use commissioned salespeople, I suggest you go with them on the first few sales calls. When you are interviewing people to represent your company, remember they are sales professionals. These people will sell you in the interview with the same tenacity that they will use on prospects in the field. Go into the relationship with your eyes wide open. Don't take anything for granted, and check the individuals out for integrity and professionalism.

DETERMINING WHERE TO ADVERTISE

Where should you advertise? Advertising in the local phone directory generates customer inquires and provides credibility for your company. Ads in the local newspaper can result in quick responses. Door-to-door pamphlets and flyers can produce satisfactory results. Radio and television ads can be very effective. Putting a slide-in ad on the video boxes at the local video rental store can give you a lot of exposure. The list of possible places to advertise is limited only by your imagination. However, some advertising media are better than others. Let's take a close-up look at some specific examples.

The phone directory

The phone directory is an excellent place to have your company name advertised. The size of your ad, however, depends on the nature of your business and the type of work you want to attract. Being listed in the phone book ads credibility to your company. Whether you are merely listed in a line listing or have a full-page display ad, you should get your company name in the phone book as soon as possible.

The size of your ad in the directory should be determined by the results you hope to achieve. Large display ads are expensive, and they might not pay for themselves in your line of work. If your business is building houses, a large display ad probably isn't necessary. When people are shopping for a builder, they are not normally in a hurry. An ad that is one column wide and an inch or two in length can pull just as many calls under these circumstances. A quick look at how your competition advertises can give you a hint as to what you should do. If all the other builders have large ads, you probably should have a large ad, too.

Over the years I have tried many experiments with directory advertising. At one time I was running a half-page ad for my business. I thought I could save money by going to a smaller ad. I did pay less for my new ad, but my business suffered. I had a noticeable drop in phone requests.

As I became more knowledgeable about business, marketing, and advertising, I continued to test the results of various directory ads. During my test marketing, I have used many types of ads for my various businesses. I found that for remodeling, real estate, and plumbing, large ads worked best. When perfecting my ads for home building, I did just as well with smaller ads. The results of ad sizes have varied geographically for me. In Virginia, I needed a bigger ad than I need in Maine.

Newspaper ads

Newspaper ads provide quick results; you either get calls or you don't. As a service contractor, my experience has shown that most people who respond to newspaper ads are looking for a bargain. If you want to command high prices, I don't think newspapers are the place to advertise. But if you are new in business, the newspaper can produce customers for you quickly.

Handouts, flyers, and pamphlets

Handouts, flyers, and pamphlets are similar to newspaper advertising. These methods seem to generate calls quickly, but the callers are usually looking for a low price on your services. Many businesses consider this form of advertising as degrading. I don't know that I would agree with that opinion, but I don't think you can receive the money you are worth with these low-cost advertising methods.

Radio advertising

Radio advertising is expensive, but it is a good way to get your name to listeners. I believe the key to radio advertising is repetition. If you can't afford to sustain a regular ad on the radio, I would advise against using this form of advertising. Most people are not going to hear your ad and run to the nearest phone to call you. However, if you can budget enough money for several radio spots for a few weeks, you can gain name recognition.

Television commercials

Television commercials can be very effective. Unlike radio, where people only hear your ad, television allows viewers to see your ad. People associate television advertisers with success. With the many cable channels available, television advertising can be an affordable and effective way to get your message out to the community. I have used ads on cable television very effectively. Television ads can increase your sales.

Direct-mail advertising

Direct-mail advertising is very effective, but it is not always cost-effective. However, due to the profit potential from building a new house, direct mail is usually a sound business decision to pursue. The cost of direct-mail advertising can easily run into thousands of dollars. Most people who use this form of advertising are happy if only 1 percent of the people they mail to become customers.

By using direct mail you can reach a targeted market. If you want to advertise to people with incomes in excess of $50,000, you can rent a mailing list of just those people. You have complete control over who sees your advertisement. This type of demographic breakdown is very effective in mailing to the best prospects.

Most mailing lists are available at prices of around $75 for each thousand names. Many list brokers require a minimum order of 3000 names. The names can be supplied to you on stick-on labels. Expect extra charges for various demographic segregations.

If you want to reduce your mailing costs, your local postmaster can set you up with a bulk-rate permit. To use the bulk-rate service, you must mail a minimum of 200 pieces of mail at a time. The cost for this type of mailing is much less than first-class postage, but you must pay one-time and annual fees up front. Talk with your local post office for full details.

Creative advertising methods

Creative advertising methods are just that—creative. You might want to rent space on a billboard to advertise your business. Perhaps you can cut a deal with a local restaurant to have your company highlighted on the menus in the restaurant. Providing uniforms for the local little league can get your name in front of a large audience. If you put your mind to it, there is almost no end to the possibilities for creative advertising.

DETERMINING A GOOD RATE OF RETURN ON ADVERTISING COSTS

What rate of return will you receive on advertising costs? The response to your advertising depends on your marketing plan and the execution of your advertising. If you are advertising in the local newspaper, you might expect about a .001 response. In other words, if the paper has 25,000 subscribers, you might get 25 responses to your ad. This projection is aggressive. In most cases your response might be much lower. If you hit the paper at the right time of the year with the right ad, 25 calls might come in. However, if you only get 10 calls, don't be surprised. In some cases you might not even get that many. Your advertising success depends entirely on how well you picked the publication and designed the ad.

Advertising a contracting business on the radio or television can seem like a waste of money. It is not uncommon for these ads to run without getting calls, but that doesn't mean the ads were not effective. Television and radio advertising builds name recognition for your company. This form of advertising works best when it is used in conjunction with some type of print advertising.

If you are running ads in the paper, distributing flyers, or doing a direct-mail campaign when the television and radio ads are on, you should see a higher response than you would without the radio and television ads.

Direct-mail advertising often provides fast results. Many people receiving ads by mail either trash them or act on them quickly. A 1-percent response on direct-mail advertising is generally considered good. For example, if you mail to 1000 houses, you should be happy if you get 10 responses. Due to the low response rate of bulk mailings, direct mail is not effective for low-priced services. However, if you are selling big-ticket items, such as new houses, direct mail can work very well.

If you target your direct-mail market, you should get a much better rate of return. For example, if you use a mixed mailing list for your advertising, you won't know what type of person is receiving your ad. But if you pick a list based on demographics, you can be sure you are reaching the type of potential customer you want.

Demographics are statistics that tell you facts about the names on your mailing list. You can rent a mailing list that comprises specific age groups, incomes, and so forth. These statistics can make a big difference in the effectiveness of your advertising.

Determining the effectiveness of advertising is a task all serious business owners must undertake. To learn which ads are paying for themselves, you need to know which ads are generating buying customers. Some ads generate a high volume of inquiries but don't result in many sales. Other ads produce fewer curiosity calls and more buying customers. You need to track the results of your advertising. Without knowing which ads and advertising mediums are working, you have no way of maximizing the return on your advertising expenses.

LEARNING TO USE ADVERTISING FOR MULTIPLE PURPOSES

Most businesses use advertising for multiple purposes. The primary use of advertising is to generate consumer interest in goods and services. But advertising can do much more for a business. As you have already seen, advertising can build name recognition for your company. Name recognition is important when trying to get the most mileage out of your advertising budget.

What else can advertising be used for? Advertising can be used to build your company image. Through advertising, you can create almost any look you like for your business. A company image can be responsible for commanding higher fees and quality customers.

Advertising can help you to build goals. If you want to be known as an expert in building replica or reproduction homes, advertising can get the job done. As you go along in business, you'll find that various forms of advertising can help you achieve success in many ways.

BUILDING NAME RECOGNITION THROUGH ADVERTISING

I have already talked about building name recognition through advertising, but now I am going to teach you how it's done. You want people to see or hear your company name and feel like they know the company. To accomplish this goal, you must use repetitive advertising.

Repetitive advertising can be used in all formats of advertising. Take radio advertising, for example. When you hear radio commercials, you normally hear the company name more than once. Pay attention the next time you hear ads on the radio. You'll probably hear the company name or the name of the product being sold at least three times.

Television uses verbal and visual repetition to ingrain a name or product in your mind. Watch a few television commercials and you'll see what I mean. During the commercials you see or hear the company name or product several times.

Not only should your name be used often in the ad, the ad should run regularly. If you advertise in the newspaper, don't run one ad and stop. Run the same ad several different times. Use your logo in the ad, and keep the ads coming on a regular schedule. This type of repetition implants your company name into the subconscious of potential customers. When these potential customers are ready to become customers, they will think of your company.

GENERATING DIRECT SALE ACTIVITY WITH ADVERTISING

The need for generating direct sale activity with advertising is the reason most people use advertising. For a service business, generating sales is possible with direct mail, radio, television, print ads, telemarketing, and other forms of creative marketing. Telemarketing and direct mail are two of the fastest ways to generate sales activity.

I've already talked about how direct mail works, but how about telemarketing? Telemarketing is a job for thick-skinned people. Calling people you don't know and asking them to use your services, buy your product, or allow you into their home for a free inspection, estimate, or whatever is not much fun. However, if you can live with rejection and are not afraid to call 100 people to get 10 sales appointments, cold calling can work.

When your motive with advertising is to generate sales activity, it helps to make your offer on a time-restricted basis. You should offer a discount for a limited time only. Create a situation in which people must act now to benefit from your advertising. Turning up the heat with time-sensitive ads can generate activity quickly.

Without advertising, the public won't know you exist. Advertising is expensive, but it is also a necessary part of doing business. If you don't spend money on advertising, the public is not going to spend more with your business. The contracting field is filled with business owners who are aggressive. These aggressive owners advertise regularly. If you don't put your name in front of people, you will be run over by the companies that do.

STAGING PROMOTIONAL ACTIVITIES

Promotional activities are an excellent way to get more sales and to build name recognition. By using special promotions, you capture public attention and create an opportunity for additional sales. Let me give you an example of how you could stage a promotional event.

For example, assume you are a contractor who specializes in new homes. You could talk to your local material supplier and develop a seminar. Have the material supplier allow you to come into the store and give a home-building or home-buying seminar to shoppers. Tell the supplier how the seminar can be good for the store's image and can increase material sales.

Advertise the free seminar for about two weeks prior to the date of your talk. The supplier might be willing to pay a portion of the ad costs; after all, the store is gaining publicity from this promotion as well.

When people begin gathering around you in the store, be sure to have your business cards, rate sheets, and other sales aids out where the shoppers can see them. After your seminar, field questions from the audience. This type of promotion can help create your image as an expert in your field.

If it is legal in your area, give away a door prize. Have the audience fill out cards with their names, addresses, and phone numbers for a prize drawing. Give away a prize; it could be a discount on remodeling services, a small appliance, or just about anything else you can think of. After the seminar, you have a box full of names and addresses to follow up on for work. This type of idea can increase your business dramatically.

STAYING BUSY IN SLOW TIMES

Every business owner wants to know how to stay busy in slow times. Until you have survived recessionary times, you don't have the experience to stay afloat in troubled waters. Since you can't always learn survival skills on a firsthand basis and survive, you must turn to the experience of others for your training.

When times are tough, you might have to alter your business procedures. But how do you do it? Will you lower your prices? Can you eliminate overhead expenses? How about cutting back on your advertising expenses—should you cut back on your ad budget? Any of these options could be the wrong thing to do.

If you lower your prices, you'll have a very hard time working your prices back up to where they used to be. Lowering prices is risky business, but sometimes it is the only way to keep food on the table. If you have to lower prices to stay in business, do so with the understanding that getting prices back to normal takes time.

Discounts might accomplish the same goal as lowering labor rates, but with less long-term effects. People expect discount offers to end. Run ads offering a discount from your regular labor rates for a limited time only. This tactic makes rebounding to regular labor rates less difficult.

Cutting unnecessary overhead expenses is a good idea at any time, but cutting needed overhead expenses can be dangerous. Even when times are tight, some overhead expenses should be maintained. If you cut back on the wrong expenses, your business might get worse. For example, if you eliminate your answering service, you might miss some of the important business calls your company needs to survive. Evaluate your expenses carefully before making cuts. Only eliminate expenses that should not affect the volume and profitability of your business.

Advertising is one of the first expenditures many business owners cut. Advertising is just as vital to your business in bad times as it is in good times. In fact, advertising might be more valuable in tough times. As your competitors cut back and shrink into obscurity, you can lunge forward with aggressive advertising. Think long and hard before you eliminate your ad in the phone book or reduce your normal advertising practices.

Business promotion should be one of your top priorities. As I've said before, without customers, you have no business. To get customers, you must always strive to find new people to build homes for. If you get too comfortable and slack off on your business promotion, you might find that your business shrivels up and blows away.

25

Hiring employees

Hiring and managing employees can be two of the largest hurdles a new business owner faces. Employees could easily be the biggest unsolved mystery of the business world. Most business owners want employees until they have them; then they often wish they didn't need employees.

Finding the proper balance for employees in your business might cause you to lose sleep at night. No question about it; employees complicate your life. However, the right employees can make you more money than you could make alone. So, how do you decide how to handle the employee issue? This chapter opens up the questions and show you the answers. You'll learn the ropes of finding, hiring, managing, and terminating employees.

Do you want employees? This a simple question, and you shouldn't need any help in answering it. You either do or don't want employees. For now, I'm not talking about your needs, I'm discussing your desires. If you want employees, you know it, but do you know why you want employees? Assuming that you do want employees, let's examine a few justifiable questions.

Why do you want employees? Ah, now there's a question that requires a little thought. Do you want employees to make more money? Do you want employees to cast a brighter public image? Would you like to have employees so that you can feel more important? Will having employees make you more successful in your own mind? Let's look at each of these questions and evaluate your answers.

WHY DO YOU WANT EMPLOYEES?

Why do you want employees? Many business owners can't provide a clear answer to this question. The owners might be quick to say they want employees, but they often can't express why they want them.

Before you begin building a list of employees, decide why you are hiring them. Look at your reasons and be sure they make sense. It is easy to hire employees, but it is not so easy to get rid of them. Don't hire people until you need them and know why you are hiring them.

To make more money

Do you want employees so you can make more money? Most employers give this as their reason for hiring employees. There is a belief, true or false, that the more employees you have, the more money you make. While this theory can hold true, it can also be dead wrong.

From my personal experience, I have seen both sides of this coin. There have been times when I had several employees and made less in net profits than I did without them. There also have been occasions when employees have been good for my financial health. What caused the differences? In my personal experiences, there were many reasons.

The economy has often affected the success of my business endeavors. Riding the cycles of the economy has been a factor in my experience with employees. In my early years, poor management contributed to my failings with employees. I have owned various types of businesses and the type of business I was in at the time seemed to have an impact on the effectiveness of employees. My selection of employees has definitely made a difference in the success and failure of the relationship.

Many factors can influence your experiences with employees, but don't get the idea that more is always better. A lot of contractors do much better financially with small crews than they do with large crews. As you move through the chapter you'll see specific examples of how employees can affect your business.

To cast a brighter public image

Do you want employees to cast a brighter public image? It is true that the public often associates the success of a company with the number of employees the company has. But, is this reason enough to hire employees? No; you must hire employees for profitable reasons, not for public opinion. Yes, I know public image has an effect on the success of your business, but don't burden yourself with employees for this one reason.

To make you feel more important

Would you like to have employees to make you feel more important? Most people won't admit this, but many of them do feel more important when they have employees. It is not a good reason to hire people; there are less costly ways to improve your self-appreciation.

To feel more successful

Will you feel more successful when you have employees? People measure success with different measuring sticks. Some people consider themselves successful when they have a lot of money. Having a number of employees can spell success for some individuals. Family health and happiness are common measurements of success for business owners.

How you measure success is up to you, but don't lean too heavily on employees to find happiness. Simply having a large number of people on your payroll doesn't mean you'll be happy. Unless you are satisfied to provide jobs for others, while getting by on meager profits, employees can be a mistake.

Does it sound like I'm against hiring employees? I can see where you might be forming that opinion, but don't judge me too quickly. I'm not against employees, but I do want to show you both sides of the issue. Many business owners never take the time to consider the bad points of hiring employees. Since I assume you have dozens of reasons why hiring employees is great, I want to expose you to some of the possibilities you might not have thought of.

DO YOU NEED EMPLOYEES TO MEET YOUR GOALS?

Do you need employees to meet your goals? The answer to this question might lie in your goals. There are limitations to what any one individual can do without help. However, you might be able to accomplish your goal with independent contractors, removing the need for regular employees. Independent contractors might seem to be more expensive than employees when their rates are first reviewed, but further investigation can prove independent professionals are less expensive. Let's look at the pros and cons of employees versus independent contractors.

Hourly rates

The hourly rates of independent contractors are normally higher than the wages you would pay an employee to perform the same function. While this is to be expected, it might not be as it seems. While the hourly rate of independents is higher, they might actually cost less. How can this be? Independents can wind up costing less for many reasons.

Since independents often work for multiple employers, they do not depend on you for their entire income. You pay only for what you need and for what you get. Employees, on the other hand, rely on their employers to pay their salaries for the whole year. While you might be paying a lower hourly rate to employees, if you are paying them for time when you don't need them, you can be wasting money.

Insurance costs

Independent contractors pay for their own insurance costs, which can amount to a substantial amount of money. If you have employees, you might have to pay for liability insurance, health insurance, dental insurance, worker's comp insurance, and disability insurance, but you won't have these same expenses with subcontractors. You still need liability insurance, but the rest of the coverages can be avoided.

Transportation

Subcontractors are responsible for their own transportation. If you hire employees, you probably need to furnish company vehicles for field personnel. When you consider the cost of acquiring, insuring, and maintaining vehicles, subcontractors who provide their own transportation are desirable options.

Payroll

Payroll for companies with many employees can be a full-time job. If you have to pay a full-time employee just to handle the payroll for your other employees, you have additional overhead that must be recovered in the prices of your services. By using subcontractors, you eliminate the need for payroll and payroll records. Of course, you still have to write checks to the subs, but this procedure is less labor-intensive than doing payroll.

Payroll taxes

Payroll taxes are another expense you eliminate by using subcontractors. When you have employees, you have additional tax deposits to make for the payroll. This tax is nonexistent when you use independent contractors.

Paid vacations

Most employees expect to receive paid vacations. If you give the average trades-person a two-week paid vacation, you are losing more than $1000 a year in giving this benefit. This loss applies to every employee who receives this benefit. If you have 10 plumbers, you lose more than $10,000 a year by being a nice boss. Sub-contractors don't expect you to give them a paid vacation. This type of savings can add up quickly.

Sick leave and other benefits

Sick leave and other benefits can be compared to paid vacations. Most employees expect these favors, but subcontractors don't.

Convenience

If convenience is a factor, employees might have an edge against subcontractors. Subcontractors can be difficult to control; after all, why do you think they are called independent contractors? If you want people at your fingertips, employees are generally more reliable than subs.

Competition

Many contractors fear that their subcontractors might become direct competition. These contractors assume the independents are just using them to get to their customers. While some subcontractors might attempt to steal your clients, most won't.

While employees are not generally looked on as competition, they can pose more of a threat than subcontractors. Subcontractors are already in business and already have customers. Employees might be looking to go into business for themselves. They might also be considering taking some of your customers with them. From the competition angle, either group is a possible threat, but I would be more concerned about the employees.

Comparisons

Comparisons between employees and subcontractors are not hard to make. Before you dash out and hire employees, consider the advantages to using subcontractors. You might find that you are happier and more prosperous using independent contractors.

HOW DO YOU FIND GOOD EMPLOYEES?

Do you know how to find good employees? If you do, you can probably make more money as a consultant to other businesses than you can in the contracting business. Business owners of all types struggle to find good employees. It is easy to hire people, but locating good employees is difficult.

Before you can hope to hire good employees, you must understand what makes a good employee. Defining the qualities you want in your staff should be easy. For most business owners, setting the criteria for a good employee is not a problem. The problem comes in trying to find such a person who is available for work. Most good employees are well taken care of, and it is hard to pry them away from their present employers.

Classified ads

Classified ads are one way of finding employees, but be prepared for many wasted hours in your search. When you run a help-wanted ad, you get responses from all types of people. Most of these people do not fit the mold you have cast.

Classified ads are the quickest way to get applicants in your office, but they generally are not the best route to finding prime candidates for your opening. However, classified ads can produce top-notch people. If you go into your employee search with the knowledge that you'll have to sift through a mass of unqualified applicants to find one good worker, classified ads can be worthwhile.

Employment agencies

Employment agencies are known for their work with executives and professionals. These agencies screen prospective applicants before giving out your name and phone number. Generally, the agency contacts the business owner and goes over the traits of a candidate before the candidate knows who the employer is. Then, if you are interested in talking with the applicant, the agency arranges an interview.

These agencies are geared more towards white-collar positions than they are to blue-collar jobs. If you are looking for people in the trades, agencies might not be of much help.

Also, you must consider the fees charged by agencies. Some agencies charge applicants when they locate a job for them. However, most agencies charge employers for finding acceptable applicants. Before you deal with an agency, be sure of what you are getting into. If you are asked to sign a contract or engagement letter, read it carefully and consider consulting your attorney before signing the document. The fees charged by some agencies are high.

Word-of-mouth

Word-of-mouth referrals for job applicants are a good way to find the best employees. If you put the word out that you are looking for help, you might drum up some applicants through your existing employees or friends. When a person applies for a job because a friend has recommended your company, you have the advantage of built-in credibility.

Unemployment office listings

Unemployment offices carry listings of job opportunities for people in need of work. If you have a job to fill, notify the local unemployment office. Your opening will be listed in their computers and on their bulletin boards. This type of listing service is free and can produce quick results.

Hand-picking

Hand-picking employees is one of the best ways to get what you want. However, you must remain ethical in your procedures as you select and solicit individuals. Stealing employees from your competition is frowned on. Let's look at how you should and shouldn't go after specific employees.

There is a right way and a wrong way to hand-pick employees. First, let's look at an example of the wrong way. Years ago, I was a project superintendent on a townhouse project. I was in charge of all the plumbers and support people for the plumbers. During these times, the economy was good and all good plumbers had jobs. However, most plumbing companies needed more plumbers to keep up with the rapid building trends.

Some companies showed no remorse in stealing good plumbers from the competition. During this project, I saw representatives from competitive companies come onto my job and offer my plumbers more money right in front of me!

The plumbers knew they could get work anywhere, and many of them changed jobs for an extra 25 cents an hour. When these people raided my job sites, they often left with my plumbers in the backs of their trucks. The plumbers didn't give any notice; they just left, following the higher hourly wages. Some plumbers even tried to get me and the other company's representative into bidding wars.

You can imagine the ill will that formed between companies under these circumstances. One company would steal a plumber on Monday, and on Friday, another company would take the plumber away to a new job. This endless turnover of plumbers hurt everyone in the business. You don't want to use these techniques in hand-picking employees.

What tactics can you use to hand-pick employees? There are many ways to make candidates aware that you are interested in offering them a position. One excellent way is to run into them at the supply house. While the two of you are standing around, waiting for your orders to be filled, start a conversation that leads to your need for help. Stress how you are looking for someone just like the person you are talking to. If the other person is interested in pursuing employment with your company, you should get some signals.

If you have your eye on a particular person for a position, don't hesitate to call or write the individual. If you don't want to look too obvious, ask the person if he knows of anyone with qualifications like his that is looking for work. By using this approach, you can give all the details you want about the position without making a direct solicitation of the individual.

You can get your point across to people who are presently employed in many tactful ways. You don't have to stoop to going public with your attempt to take the employee away from the existing employer. Be discreet and keep your dealings fair.

WHAT PAPERWORK IS INVOLVED?

Let me give you a few pointers to provide a starting point for obeying the law in your employee paperwork. Don't take these pointers as the last word and don't consider them conclusive. Look at them as a guide to the questions to ask your attorney and tax professional. Remember, you are trying to keep as much of your cash as you can, and losing a portion of it to fines and lawsuits is not a good use of your resources.

Employment applications

You should have all prospective employees complete an approved employment application. These applications tell you something about the people you are considering hiring, and they provide a physical record for your files. Be sure the application forms you use are legal and don't ask questions you are prohibited from asking.

W-2 forms

W-2 forms are used to notify employees of what their earnings were for the past year and how much money was withheld for taxes. The forms must be mailed or given to employees no later than the last day of January following the taxable year.

W-4 forms

W-4 forms are government forms that should be completed and signed by all employees when they are hired. These forms tell the employer how much tax to withhold from an employee's paycheck. Once the form is filled out and signed, keep it in the employee's employment file. New W-4 forms should be completed and signed by employees each year.

I-9 forms

I-9 forms are employment-eligibility verification forms. These forms must be completed by all employees hired after November 7, 1987. I-9 forms must be completed within the first three days of employment.

Employers are required to verify an employee's identity and employment eligibility. Employees must provide the employer with documents to substantiate these facts. Some acceptable forms of identification are birth certificates, driver's licenses, and U. S. passports. If an employer fails to comply with the I-9 requirements, stiff penalties can result.

1099 forms

1099 forms are used to report money you pay to subcontractors. These forms must be completed and mailed at the end of each year. You send one copy to the subcontractor, one copy to the tax authorities, and you retain a copy for your files.

Since subcontractors come and go, they can be difficult to locate when the time comes to send out the 1099 forms. Insist on having current addresses on all of your subcontractors at all times.

Employee tax withholdings

Employee tax withholdings cannot be ignored. When you do payroll, you must withhold the proper taxes from an employee's check. The Internal Revenue Service provides you with a guide to explain how to figure the income tax withholdings.

In addition to withholding for income taxes, you also have to deduct for Social Security. Again, you can use a tax guidebook available from the IRS to figure the deductions.

Once you have computed the income and FICA withholdings, deduct them from the gross amount due the employee for wages. After doing this, enter the amount of withholdings in your bookkeeping records. While the money withheld is still in your bank account, it doesn't belong to you; don't spend it.

As an employer, you must contribute to the Social Security fund for your employees. Your contribution must equal the amount withheld from the employee. This is just another hidden expense to having employees. Your portion of the funding does you no good; it is only to benefit the employee.

Employer ID number

When you establish your business, you should apply to the Internal Revenue Service for an employer ID number. Some small business owners use their Social Security numbers as an employer ID number, but it is better to receive a formal ID number from the IRS. You use this identification number when making payroll-tax deposits.

Payroll-tax deposits

Payroll-tax deposits are required by companies with employees. The IRS provides you with a book of deposit coupons for making these deposits. The deposits can be made at the bank you do business with.

Remember the money you withheld from your employee's paycheck? That's the money you use to make the payroll-tax deposit. The requirements for when these deposits are made fluctuate from business to business. Consult your CPA for precise instructions on making your payroll-tax deposits.

As a business owner, you can be held personally responsible for unpaid payroll taxes. Even if you sell or close your business with payroll taxes left unpaid, the tax authorities can come after your personal assets to settle the debt. Don't play around with the money owed on payroll taxes.

Federal unemployment tax

The federal unemployment tax is also know as FUTA. Your requirements for making FUTA deposits depend on the gross amount of wages paid in a given period of time. Talk to your accountant for full details on how FUTA affects your business.

Self-employment tax

As your own boss, you must pay self-employment taxes. As a self-employed individual, your Social Security tax rate can be nearly double what it was when you were an employee to make up for the fact that you don't have an outside employer contributing to your Social Security fund.

State taxes

State taxes are another consideration for your business. Different states have various rules on their tax requirements. To be safe, check with your CPA or local tax authority to establish your requirements under local tax laws.

Labor laws

The labor laws control such areas as minimum-wage payments, overtime wages, child labor, and similar requirements. As an employer, you must adhere to the rulings set forth in these laws. The government can provide you with information on your responsibilities to these laws. Call the Department of Labor to request that details be mailed to you.

OSHA

The Occupational Safety and Health Act (OSHA) controls safety in the workplace. Your business can be affected by OSHA in many ways. To learn the requirements of OSHA, contact the Department of Labor.

HOW DO YOU TERMINATE EMPLOYEES?

Terminating employees can become a sticky situation. With so many employee rights that might be violated, you must be careful when you are forced to fire an employee. Since you never know when termination might be your only option in dealing with a troublesome employee, you should assume all employees are possible targets for termination. By this, I mean you should create and maintain a paper trail on each employee's activities.

As soon as you hire an employee, start an employment file on the individual (FIG. 25-1). The file will grow to contain all documentation you have on the employee. Examples of the file contents include tax forms, employment application, income records, performance reviews, attendance records, and disciplinary actions and warnings. If a time comes when you must dismiss an employee, these records of employment history (FIG. 25-2) might come in handy.

Before you lose your temper and fire an employee, consider the costs you'll incur replacing the worker. Give yourself time to think about the offense; is it really necessary to fire the individual? If the circumstances demand termination, do so with care. Consult with your attorney in advance to be certain of your requirements in letting an employee go.

HOW CAN YOU KEEP GOOD EMPLOYEES?

After you have hired your people, you need to learn how to keep good employees. If employees are worth having, someone else will want them. You always have the

Employee-File Checklist

Employee name: _____

Employee ss # _____

Item	In file	Need	Notes
I-9 Form			
W-4 Form			
Application			
Tax Info			
Insurance Info			
Reviews			
Warnings			
Attendance			

25-1 Employee file checklist.

risk of someone trying to persuade your best employees to leave your company for theirs. As another risk, you must be concerned about good employees going into business for themselves and into competition against you. You can't really blame the employees for wanting their own business; after all, you wanted your own business. What you have to do is make the working conditions so good that the employees won't want to leave.

Weekly Work History

Employee: _____

Payroll number: _____

Date	Work phase & Job name	Time in	Time out	Total time

25-2 Weekly work history.

Many factors can influence employees to stick with your company. Some of these factors are:

- Comfortable wages
- Health insurance
- Dental insurance
- Paid vacations
- Sick leave
- Company vehicles
- Retirement plans
- Good working conditions
- A friendly atmosphere
- Competent co-workers
- Fair supervisors
- Pride in the company

If you establish a good environment for your employees, they have no reason to leave your employ. Bonus plans and other incentives can even remove much of the risk of having the employees go into business for themselves. It's up to you to communicate with your employees and to create circumstances to keep them happy. If you have valuable employees, they are worth the extra effort.

HOW CAN YOU CONTROL EMPLOYEE THEFT?

Controlling employee theft is a consideration most business owners prefer not to think about. All business owners would like to believe that their employees aren't thieves, but some are, and you have to protect yourself and your good employees from the few bad ones. Most employee theft is petty, but that doesn't mean it is not serious. Stealing is stealing, and you can't afford to have criminals for employees.

How you run your business can have a bearing on how much employee theft you experience. If you screen all of your job applicants thoroughly, you can cut down on the chances of hiring a crook. By keeping tight control on your inventory and making your employees aware of your antitheft policies, you can reduce your risks even more. If you eliminate temptation, you eliminate most casual theft.

HOW CAN YOU EXERCISE QUALITY CONTROL?

By exercising quality control over employees you can build a better business. What is meant by quality control? Quality control is just what it sounds like—controlling quality. The qualities you control could be numerous.

The qualities most business owners are interested in controlling are:

- Punctuality
- Good work habits
- Customer service
- Work quality
- Dependability
- Loyalty

There are, of course, other qualities you might wish to keep an eye on. You might request your employees to maintain ongoing continuing education programs. Having your employees expand their capabilities into other work areas could be one of your pet projects. Once you know what you want from your employees, work with the employees to meet your goals.

DO YOU WANT TO TRAIN EMPLOYEES?

Training employees to do the job used to be standard procedure, but it's not today. Today, most employers are looking for experienced people who can step into a position and be productive. The days of training apprentices is all but gone.

Why has this shift in the workplace occurred? One reason is money; it costs money to train employees. Even if you do the training yourself, it costs money. The time that you spend away from your routine duties is lost income. To train employees, you must look upon the training as an investment. This is another reason why employers don't spend the time training new employees.

In the old days, employees stuck with their employers for a long time. If an employer trained employees, the business owner could be reasonably confident the investment would pay off. Today, employees change jobs in the blink of an eye. Employers know this and are reluctant to train employees who might run to another employer if the opportunity arises. It is sad, but the traditional values that once existed have been eroded with the increased demand for the mighty dollar.

The quest for money is not only in the minds of employees. Many employers don't want to hire inexperienced help because they know the new employees won't make as much money for them as an experienced employee. With everyone being in a hurry to grab the brass ring, no one has time to invest in building a stable business or career. Everyone seems to take the shortest path they can find to potential riches.

These circumstances are changing the business world. Since the old masters are not passing their knowledge down to apprentices, the crop of qualified tradespeople is shrinking. In time, if this pattern continues, the artful craftsmanship of the past might be only a memory.

Should you hire experienced help or train new people to do the job your way? I guess it makes more sense to hire people that can jump in and start turning a dollar. But if you do train employees to produce the type of work you want, you might be happier. There is a certain satisfaction to be gained from watching a rookie mature into a journeyman. The choice is yours, but be advised, trainees are likely to look for a higher-paying job once you have trained them.

SHOULD YOU TRAIN EMPLOYEES TO DEAL WITH CUSTOMERS?

I believe training employees to deal with customers is the responsibility of every business owner. Every business is run differently, and even experienced mechanics have to be taught to treat customers the way you want them treated.

You might choose not to provide on-the-job training for work skills, but don't forego training your employees to deal with customers. Your customers are your business; if you alienate them, you lose your business. Employees are representatives of your company. If they act improperly around customers, it is a reflection on your business.

It is a good idea to develop a policy manual on how you want customers treated. Issue the manual to each of your employees and require them to commit it to memory. If necessary, test the employees' knowledge of the manual. Before you put people in touch with your customers, make sure they will behave in a suitable manner.

HOW DO YOU ESTABLISH THE COST OF EACH EMPLOYEE TO YOUR COMPANY?

Establishing the cost of each employee to your company is a necessary part of setting your pricing structure. Many hidden costs are involved with employees, and each employee might have a different set of circumstances. Before you set your prices, know what each of your employees is costing you.

What costs should you look for? The most obvious cost is the hourly wage, but there are other factors. Some employees might receive more benefits than others. If one employee gets a two-week paid vacation and another employee gets a one-week paid vacation, the cost of the employees will be different by an amount equal to the extra week of vacation pay.

As employees build seniority, they normally gain additional benefits. You must consider all of these costs when determining the overall cost of an employee. Whether it is health insurance, dental insurance, or paid leave, you must factor the cost into your projections.

Bonus pay is another item that can influence the cost of your employees. If you are in the habit of giving each employee a bonus during the holiday season, you must count this money in the cost of the employee.

Turn over every rock when looking for hidden expenses. Don't let any part of the expenses for your employees go unnoticed. Once you have all the figures, chart the hourly differences. When you bid a job, bid it based on your most expensive employees. Then, if you can put employees who cost less on the job, you make more money. But if your least expensive labor is not available, you won't lose anything by putting your top-paid people on the job.

HOW CAN YOU DEAL WITH PRODUCTION DOWNTIME?

Dealing with production downtime when paying employees can be frustrating and costly. You have already seen how you can lose money if your crews must stop to run for materials, but that is not the only way you can lose money to downtime. Some causes for these losses are beyond your control, but many of them can be avoided with strong management skills.

Bad weather

Bad weather can often shut a contractor down. While you can't control the weather, you can plan for its effects on your business. If you have a business that involves some inside work, try to save this work for days when the weather won't allow your normal outside operations.

If you are starting a job on which the weather might cause delays, plan ways to circumvent the lost time. This might involve using tarps to cover the work area or renting heaters to keep the job comfortable. Look for ways to keep production up during any weather conditions.

When the circumstances cannot be overcome, use your best judgement as to what to do with your employees. Most employers send the employees home without pay. On the surface this saves money, but it might cause you to lose your employees. Good employees are hard to find, and a turnover in employees is expensive. You might be better off to create some busy work for the crews, even if it is not cost-effective.

Some ideas for busy work could be counting inventory, taking trucks in for service, or performing maintenance on equipment. While these tasks might not warrant the use of highly paid personnel, they must be done, and if this keeps your employees in place, you might be better off in the long run.

Past-due deliveries

Past-due deliveries can bring your crews to a halt. If you are a good manager, you won't let this happen often. However, at times a delivery won't be made, and you must find work for your crews. Be prepared for these times with some back-up plans. If you send the crews home, they might not be happy. On the other hand, maybe they would enjoy having the day off, even if they aren't getting paid. Give them the option of taking the day off or doing fill-in work.

Code-enforcement rejections

Code-enforcement rejections can bring a sudden stop to work. It is difficult to think of a suitable excuse for this type of downtime. If you or your field supervisors are supervising the work, it should not fail an inspection. If you start to have recurrent problems of this nature, you need tighter control on your field supervision.

Disabled vehicles

Every contractor is going to have problems with disabled vehicles from time to time. The most you can do to prevent these problems is regular maintenance. When a truck breaks down and is going to be out of service for an extended time, try to double your crews up. There isn't much else you can do.

A lull in work

Sooner or later, a lull in work will cause you downtime with your crews. These lulls can be projected at certain times of year. Typical times include holidays, summer vacation seasons, tax seasons, and school start-up seasons.

Proper preparation can help to overcome these slow-downs. Line work up in advance for the slow times. Be aggressive in advertising, and offer discounts, if necessary, to keep your people busy. Avoid laying your people off; once they are gone, you might not get them back.

HOW CAN YOU REDUCE EMPLOYEE CALL-BACKS AND WARRANTY WORK?

By reducing employee call-backs and warranty work, you can increase your profits. Customers won't pay you to do the same work twice, but you have to pay your employees for their time. This can get expensive, fast. If you have sloppy workers who frequently cause call-backs, you must take action.

Everyone is going to make mistakes, but professionals shouldn't make many. Call-backs are generally the result of negligence; either the mechanic did the job too quickly, too poorly, or didn't check the work before leaving the job. You can and must control this type of behavior.

Call-backs and warranty work hurt your business in two big ways. The first hurt is financial; you lose money on this type of work. The second problem is the confidence your customers lose in the quality of your work. You cannot afford either of these results.

How can you stop call-backs and warranty work? You have several options available for controlling these costly occurrences. Let's look at some of the ways that have worked for others and that might work for you.

Call-back boards

Call-back boards can reduce your call-backs if you have multiple employees. Hang a call-back board in a part of your office that all employees can see. When a mechanic has a call-back, the mechanic's name is put on the board. The board is cleared each month, but people with call-backs must see their name on the board for up to a full month.

Generally, there is a certain competitiveness among tradespeople. If a mechanic's name is on the call-back board, he or she might be embarrassed. This simple tactic can have a profound effect on your call-back ratio.

Employee participation

Employee participation in the financial losses of call-backs is another option. However, your employees must agree to this plan without being pressured. For your protection, have all employees agree to the policy in writing.

Under the employee-participation program, employees agree to handle their call-backs on their own time. You pay for materials and the employees absorb the cost of the labor.

As a variation of this program, you can agree to pay the employee for the first two call-backs in a given month, with the employee taking any additional call-backs without pay. But, before you implement either of these programs, confirm their legality in your area and have employees agree to your employment terms in writing.

Bonus incentives

Bonus incentives are another way to curtail call-backs. Call-backs are expensive and detrimental to your business image. If you can eliminate warranty work by offering bonuses, do it. You won't lose any more money and you won't lose any credibility with your customers.

If you don't like the idea of giving employees bonuses for doing a job the way they should in the first place, hedge your bets. Determine what the maximum annual bonus for any employee might be, and adjust your starting wages to build in a buffer for the bonuses. The employee will feel rewarded with the bonus, and you won't be paying extra for services you expect to get out of a fair day's work.

HOW DO YOU MANAGE OFFICE EMPLOYEES?

Office employees are a little easier to manage than field employees. People working in your office are easy to find and keep an eye on. If you are an office-based owner, your office employees feel compelled to stay busy; they know you are watching their performance. However, you must not abuse the power your presence presents.

Office employees can become intimidated by having the boss close at hand. If this happens, production drops off or mistakes multiply. You should hire the best help you can find and then let them do their jobs. If you are constantly looking over their shoulders, you are doing more harm than good.

A common mistake made by first-time bosses is their involvement with office help. If you get bored, don't start bending the ear of your office help. When you distract the office workers, your work is not getting done. You must set an example for your employees. If they see you hanging around the coffee pot swapping stories, they will feel cheated that they don't have the same privileges. If you want to goof off, do it behind closed doors.

When you establish your office employees, don't neglect their needs and desires. If you have a good employee who wants a new chair, buy a new chair. When your workers want a coffee maker, buy a coffee maker. If the requests of your help are reasonable, attend to them. Happy employees are more productive, not to mention nicer to be around.

HOW DO YOU MANAGE FIELD EMPLOYEES?

Field employees present more management challenges than office help. These employees are mobile and can be difficult to keep up with. Since the employees are not under foot, their actions are more difficult to monitor. You might know about every trip your secretary makes to the snack area, but you'll be hard-pressed to keep up with how many times your field crews take a break.

By monitoring job production, you can keep tabs on your crews. If the work is getting done on time, what difference does it make if the crew took three breaks instead of two? If you have good employees who are turning out strong production, leave them alone.

Too much employer presence is not good. You are a boss, not a babysitter. When you hire professionals, you should expect them to be competent workers. If you make the decision to hire people, you are going to have to trust them, at least to some extent.

If you are concerned about your field crews, talk to your customers. Customers are generally very aware of how crews are acting. Make some unannounced visits to the job sites. Don't let the crews get too comfortable, but don't crowd them either.

HOW CAN YOU MOTIVATE EMPLOYEES?

You can use employee motivation tactics to increase the profits of your business. There are many ways to shape employees into a mold that suits you. There are books written for the express purpose of showing employers how to motivate their employees. A creative employer can always find ways to influence employees to do better. Let me give you just a few suggestions that might work with your company.

Awards

Awards are welcomed by everyone. You can issue award certificates for everything from perfect attendance to outstanding achievements. These inexpensive pieces of paper can make a world of difference in the ways employees act.

An employee who knows she will get a certificate for coming to work every day might think twice before calling in sick when she isn't. While the award might not have a financial value, it becomes a goal. Employees who are working towards a goal work better.

Money

Money is a great motivator. Since most people work for money, it stands to reason that they might work a little harder for extra pay. Any type of bonus program can be beneficial to the production rate of your employees.

A day off with pay

Sometimes a day off with pay is worth much more to an employee than the value of the wages. This special treat could become a coveted goal. One idea would be to hold a contest in which the most productive employee of the month gets a day off with pay. Sure, you'll lose the cost of a day's pay, but how much will you gain from all of your employees during the competition?

Performance ratings

Performance ratings can be compared to awards. If employees know they are rated on their performance, they might work harder. These ratings should be put in writing and kept in the employee files.

Titles

Wise business owners know that a lot of people would rather have a fancy title than extra money. In fact, many companies promote people into new titles to avoid giving higher raises. Even if your company is small, you can hand out some impressive titles. For example, instead of calling your field supervisor a foreman, call him a field coordinator. Instead of having a secretary, have an office manager. When you have someone who enters data in a computer all day, change the title from data entry clerk to computer operations specialist. Titles make employees feel better about themselves, and they don't cost you anything.

If you decide to hire employees, be prepared for some rocky spots in the road. Putting people on your payroll can be very beneficial, but it is not a job that can be done without some sacrifices. Take your time and plan carefully. If you are careful and meticulous, you can avoid most of the problems that so many people have with employees.

26
Planning for your future

Planning for your future is one of the first things you should do as a new business owner. This job entails thinking about insurance, benefits, and retirement plans, which can be very perplexing. These areas of your business are not simple, and the responsibilities for you as a business owner are much more imposing than they were as an employee.

As a business owner, when you think of insurance, you must consider all aspects of the issue. If your mind is on health insurance, you must acknowledge the fact that you no longer have deductions taken from your paycheck and coverage provided by your employer. You must establish your own insurance program, pay all the costs, consider tax consequences, and determine what impact employees might have on the program you choose.

If you have employees or plan to hire employees, benefit packages are a serious consideration. If you don't offer employees benefits, you might not get or keep the best employees. It has become standard practice for employers to provide their workers with benefits.

The task of establishing and administering benefit packages can get complicated. Many options are available, each with its own advantages and disadvantages. If you are not knowledgeable of how the laws and rules regulate your actions as they pertain to benefits, you can get in a lot of trouble.

Whether you are looking at retirement plans for yourself or your employees, the possibilities can be mind-boggling. Setting up a plan for yourself is one thing; establishing programs for employees is another.

Many business owners are not aware of the alternatives and combinations available for insurance, benefits, and retirement plans. When they were employees, new business owners didn't need to know the ins and outs of these business components; their employers handled the details. But now that the employees are employers, the responsibility is on their shoulders, and it can be quite a burden.

If you are the only employee of your company, your choices are easier to make. However, if you employ others to work for your company, you might have some studying to do. This chapter can help prepare you for the kinds of decisions you have to make when planning for the future of your business and your own long-range requirements.

GETTING COMPANY-PROVIDED INSURANCE FOR YOURSELF

How hard is it to establish company-provided insurance for yourself? Putting an insurance program in place for yourself when no other employees are involved is not difficult. However, choosing the right plans takes some research. Is the cost of your insurance a deductible expense? If your business is structured as a corporation, you should be able to get some tax relief from the insurance premiums the company pays. If your business is not a corporation, you probably won't be able to deduct your personal insurance premiums.

What types of personal insurance coverage do you want your company to provide for you? Health insurance is almost a given; everyone should try to maintain this type of coverage. Dental insurance is not as critical as health insurance, but it does provide some additional peace of mind. Disability insurance is often ignored, but it can be very appreciated if you are injured or become seriously ill. Life insurance might not be too important if you don't have a family, but if you do, life insurance should be considered a necessity. Key-man insurance isn't needed for a mom-and-pop business, as long as you have enough life insurance in force. Let's take a closer look at each of these types of coverage to see how they fit into your business plans.

Health insurance

Health insurance is expensive and the plans are complex. Deciding on what type of insurance to get requires research and thought. What should you look for in health insurance programs?

Pre-existing conditions

Pre-existing conditions can be a major factor in your choice of health insurance. Most insurance companies do not cover expenses related to a pre-existing condition. For example, if you have problems with your back when you obtain your new insurance, the insurance company might refuse to cover medical expenses related to these back problems. If you have had a pregnancy that involved surgery or medical attention beyond the normal childbirth requirements, a reoccurrence of these circumstances might not be covered by your new policy.

It is possible to obtain insurance coverage where pre-existing conditions are not eliminated from coverage. The price for these policies might be higher, but the protection could be worth the additional cost.

Deductible payments

The deductible payments for insurance plans vary. Typically, the more you have to pay in deductible expenses, the lower your monthly premiums are. A plan with a $200 deductible costs more on a monthly basis than a plan with a $500 deductible. It is generally considered wise to choose a plan with a higher deductible and lower installment payments.

Limits of standard coverage

Before you buy any insurance plan, know what the limits of standard coverage are. Not all policies cover all possible circumstances. Read policies closely and ask questions. The insurance company might not have to disclose facts to you unless you ask direct questions.

Waiting period

Some insurance policies requires a waiting period. These waiting periods stipulate that a specific amount of time must pass before a procedure is covered. For example, most insurance would not cover the costs of a pregnancy until after a waiting period was passed. Since the insured might have been pregnant when the policy was taken out, the waiting period eliminates the risk to the insurance company. Determine if the policy you are considering has a waiting period and if so, what conditions apply to the rules of the waiting period.

Copayments

Average health plans call for the insured to make copayments. This means that you are responsible for paying a portion of your own medical expenses even though you are insured. A common copayment amount is 20 percent of the costs incurred. You pay 20 percent and the insurance company pays 80 percent. However, the split on how much each party pays can vary. You might find that you are responsible for 30 percent of the bills.

Verify how your intended policy deals with copayments. Some types of coverage are much more generous and pay nearly the entire cost of your medical expenses. For example, you might only pay a few dollars for each office visit to your doctor. These pay-all policies cost more, but they provide excellent coverage, and you do not have to come up with large sums of out-of-pocket cash unexpectedly.

Dependent coverage

If you have dependents, you'll be interested to know how a policy deals with dependent coverage. Will your dependents receive the same coverage as you do? Will the premiums be set at reduced rates for the additional coverage? Are there limits on dependent coverage? Is there an age limit on the coverage extended to your dependents? You need to ask all of these questions about dependent coverage.

Rate increases

Rate increases are a fact of life with insurance. However, some insurance policies are more prone to rate increases than others. Ask how often the insurance company is allowed to raise its rates. Will you be faced with increases quarterly, semiannually, or annually? Inquire about caps on the amount of increase at any one interval. For example, if your rates are subject to an increase on an annual basis, how much is the maximum the rate can be elevated? With insurance, you can never ask too many questions.

Group advantages

As a business owner, you might be eligible for group advantages. Some insurance companies take small groups of customers and create a large group. This type of grouping is designed to offer coverage at lower rates. Normally, your company needs at least two employees for this type of coverage, but the savings might make it worth putting your spouse on the payroll. Check with your insurance representative for the requirements of joining a group plan.

Dental insurance

Dental insurance is a blessing for people with bad teeth. If you have paid for crowns or root canals lately, you know they aren't cheap. Should you buy dental insurance?

This type of insurance is shunned by some and coveted by others. The decision is yours, but dental insurance can be well worth its cost for the right people.

When you shop for dental insurance, you can ask nearly the same questions you ask about health insurance. Like health insurance, dental insurance comes in many forms. Choosing the right policy is a matter of your personal needs.

If you decide to buy dental insurance, expect to go through a waiting period for major-expense coverage. While some policies pick up routine maintenance of your teeth immediately, you might have to wait for those needed crowns and caps. The waiting period for major work is usually one year.

Once you are covered and eligible for payments on major work, don't expect the coverage to pick up the whole tab. Many dental plans pay no more than one half of your major expenses. For example, if you are getting a $500 crown, your insurance might only pay $250.

Disability insurance

Disability insurance provides protection against lost income due to injuries and sudden disabilities. Short-term disability policies are designed to provide assistance for a short period of time. Long-term disability insurance continues to make payments for an extended period of time.

Disability policies provide a percentage of your normal income to you while you are unable to work. The percentage of your income that is paid depends on your policy. These policies might also be loaded with a pre-existing condition waiver. Let me give you examples of how each type of disability plan might work.

Short-term disability

Short-term disability polices set a limit on the amount of time you can receive benefits. Six months is a common benchmark for the maximum period of time you can collect from a short-term policy.

You usually have a short waiting period before the disability income (DI) kicks in. In most cases, you have to be out of work for at least a week before you can collect on your DI. The amount you can collect is a percentage of your normal income. A plan that pays up to 50 percent of your income is not unusual. However, there are generally limits on the maximum amount you can collect.

For example, your policy might pay 50 percent of your normal weekly pay, but it might stipulate that the maximum you can receive in any given week is $150. Obviously, if you make more than $300 a week, and most contractors do, you won't be getting half of your income in benefits. You need to watch out for these little stingers when you are shopping for disability insurance.

Long-term disability

Long-term disability works on a principle similar to short-term disability. These plans might pay a higher percentage of your income than short-term DI. There are limits on the minimum and the maximum monthly payments, but the length of time you can collect payments is frequently unlimited.

Life insurance

Life insurance doesn't seem very important until you have dependents. While you are single, you don't need to worry about how people can get along without your income when you die. Your mind isn't filled with questions of how your bills will be paid after you are gone or how your child will grow up and be educated. However, when you have people you care for who will be left behind after your death, life insurance becomes important.

How much life insurance do you need? What type of life insurance suits your needs best? These are the two most common questions asked about life insurance. Every person could have a different answer for each question. Life insurance must be tailored to your personal requirements.

Amount of life insurance

How much life insurance is enough? The amount of life insurance coverage you need depends on several factors. The first factor is the number of dependents you might leave behind. A person with only a spouse needs less insurance than a person with a spouse and two children. Another factor is your income.

Many people suggest buying insurance coverage based on a multiple of your annual income. Some people say insurance benefits equal to your annual salary is enough. More people are inclined to believe it is better to have coverage equal to three years of income. Your spouse's employment conditions might influence this decision.

If your spouse isn't working and hasn't worked for some time, it might be difficult for him or her to find a job. If you have been the sole provider, your spouse has to grieve, adjust to your death, find work, and establish a new life. This is not only stressful, it takes time. Can all of this take place in one year? It could, but it would be a strain. So if you leave behind only one year's worth of benefits, the spouse is under extra pressure, and don't forget, there are burial expenses and other related expenses to be paid out of the benefits you bequeath.

Is your spouse capable of being self-supportive? If you are leaving behind a spouse and children, your spouse might not have the earning ability to support the remaining family members. If this possibility exists, you should carry enough insurance to allow for investments and long-term support.

If you died today, how many personal and business debts would your spouse be left with? This consideration must be weighed in setting an amount on your life insurance.

As you can see, you have to consider a number of factors in determining the face amount of your life insurance. Some people look at life insurance as a one-time shot in the arm for the distressed family members. These people assume leaving their spouses $100,000 in cash is more than adequate. In this mindset, the spouse is expected to live off the $100,000 until he or she builds a new life. This isn't a bad plan, but consider another perspective.

In my estate planning, I have structured a way for my wife and children to derive most of their annual income needs from the interest of my life insurance dividends. When I die, if the proceeds from my life insurance are invested wisely, the passive income generated can be substantial. This passive income can support my family, without them having to deplete the lump sum of the premium payoff.

Setting this plan into motion allows my wife and children to be well cared for, and the money paid by the insurance company remains virtually untouched. Then, when my wife passes on, her life insurance dividends can be handled in a similar way. The end result for our children should be a comfortable income from investments and a sizable nest egg in cash.

Of course, to generate this type of insurance payoff, you have to carry some steep premiums. Not all people are willing to invest their money in life insurance, and I'm not saying you should. I believe you should buy as much life insurance as you feel you need, and not a penny more. Now, let's look at the various types of insurance.

Term life insurance

Term life insurance is one of the least expensive forms of life insurance you can buy. While it is the cheapest, it might not be the best value. Many types of term policies are available. Some of the programs feature premiums that increase annually and others have face amounts that are reduced each year; some might do both.

Term insurance is fine as a supplemental life insurance, but it might not be the best choice as a primary insurance. When you are in your prime earning years and building assets, term policies can protect your family from incurring your debts. For example, if you are buying a house with a 30-year mortgage and die while there is a substantial loan outstanding, what will your spouse do? If the spouse can't afford the house payments alone, the house will have to be sold. However, if you have term life insurance, the proceeds from the policy could be used to satisfy the mortgage on the house. As time passes, the amount you owe on the home is reduced, so the reducing term insurance is not such a bad deal. You are paying only for the insurance you need while you need it.

If you depend on term life insurance as your only life insurance, you might be distressed in later life. As you grow older, the premiums go up and the value goes down. If you live a normal life span, the policy might not be worth much at the time of your death.

Whole-life policies

Whole-life policies are more expensive than term insurance, but they are more dependable. The face amount of these policies doesn't decrease and the premiums don't go up.

There are other advantages to a whole-life policy. As you make your monthly payments, you build a cash value in the policy. In effect, you are creating a savings account of a sort. Later in life, if you need some quick cash, you can borrow against your built-up cash value. Interest rates on these loans are usually very low, and you can pay back the money at your discretion.

If you reach a point in life where you no longer want to maintain your life insurance, you can cash in a whole-life policy and receive the money from the cash value. These policies are considered one of the best available for the long haul.

Universal and variable policies

Universal and variable life policies are variations of whole-life policies. These policies feature investment angles for your premium dollars. As you pay your premiums, you build cash value and your account earns interest. The interest you are earning is rolled over and is not taxable unless it is withdrawn. Many business owners choose these policies.

Key-man insurance

If you have been in business, you have probably heard of key-man insurance. This form of life insurance protects a company against the death of a vital employee.

Normally, the employee is insured by the employer and the employer pays the insurance premiums. If the employee dies, the proceeds of the insurance goes to the employing company, allowing the company to have a cash buffer until the key employee can be replaced.

Unless you are in a partnership or a corporation with other stockholders, you shouldn't need key-man insurance. Regular life insurance can protect your family and cover your business debts. However, if you have a partner whom you depend upon heavily, you might want to set up a key-man plan.

Other options for life insurance

There are many other options for life insurance. The abundance of possibilities in structuring plans is almost overwhelming. Riders can be added to standard policies, and terms and conditions can be adjusted to meet every conceivable need. Due to the complexity of insurance programs, you should talk to several insurance professionals before making a buying decision.

CHOOSING AN INSURANCE COMPANY

Choosing an insurance company is no easy job, but it can be one of the most critical aspects of your insurance planning. No one wants to pay premiums on insurance for years only to have the insurance company go out of business. Not all insurance companies have the same financial strength. The investment abilities of some companies are much better than those of other companies.

Choose your insurance firm carefully. Research the company and attempt to establish its financial power and track record. By talking with your state agencies and going to major libraries, you should be able to find performance ratings on the various companies. Dig deep into a company's background before you depend on it to protect you.

OFFERING EMPLOYEE BENEFITS

Employee benefits can be even harder to decipher than your own benefits. The rules and regulations that go with providing benefits to your employees make the chore more challenging. As an employer, you are responsible for the compliance of some rather strict laws and regulations. If you fail to execute your duties in the proper manner, you can wind up in serious trouble. With that said, let's move on to some of the benefits you might consider for your employees.

The benefits you offer to your employees might include any of the insurance coverages I have already discussed. However, when you are setting up plans for employees, you have to follow some additional guidelines.

Most companies use an employment manual to explain company policies to their employees. These policy manuals tell the employees what benefits they might be eligible for and when their eligibility begins. It is important that you treat all of your employees equally. You should not provide benefits for some employees and deny the same offering to other employees. If you do this, you are asking for trouble. The policy manual makes it easy for you to set and maintain protocol.

As you are shopping for benefit plans to offer your employees, you should find a host of them. Every company offering benefit services will tell you their benefits are the best. Wading through the myriad possibilities takes some time.

Many employers choose an insurance company that offers multiple benefits in a single plan. The benefits can include coverage for medical, dental, life, disability, and accident insurance. This type of employee package can be cost-prohibitive, but it is an attractive feature when you are trying to hire and keep top-notch employees.

Some of these multiple plans allow employees to make some of their own choices in the types of coverage they want. The employer gives each employee a set allowance to allocate to various types of coverage, and the employee is free to customize his or her individual plan. This type of employee package is often referred to as a flexible benefit package and is sometimes called a cafeteria plan. The reasoning behind calling the program a cafeteria plan is the employees' ability to choose from various benefits.

Other benefits you might offer your employees include the standards of paid sick leave, paid personal days, paid vacation, retirement plans, and bonus programs. Retirement plans for you and your employees are discussed in detail later in the chapter.

Before you make a decision on giving benefits to your employees, research the rules and regulations you must follow. Talk to your attorney, your insurance agent, and your state agencies. By talking to these professionals, you should be able to obtain all the information needed to stay on the right side of the law.

GETTING LIABILITY INSURANCE

Liability insurance is one type of insurance coverage no business can afford to be without. The extent of coverage needed varies, but all business ventures should be protected with liability insurance.

Most business owners are aware of what liability coverage is and why they need it. However, some readers might not be familiar with this type of insurance. Allow me to explain how liability insurance works.

General liability insurance protects its holder from claims arising from personal injury or property damage. When a company has a current general liability policy, all representatives of the company are typically covered under the policy while performing company business.

The cost of liability insurance is determined by the nature of your business. Rates are lower for someone engaged in relatively safe endeavors compared to those assessed against businesses dealing in high-risk ventures. For example, if you own a blasting company and work with explosives, your premiums will be higher than those of someone who installs interior trim molding.

Without adequate coverage against liability claims, you could lose your business and all of your other assets. Contractors are in particular need of this type of insurance. With so many possibilities for accidents on the job site, you can't afford to do business without it.

PAYING WORKER'S COMPENSATION INSURANCE

Worker's compensation insurance is insurance that is generally required by individual states for companies having employees who are not close family members. The cost of this insurance can be crippling, but it is a necessity for most businesses with nonfamily employees.

Worker's comp insurance benefits your employees. If employees are injured in the performance of their duties on your payroll, this insurance helps them financially. The employees might receive payment for their medical expenses that are related to the injury. If employees are disabled, they might receive partial disability income from the program. Other events, such as a fatal injury, could result in similar benefits to the heirs of the employees.

The cost of workmen's compensation insurance is based on a company's total payroll expenses and the types of work performed by various employees. The rate for a secretary is much lower than the rate for a roofer. Each employee is put into a job classification and rated for a degree of risk. Once the risk of injury and other factors are assessed, an estimated premium is established.

At the end of the year, the insurance carrier conducts an audit of the insured company's payroll expenses to determine how much was actually paid out in payroll and to what job classifications the wages were paid. At this time, the insurance company renders an accurate accounting of what is owed or due to the company. Since some preliminary annual estimates are high, it is possible a company might receive a refund. However, if the original estimate was low, the company must pay the additional premium requirements.

Worker's comp is at best a bad experience for companies that have personnel injured. If your company is accident prone, you pay for it in higher premiums.

Worker's comp for subcontractors

Worker's comp for subcontractors can give you a nasty surprise. When you engage a subcontractor to work for your company, you might be held responsible for the cost of workmen's compensation insurance on that sub. You can avoid this by requiring subcontractors to furnish you with a certificate of insurance before allowing them to do any work.

The certificate of insurance comes from the company issuing the insurance. Don't accept a copy of an insurance certificate that a subcontractor hands you; the policy might not still be in force. When you receive the certificate of insurance, check it for coverage and expiration. When you are satisfied that the sub has proper insurance, file the certificate for future proof of insurance.

When your insurance company audits you at the end of the year, you might need to produce certificates of insurance on all of your independent contractors. You cannot afford to let your guard down on this one. Paying premiums for insurance that subcontractors should be responsible for can cause you great grief.

Some contractors deduct money from payments due subcontractors when the subs don't carry the necessary insurance. The money is used at the end of the year when the contractors must settle up with their insurance companies. While this has been done for years, I don't recommend it. It is best to require the subcontractors to carry and provide proof of their own insurance.

MAKING PLANS FOR YOUR LATER YEARS

You should start making plans for your later years now. There is no time like the present to prepare for the future. In planning your future, you must define the paths you want to take with your business. Will your business be handed down to your children? Will the business be run by an employee when you retire? Are you interested in selling your business at some point in the future? These are only some of the questions you should start asking yourself now.

Passing the business on to your children

Passing the business on to your children is a fine way to keep your company going when you are tired of the daily grind. However, some children have no desire to own or operate the business you spent years building. They might have their own dreams to fulfill.

If you have hopes of one day giving your business, or the management of it, to your children, discuss it with them as soon as possible. The sooner the kids become involved in the business, the better they are prepared to handle the responsibilities when you step down.

Don't count on your children being overly enthusiastic about taking the reins, and don't become angry with them if they want to pursue other goals. After all, you wanted to build your own business; perhaps they want the same freedom. Taking over the family business can put a lot of strain on devoted children. If they do to well, you might be offended that they are more capable in business than you were. If they perform poorly, they might feel they have let you down. Respect the wishes of your children.

Allowing employees to manage your business

Allowing employees to manage your business can be a hard pill to swallow, even when it's only for a few days. If you have employees now, would you trust them to mind the store while you took a vacation? Does the thought of having someone else at the helm of your business send shivers down your back?

Putting your business into the hands of employees might take some getting used to. If your plans call for having employees manage your business, start testing the waters now. Delegate duties to your best people and see how they handle them. When you're comfortable with their performance under your watchful eye, take a short vacation.

You never know how managers function under pressure until you let them take control. If you are standing behind them every step of the way, the managers might be nervous and not perform to their best abilities. If you're too close at hand, the managers might rely on you to make the tough calls. Get away from the business and let them have a go at it. If something does go wrong, you can step back in and pick up the pieces quickly. This is the only way you are going to be able to assess fully the abilities of your chosen few.

If while you're gone the business runs smoothly, give the managers a little more rope. Keep testing the employees with additional responsibilities. If you have the right people, you can enjoy life more and rest comfortably, knowing you have good people to back you up.

Grooming your business for sale

Grooming your business for sale is an important step towards liquidation. If you wake up one morning and decide to sell your business immediately, you are going to make mistakes. When your long-range plans call for the sale of the business, begin your preparations early. When the time comes to put the business on the auction block, you'll be ready to make your best deal.

Closing the doors

Closing the doors on a business you have invested your life in can be traumatic. You might feel you are throwing away a part of yourself. If shutting down is the ultimate fate for your business, prepare yourself mentally for the final days.

Reducing your workload

As an alternative to closing the doors, you might consider reducing your workload. Going into semiretirement might be the ideal answer to your problems. You can be selective in the work you do, and you can enjoy some additional income. This option is very appealing to a lot of contractors. Again, proper advance planning is the key to making your desires reality.

CHOOSING RETIREMENT PLAN OPTIONS

The number of retirement plan options that exist is amazing. Whether you're looking for a plan for yourself or a plan for your employees, you have many to choose from. To prove this point, let's look at some of the most common methods of building retirement capital.

Rental properties

Rental properties can be an ideal source of retirement income for yourself. Real estate is one of the best ways to keep up with the rising rates of inflation. Inflation is one of your biggest enemies when planning for retirement. With some investments, the money earned from the investment won't amount to a hill of beans when you retire. Real estate has the edge in these circumstances because of its typical pattern of appreciation.

Rental real estate can be advantageous to you now and later. When you first buy income properties, the net rental income might not turn a profit for you, but the tax advantages can be significant. Even though the 1986 changes in tax laws dealt a deadly blow to real estate investors, there is still room to capitalize on deductions.

To make the most of your tax advantages, you must maintain an active interest in the management of your rental properties. If you are merely a passive investor, you miss out on the bulk of the tax savings. However, being a contractor, you should be well suited to being a landlord. You have the ability or the contacts to keep maintenance costs at a minimum.

If you own rental property that breaks even, you're doing fine in your retirement plans. While you are not turning a profit, you are paying for the real estate. If you start your real estate investing early, when you retire your rental income can come to you, not to the mortgage holder.

Income-producing real estate allows you to win three ways. The first way is in the form of routine cash flow; the rents you collect allow you to have some spending money.

The second way rental properties help you is in your net worth. As your buildings are paid off, you gain equity. This equity can be used as leverage to borrow money against your properties. Since the money is borrowed, you don't have to pay taxes on it. When you have enough equity in rental real estate, you can literally live on borrowed money for the rest of your life.

The third option you have with real estate is selling it. As you have been paying off the mortgages over the years, your real estate should have increased in value. If you don't want to be an active landlord in your later years, you can sell the property for a handsome profit and live off the proceeds. Since you will be selling the real estate at current prices, you won't losing ground to the effects of inflation.

If you have the temperament and time to be a landlord, rental properties are one of the best retirement plans you can come up with. However, you might have some problems along the way. Running rental properties can be a thankless job. Tenants are not always the easiest people to get along with, and you must comply with many laws and regulations. Some people simply are not cut out to be property managers. If you don't think real estate is your way to retirement riches, let's examine some other options.

Keogh plans

Keogh plans for self-employed people can get a little complicated. If you are self-employed, you can contribute up to 25 percent of your earnings to the fund. However, the maximum dollar contribution is capped at $30,000. This formula gets a little trickier.

The 25 percent you are allowed to contribute is not computed on your gross earnings alone. When you determine how much you are going to put into your Keogh, you must subtract that amount from your earnings. Then you can contribute up to 25 percent of what is left of your earnings. In effect, you can only fund 20 percent of your total earnings. Let me give you an example.

Let's say you had a great year and earned $100,000. You want to contribute $20,000 to your retirement plan. After subtracting the $20,000 from your earnings, you are left with $80,000. Twenty-five percent of $80,000 is $20,000, the maximum you can invest.

For older people, setting up a defined-benefit plan can allow larger contributions. However, these plans are expensive to establish and maintain. If you have employees, these plans become even more confusing. As the employer, you not only must deduct the contribution to your personal plan before arriving at the earnings figure used to factor your maximum contribution, you must also deduct the contributions you made as your part of the employees' contributions. Other rules apply to these plans, and, as you can see, the plan can be confusing.

When setting up a Keogh plan, you must name a trustee. The trustee is usually a financial institution. Before attempting to establish and use your own Keogh plan, consult with an attorney who is familiar with the rules and regulations.

Pension plans

Pension plans for your employees must be funded in good years and bad years. If you are hiring older employees, they might prefer a pension plan over a profit-sharing plan. Pension plans provide a consistent company contribution to the employee's retirement plan.

Pension plans are termed as qualified plans, which means they meet the requirements of Section 401 of the Internal Revenue Code and qualify for favorable tax advantages. These tax advantages help you and your employees.

If you decide to use a qualified pension plan, you must cover at least 70 percent of your average employees. The features and benefits of these plans are extensive. For complete details on forming and using such a plan, consult with a qualified professional.

Profit-sharing plans

Profit-sharing plans can also be termed as qualified plans. One advantage to you, as the employer, of a profit-sharing plan is that no regulation requires you to fund the plan in bad economic years.

A formula is established to identify the amount of contributions that will be made to profit-sharing plans. The plan also details when contributions are made. Many new companies prefer profit-sharing plans because there is no mandatory funding in years in which the company doesn't make a profit.

Social security

Did you know that social security benefits are taxable? Well, they can be. If an individual's adjusted gross income, tax-exempt interest, and one-half of the individual's social security benefits exceed $25,000, the social security benefits can be taxed. The maximum tax is one-half of the social security benefits.

Annuities

Annuities can be good retirement investments. These investments are safe, pay good interest rates, and the interest you earn is tax-deferred until you cash the annuity. However, if you need access to your money early, you have to pay a penalty for early withdrawal. When you plan to let your money work for you in the annuity for 7 to 10 years, annuities are a safe bet.

If you decide to put your money in annuities, shop around; there are a multitude of programs open to you. If you want to investigate annuity plans for employees, talk with professionals in the field. Again, many options are available for these programs, but there are also rules to be followed.

Some considerations

Investing for your future can involve a variety of strategies. Bonds, art, antiques, diamonds, gold, silver, rare coins, stocks, and mutual funds are all conceivable retirement investments.

Any of these forms of investments can return a desirable rate of return. However, many of these investments require a keen knowledge of the market. For example, if you are not an experienced coin buyer, your rare coin collection might wind up being worth little more than its face value.

For most business owners, sticking to conservative investments is best for retirement. If you have some extra money you can afford to play with, you might diversify your conservative investments with some of the more exciting opportunities available. However, when you are betting on your golden years, play your cards carefully.

Some final words on retirement plans

Allow me to give you some final words on retirement plans. Retirement plans for you and your employees can be quite sophisticated. With the complexity of the circumstances surrounding these plans, you should always consult experts before making decisions. Make yourself aware of your responsibilities to your employees.

Many factors affect how you must treat each employee fairly. Don't assume that part-time employees are not the same as full-time employees under your benefits package. While there are probably exceptions to part-time help, don't make that assumption; don't assume anything. Employees' rights under the law are too important. Consult professionals and maintain your integrity as an employer.

CONCLUSION

In closing, I would like to thank you for taking this time to become a better contractor. With so much publicity on the bad contractors, we all have to work to keep our public image shining. I hope your investment in this book proves very beneficial to you and your company. Again, thank you, and good luck in all your endeavors.

Appendix:
Federal tax forms

Form **8829**	**Expenses for Business Use of Your Home**	OMB No. 1545-1266

Form **8829**

Department of the Treasury
Internal Revenue Service   (99)

Expenses for Business Use of Your Home

▶ File only with Schedule C (Form 1040). Use a separate Form 8829 for each home you used for business during the year.

▶ See separate instructions.

OMB No. 1545-1266

1995

Attachment
Sequence No. **66**

Name(s) of proprietor(s)

Your social security number

Part I   **Part of Your Home Used for Business**

1	Area used regularly and exclusively for business, regularly for day care, or for inventory storage. See instructions	**1**		
2	Total area of home	**2**		
3	Divide line 1 by line 2. Enter the result as a percentage	**3**		%

     • **For day-care facilities not used exclusively for business, also complete lines 4–6.**

     • **All others, skip lines 4–6 and enter the amount from line 3 on line 7.**

4	Multiply days used for day care during year by hours used per day .	**4**		hr.
5	Total hours available for use during the year (365 days × 24 hours). See instructions	**5**	8,760 hr.	
6	Divide line 4 by line 5. Enter the result as a decimal amount . . .	**6**	.	
7	Business percentage. For day-care facilities not used exclusively for business, multiply line 6 by line 3 (enter the result as a percentage). All others, enter the amount from line 3 . . . . ▶	**7**		%

Part II   **Figure Your Allowable Deduction**

8	Enter the amount from Schedule C, line 29, **plus** any net gain or (loss) derived from the business use of your home and shown on Schedule D or Form 4797. If more than one place of business, see instructions			**8**	
	See instructions for columns (a) and (b) before completing lines 9–20.	**(a)** Direct expenses	**(b)** Indirect expenses		
9	Casualty losses. See instructions	**9**			
10	Deductible mortgage interest. See instructions .	**10**			
11	Real estate taxes. See instructions	**11**			
12	Add lines 9, 10, and 11.	**12**			
13	Multiply line 12, column (b) by line 7		**13**		
14	Add line 12, column (a) and line 13.			**14**	
15	Subtract line 14 from line 8. If zero or less, enter -0- .			**15**	
16	Excess mortgage interest. See instructions . .	**16**			
17	Insurance	**17**			
18	Repairs and maintenance	**18**			
19	Utilities	**19**			
20	Other expenses. See instructions	**20**			
21	Add lines 16 through 20	**21**			
22	Multiply line 21, column (b) by line 7		**22**		
23	Carryover of operating expenses from 1994 Form 8829, line 41 . .	**23**			
24	Add line 21 in column (a), line 22, and line 23			**24**	
25	Allowable operating expenses. Enter the **smaller** of line 15 or line 24			**25**	
26	Limit on excess casualty losses and depreciation. Subtract line 25 from line 15			**26**	
27	Excess casualty losses. See instructions	**27**			
28	Depreciation of your home from Part III below	**28**			
29	Carryover of excess casualty losses and depreciation from 1994 Form 8829, line 42	**29**			
30	Add lines 27 through 29			**30**	
31	Allowable excess casualty losses and depreciation. Enter the **smaller** of line 26 or line 30 . .			**31**	
32	Add lines 14, 25, and 31			**32**	
33	Casualty loss portion, if any, from lines 14 and 31. Carry amount to **Form 4684**, Section B . .			**33**	
34	Allowable expenses for business use of your home. Subtract line 33 from line 32. Enter here and on Schedule C, line 30. If your home was used for more than one business, see instructions ▶			**34**	

Part III   **Depreciation of Your Home**

35	Enter the **smaller** of your home's adjusted basis or its fair market value. See instructions . .	**35**		
36	Value of land included on line 35	**36**		
37	Basis of building. Subtract line 36 from line 35	**37**		
38	Business basis of building. Multiply line 37 by line 7	**38**		
39	Depreciation percentage. See instructions	**39**		%
40	Depreciation allowable. Multiply line 38 by line 39. Enter here and on line 28 above. See instructions	**40**		

Part IV   **Carryover of Unallowed Expenses to 1996**

41	Operating expenses. Subtract line 25 from line 24. If less than zero, enter -0-	**41**	
42	Excess casualty losses and depreciation. Subtract line 31 from line 30. If less than zero, enter -0- .	**42**	

For Paperwork Reduction Act Notice, see page 1 of separate instructions.   Cat. No. 13232M   Form **8829** (1995)

A-1 Form 8829.

SCHEDULE SE
(Form 1040)

Department of the Treasury
Internal Revenue Service   (99)

Self-Employment Tax

▶ See Instructions for Schedule SE (Form 1040).

▶ **Attach to Form 1040.**

OMB No. 1545-0074

Attachment
Sequence No. **17**

Name of person with **self-employment** income (as shown on Form 1040)	Social security number of person with **self-employment** income ▶	

Who Must File Schedule SE

You must file Schedule SE if:

- You had net earnings from self-employment from **other than** church employee income (line 4 of Short Schedule SE or line 4c of Long Schedule SE) of $400 or more, **OR**
- You had church employee income of $108.28 or more. Income from services you performed as a minister or a member of a religious order **is not** church employee income. See page SE-1.

Note: *Even if you have a loss or a small amount of income from self-employment, it may be to your benefit to file Schedule SE and use either ™optional method∫ in Part II of Long Schedule SE. See page SE-3.*

Exception. If your only self-employment income was from earnings as a minister, member of a religious order, or Christian Science practitioner **and** you filed Form 4361 and received IRS approval not to be taxed on those earnings, **do not** file Schedule SE. Instead, write ™Exempt±Form 4361∫ on Form 1040, line 47.

May I Use Short Schedule SE or MUST I Use Long Schedule SE?

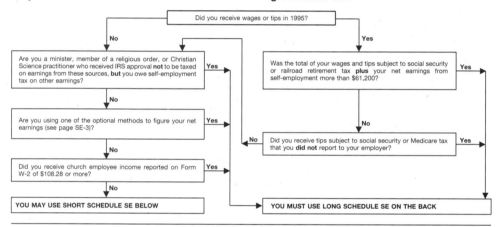

Section A–Short Schedule SE. Caution: *Read above to see if you can use Short Schedule SE.*

1	Net farm profit or (loss) from Schedule F, line 36, and farm partnerships, Schedule K-1 (Form 1065), line 15a	**1**	
2	Net profit or (loss) from Schedule C, line 31; Schedule C-EZ, line 3; and Schedule K-1 (Form 1065), line 15a (other than farming). Ministers and members of religious orders see page SE-1 for amounts to report on this line. See page SE-2 for other income to report	**2**	
3	Combine lines 1 and 2	**3**	
4	**Net earnings from self-employment.** Multiply line 3 by 92.35% (.9235). If less than $400, **do not** file this schedule; you do not owe self-employment tax . . . . ▶	**4**	
5	Self-employment tax. If the amount on line 4 is: • $61,200 or less, multiply line 4 by 15.3% (.153). Enter the result here and on **Form 1040, line 47.** • More than $61,200, multiply line 4 by 2.9% (.029). Then, add $7,588.80 to the result. Enter the total here and on **Form 1040, line 47.**	**5**	
6	**Deduction for one-half of self-employment tax.** Multiply line 5 by 50% (.5). Enter the result here and on **Form 1040, line 25**	**6**	

For Paperwork Reduction Act Notice, see Form 1040 instructions.   Cat. No. 11358Z   **Schedule SE (Form 1040) 1995**

A-2a Schedule SE—page 1.

Name of person with **self-employment** income (as shown on Form 1040)	Social security number of person with **self-employment** income ▶		

Section B–Long Schedule SE

Part I   Self-Employment Tax

Note: *If your only income subject to self-employment tax is* **church employee income,** *skip lines 1 through 4b. Enter -0- on line 4c and go to line 5a. Income from services you performed as a minister or a member of a religious order* **is** *not church employee income. See page SE-1.*

A   If you are a minister, member of a religious order, or Christian Science practitioner **and** you filed Form 4361, but you had $400 or more of **other** net earnings from self-employment, check here and continue with Part I. . . . . . ▶ ☐

1	Net farm profit or (loss) from Schedule F, line 36, and farm partnerships, Schedule K-1 (Form 1065), line 15a. **Note:** *Skip this line if you use the farm optional method. See page SE-3* . .	**1**	
2	Net profit or (loss) from Schedule C, line 31; Schedule C-EZ, line 3; and Schedule K-1 (Form 1065), line 15a (other than farming). Ministers and members of religious orders see page SE-1 for amounts to report on this line. See page SE-2 for other income to report. **Note:** *Skip this line if you use the nonfarm optional method. See page SE-3.*	**2**	
3	Combine lines 1 and 2	**3**	
4a	If line 3 is more than zero, multiply line 3 by 92.35% (.9235). Otherwise, enter amount from line 3	**4a**	
b	If you elected one or both of the optional methods, enter the total of lines 15 and 17 here . .	**4b**	
c	Combine lines 4a and 4b. If less than $400, **do not** file this schedule; you do not owe self-employment tax. **Exception.** If less than $400 and you had **church employee income,** enter -0- and continue ▶	**4c**	

5a   Enter your **church employee income** from Form W-2. **Caution:** *See page SE-1 for definition of church employee income* . . . . . .   **5a** ☐

b	Multiply line 5a by 92.35% (.9235). If less than $100, enter -0-	**5b**	
6	**Net earnings from self-employment.** Add lines 4c and 5b	**6**	
7	Maximum amount of combined wages and self-employment earnings subject to social security tax or the 6.2% portion of the 7.65% railroad retirement (tier 1) tax for 1995	**7**	61,200 00

8a   Total social security wages and tips (total of boxes 3 and 7 on Form(s) W-2) and railroad retirement (tier 1) compensation . . . .   **8a** ☐

   b   Unreported tips subject to social security tax (from Form 4137, line 9)   **8b** ☐

c	Add lines 8a and 8b	**8c**	
9	Subtract line 8c from line 7. If zero or less, enter -0- here and on line 10 and go to line 11 . ▶	**9**	
10	Multiply the **smaller** of line 6 or line 9 by 12.4% (.124)	**10**	
11	Multiply line 6 by 2.9% (.029).	**11**	
12	**Self-employment tax.** Add lines 10 and 11. Enter here and on **Form 1040, line 47**	**12**	

13   **Deduction for one-half of self-employment tax.** Multiply line 12 by 50% (.5). Enter the result here and on **Form 1040, line 25** . . . .   **13** ☐

Part II   Optional Methods To Figure Net Earnings (See page SE-3.)

Farm Optional Method. You may use this method **only if:**
- Your gross farm income[1] was not more than $2,400, **or**
- Your gross farm income[1] was more than $2,400 and your net farm profits[2] were less than $1,733.

14	Maximum income for optional methods	**14**	1,600 00
15	Enter the **smaller** of: two-thirds (²⁄₃) of gross farm income[1] (not less than zero) **or** $1,600. Also, include this amount on line 4b above	**15**	

Nonfarm Optional Method. You may use this method **only if:**
- Your net nonfarm profits[3] were less than $1,733 and also less than 72.189% of your gross nonfarm income,[4] **and**
- You had net earnings from self-employment of at least $400 in 2 of the prior 3 years.

Caution: *You may use this method no more than five times.*

16	Subtract line 15 from line 14	**16**	
17	Enter the **smaller** of: two-thirds (²⁄₃) of gross nonfarm income[4] (not less than zero) **or** the amount on line 16. Also, include this amount on line 4b above	**17**	

[1]From Schedule F, line 11, and Schedule K-1 (Form 1065), line 15b.   [3]From Schedule C, line 31; Schedule C-EZ, line 3; and Schedule K-1 (Form 1065), line 15a.
[2]From Schedule F, line 36, and Schedule K-1 (Form 1065), line 15a.   [4]From Schedule C, line 7; Schedule C-EZ, line 1; and Schedule K-1 (Form 1065), line 15c.

A-2b Schedule SE—page 2.

Capital Gains and Losses

► Attach to Form 1040.   ► See Instructions for Schedule D (Form 1040).

► Use lines 20 and 22 for more space to list transactions for lines 1 and 9.

OMB No. 1545-0074

1995

Attachment
Sequence No. **12**

Name(s) shown on Form 1040

Your social security number

Part I   Short-Term Capital Gains and Losses–Assets Held One Year or Less

(a) Description of property (Example: 100 sh. XYZ Co.)	**(b)** Date acquired (Mo., day, yr.)	**(c)** Date sold (Mo., day, yr.)	**(d)** Sales price (see page D-3)	**(e)** Cost or other basis (see page D-3)	**(f)** LOSS If (e) is more than (d), subtract (d) from (e)	**(g)** GAIN If (d) is more than (e), subtract (e) from (d)
1						

2 Enter your short-term totals, if any, from line 21 . . . . **2**

3 **Total short-term sales price amounts.** Add column (d) of lines 1 and 2 . . . **3**

4 Short-term gain from Forms 2119 and 6252, and short-term gain or loss from Forms 4684, 6781, and 8824 . . . . **4**

5 Net short-term gain or loss from partnerships, S corporations, estates, and trusts from Schedule(s) K-1 . . . . **5**

6 Short-term capital loss carryover. Enter the amount, if any, from line 9 of your 1994 Capital Loss Carryover Worksheet **6**

7 Add lines 1 through 6 in columns (f) and (g) . . **7** ( )

8 **Net short-term capital gain or (loss).** Combine columns (f) and (g) of line 7 . . . . . . ► **8**

Part II   Long-Term Capital Gains and Losses–Assets Held More Than One Year

9						

10 Enter your long-term totals, if any, from line 23 . . . . **10**

11 **Total long-term sales price amounts.** Add column (d) of lines 9 and 10 . . . **11**

12 Gain from Form 4797; long-term gain from Forms 2119, 2439, and 6252; and long-term gain or loss from Forms 4684, 6781, and 8824 . . . . . **12**

13 Net long-term gain or loss from partnerships, S corporations, estates, and trusts from Schedule(s) K-1 . . . . **13**

14 Capital gain distributions **14**

15 Long-term capital loss carryover. Enter the amount, if any, from line 14 of your 1994 Capital Loss Carryover Worksheet **15**

16 Add lines 9 through 15 in columns (f) and (g) . . . . **16** ( )

17 **Net long-term capital gain or (loss).** Combine columns (f) and (g) of line 16 . . . . . ► **17**

Part III   Summary of Parts I and II

18 Combine lines 8 and 17. If a loss, go to line 19. If a gain, enter the gain on Form 1040, line 13.
   Note: If both lines 17 and 18 are gains, see the **Capital Gain Tax Worksheet** on page 24 . **18**

19 If line 18 is a loss, enter here and as a (loss) on Form 1040, line 13, the **smaller** of these losses:
   a The loss on line 18; **or**
   b ($3,000) or, if married filing separately, ($1,500) . . . . **19** ( )
   Note: See the **Capital Loss Carryover Worksheet** on page D-3 if the loss on line 18 exceeds the loss on line 19 **or** if Form 1040, line 35, is a loss.

For Paperwork Reduction Act Notice, see Form 1040 instructions.   Cat. No. 11338H   **Schedule D (Form 1040) 1995**

Watermark: May change this form. See page 221 or instructions. Pending tax law.

A-3a Schedule D—page 1.

Name(s) shown on Form 1040. Do not enter name and social security number if shown on other side.   **Your social security number**

Part IV   **Short-Term Capital Gains and Losses–Assets Held One Year or Less**   *(Continuation of Part I)*

(a) Description of property (Example: 100 sh. XYZ Co.)	(b) Date acquired (Mo., day, yr.)	(c) Date sold (Mo., day, yr.)	(d) Sales price (see page D-3)	(e) Cost or other basis (see page D-3)	(f) LOSS If (e) is more than (d), subtract (d) from (e)	(g) GAIN If (d) is more than (e), subtract (e) from (d)
20						
21 Short-term totals. Add columns (d), (f), and (g) of line 20. Enter here and on line 2 . **21**						

Part V   **Long-Term Capital Gains and Losses–Assets Held More Than One Year**   *(Continuation of Part II)*

22						
23 Long-term totals. Add columns (d), (f), and (g) of line 22. Enter here and on line 10 . **23**						

A-3b Schedule D—page 2.

SCHEDULE C
(Form 1040)

Department of the Treasury
Internal Revenue Service (99)

Profit or Loss From Business
(Sole Proprietorship)
▶ Partnerships, joint ventures, etc., must file Form 1065.
▶ **Attach to Form 1040 or Form 1041.** ▶ **See Instructions for Schedule C (Form 1040).**

OMB No. 1545-0074

1995

Attachment
Sequence No. **09**

Name of proprietor	Social security number (SSN)

A Principal business or profession, including product or service (see page C-1)

B Enter principal business code
(see page C-6) ▶

C Business name. If no separate business name, leave blank.

D Employer ID number (EIN), if any

E Business address (including suite or room no.) ▶ ....
City, town or post office, state, and ZIP code

F Accounting method: **(1)** ☐ Cash **(2)** ☐ Accrual **(3)** ☐ Other (specify) ▶ ....

G Method(s) used to value closing inventory: **(1)** ☐ Cost **(2)** ☐ Lower of cost or market **(3)** ☐ Other (attach explanation) **(4)** ☐ Does not apply (if checked, skip line H) | Yes | No |

H Was there any change in determining quantities, costs, or valuations between opening and closing inventory? If ™Yes,ʃ attach explanation . . . .

I Did you ™materially participateʃ in the operation of this business during 1995? If ™No,ʃ see page C-2 for limit on losses . . .

J If you started or acquired this business during 1995, check here . . . . ▶ ☐

Part I Income

1	Gross receipts or sales. **Caution:** If this income was reported to you on Form W-2 and the ™Statutory employeeʃ box on that form was checked, see page C-2 and check here . . . . ▶ ☐	**1**	
2	Returns and allowances	**2**	
3	Subtract line 2 from line 1	**3**	
4	Cost of goods sold (from line 40 on page 2)	**4**	
5	**Gross profit.** Subtract line 4 from line 3	**5**	
6	Other income, including Federal and state gasoline or fuel tax credit or refund (see page C-2) . . .	**6**	
7	**Gross income.** Add lines 5 and 6 . . . . ▶	**7**	

Part II Expenses. Enter expenses for business use of your home **only** on line 30.

8	Advertising	**8**		**19**	Pension and profit-sharing plans	**19**	
9	Bad debts from sales or services (see page C-3) . .	**9**		**20**	Rent or lease (see page C-4):		
				a	Vehicles, machinery, and equipment .	**20a**	
10	Car and truck expenses (see page C-3)	**10**		**b**	Other business property . .	**20b**	
11	Commissions and fees . . .	**11**		**21**	Repairs and maintenance . .	**21**	
12	Depletion	**12**		**22**	Supplies (not included in Part III) .	**22**	
13	Depreciation and section 179 expense deduction (not included in Part III) (see page C-3) . .	**13**		**23**	Taxes and licenses	**23**	
				24	Travel, meals, and entertainment:		
14	Employee benefit programs (other than on line 19) . . .	**14**		**a**	Travel	**24a**	
15	Insurance (other than health) .	**15**		**b**	Meals and entertainment .		
16	Interest:			**c**	Enter 50% of line 24b subject to limitations (see page C-4) .		
a	Mortgage (paid to banks, etc.) .	**16a**					
b	Other	**16b**		**d**	Subtract line 24c from line 24b .	**24d**	
17	Legal and professional services	**17**		**25**	Utilities	**25**	
				26	Wages (less employment credits) .	**26**	
18	Office expense	**18**		**27**	Other expenses (from line 46 on page 2)	**27**	

28	**Total expenses** before expenses for business use of home. Add lines 8 through 27 in columns . . ▶	**28**	
29	Tentative profit (loss). Subtract line 28 from line 7	**29**	
30	Expenses for business use of your home. Attach **Form 8829**	**30**	
31	**Net profit or (loss).** Subtract line 30 from line 29.		
	● If a profit, enter on **Form 1040, line 12,** and ALSO on **Schedule SE, line 2** (statutory employees, see page C-5). Estates and trusts, enter on Form 1041, line 3.	**31**	
	● If a loss, you MUST go on to line 32.		
32	If you have a loss, check the box that describes your investment in this activity (see page C-5).		
	● If you checked 32a, enter the loss on **Form 1040, line 12,** and ALSO on **Schedule SE, line 2** (statutory employees, see page C-5). Estates and trusts, enter on Form 1041, line 3.	**32a** ☐ All investment is at risk.	
	● If you checked 32b, you MUST attach **Form 6198.**	**32b** ☐ Some investment is not at risk.	

For Paperwork Reduction Act Notice, see Form 1040 instructions.   Cat. No. 11334P   Schedule C (Form 1040) 1995

A-4a Schedule C—page 1.

Part III   **Cost of Goods Sold** (see page C-5)

33	Inventory at beginning of year. If different from last year's closing inventory, attach explanation . .	**33**	
34	Purchases less cost of items withdrawn for personal use	**34**	
35	Cost of labor. Do not include salary paid to yourself	**35**	
36	Materials and supplies	**36**	
37	Other costs	**37**	
38	Add lines 33 through 37	**38**	
39	Inventory at end of year	**39**	
40	**Cost of goods sold.** Subtract line 39 from line 38. Enter the result here and on page 1, line 4 . .	**40**	

Part IV   **Information on Your Vehicle.** Complete this part **ONLY** if you are claiming car or truck expenses on line 10 and are not required to file Form 4562 for this business. See the instructions for line 13 on page C-3 to find out if you must file.

41   When did you place your vehicle in service for business purposes? (month, day, year) ▶ ..../........./....... .

42   Of the total number of miles you drove your vehicle during 1995, enter the number of miles you used your vehicle for:

a   Business ....   b   Commuting   c   Other ....

43   Do you (or your spouse) have another vehicle available for personal use? . . . .   ☐ **Yes**   ☐ **No**

44   Was your vehicle available for use during off-duty hours? . . . .   ☐ **Yes**   ☐ **No**

45a   Do you have evidence to support your deduction? . . . .   ☐ **Yes**   ☐ **No**
  b   If "Yes," is the evidence written? . . . .   ☐ **Yes**   ☐ **No**

Part V   **Other Expenses.** List below business expenses not included on lines 8±26 or line 30.

....		
....		
....		
....		
....		
....		
....		
....		
....		
46 **Total other expenses.** Enter here and on page 1, line 27	**46**	

A-4b Schedule C—page 2.

SCHEDULE C-EZ
(Form 1040)

Department of the Treasury
Internal Revenue Service

Net Profit From Business
(Sole Proprietorship)

▶ Partnerships, joint ventures, etc., must file Form 1065.

▶ Attach to Form 1040 or Form 1041. ▶ See instructions on back.

OMB No. 1545-0074

1995

Attachment
Sequence No. **09A**

Name of proprietor

Social security number (SSN)

Part I   **General Information**

**You May Use
This Schedule
Only If You:**

- Had gross receipts from your business of $25,000 or less.
- Had business expenses of $2,000 or less.
- Use the cash method of accounting.
- Did not have an inventory at any time during the year.
- Did not have a net loss from your business.
- Had only one business as a sole proprietor.

And You:

- Had no employees during the year.
- Are not required to file **Form 4562,** Depreciation and Amortization, for this business. See the instructions for Schedule C, line 13, on page C-3 to find out if you must file.
- Do not deduct expenses for business use of your home.
- Do not have prior year unallowed passive activity losses from this business.

A Principal business or profession, including product or service

B Enter principal business code
(see page C-6) ▶

C Business name. If no separate business name, leave blank.

D Employer ID number (EIN), if any

E Business address (including suite or room no.). Address not required if same as on Form 1040, page 1.

City, town or post office, state, and ZIP code

Part II   **Figure Your Net Profit**

1   **Gross receipts.** If more than $25,000, you **must** use Schedule C.
Caution: *If this income was reported to you on Form W-2 and the ™Statutory employee⌋ box on that form was checked, see **Statutory Employees** in the instructions for Schedule C, line 1, on page C-2 and check here* . . . . ▶ ☐ | **1** |

2   **Total expenses.** If more than $2,000, you **must** use Schedule C. See instructions . . . . . | **2** |

3   **Net profit.** Subtract line 2 from line 1. If less than zero, you **must** use Schedule C. Enter on **Form 1040, line 12,** and ALSO on **Schedule SE, line 2.** (Statutory employees **do not** report this amount on Schedule SE, line 2. Estates and trusts, enter on Form 1041, line 3.) . . . . . | **3** |

Part III   **Information on Your Vehicle.** Complete this part **ONLY** if you are claiming car or truck expenses on line 2.

4   When did you place your vehicle in service for business purposes? (month, day, year) ▶ ..... / ..... / ......

5   Of the total number of miles you drove your vehicle during 1995, enter the number of miles you used your vehicle for:

a Business .... **b** Commuting .... **c** Other ....

6   Do you (or your spouse) have another vehicle available for personal use? . . . . ☐ **Yes** ☐ **No**

7   Was your vehicle available for use during off-duty hours? . . . . ☐ **Yes** ☐ **No**

8a Do you have evidence to support your deduction? . . . . ☐ **Yes** ☐ **No**

b If ™Yes,⌋ is the evidence written? . . . . ☐ **Yes** ☐ **No**

For Paperwork Reduction Act Notice, see Form 1040 instructions.   Cat. No. 14374D   **Schedule C-EZ (Form 1040) 1995**

A-5a Schedule C-EZ—page 1.

Instructions

You may use Schedule C-EZ instead of Schedule C if you operated a business or practiced a profession as a sole proprietorship and you have met all the requirements listed in Part I of the form.

Line A

Describe the business or professional activity that provided your principal source of income reported on line 1. Give the general field or activity and the type of product or service.

Line B

Enter on this line the four-digit code that identifies your principal business or professional activity. See page C-6 for the list of codes.

Line D

You need an employer identification number (EIN) only if you had a Keogh plan or were required to file an employment, excise, estate, trust, or alcohol, tobacco, and firearms tax return. If you need an EIN, file **Form SS-4,** Application for Employer Identification Number. If you don't have an EIN, leave line D blank. **Do not** enter your SSN.

Line E

Enter your business address. Show a street address instead of a box number. Include the suite or room number, if any.

A-5b Schedule C-EZ—page 2.

Line 1–Gross Receipts

Enter gross receipts from your trade or business. Be sure to include any amount you received in your trade or business that was reported on Form(s) 1099-MISC. You must show all items of taxable income actually or constructively received during the year (in cash, property, or services). Income is constructively received when it is credited to your account or set aside for you to use. Do not offset this amount by any losses.

Line 2–Total Expenses

Enter the total amount of all deductible business expenses you actually paid during the year. Examples of these expenses include advertising, car and truck expenses, commissions and fees, insurance, interest, legal and professional services, office expense, rent or lease expenses, repairs and maintenance, supplies, taxes, travel, 50% of business meals and entertainment, and utilities (including telephone). For details, see the instructions for Schedule C, Parts II and V, on pages C-2 through C-5.

If you claim car or truck expenses, be sure to complete Part III.

Form **4684**

Department of the Treasury
Internal Revenue Service

Casualties and Thefts

▶ See separate instructions.
▶ Attach to your tax return.
▶ Use a separate Form 4684 for each different casualty or theft.

OMB No. 1545-0177

Attachment
Sequence No. **26**

Name(s) shown on tax return

Identifying number

SECTION A–Personal Use Property (Use this section to report casualties and thefts of property **not** used in a trade or business or for income-producing purposes.)

1 Description of properties (show type, location, and date acquired for each):

Property **A** ....

Property **B** ....

Property **C** ....

Property **D** ....

		Properties (Use a separate column for each property lost or damaged from one casualty or theft.)			
		A	**B D**	**C**	
2 Cost or other basis of each property	**2**				
3 Insurance or other reimbursement (whether or not you filed a claim). See instructions	**3**				
Note: *If line 2 is **more than** line 3, skip line 4.*					
4 Gain from casualty or theft. If line 3 is **more than** line 2, enter the difference here and skip lines 5 through 9 for that column. See instructions if line 3 includes insurance or other reimbursement you did not claim, or you received payment for your loss in a later tax year	**4**				
5 Fair market value **before** casualty or theft . . .	**5**				
6 Fair market value **after** casualty or theft	**6**				
7 Subtract line 6 from line 5	**7**				
8 Enter the **smaller** of line 2 or line 7	**8**				
9 Subtract line 3 from line 8. If zero or less, enter -0-	**9**				

10 Casualty or theft loss. Add the amounts on line 9. Enter the total	**10**	
11 Enter the amount from line 10 or $100, whichever is **smaller**	**11**	
12 Subtract line 11 from line 10	**12**	
Caution: *Use only one Form 4684 for lines 13 through 18.*		
13 Add the amounts on line 12 of all Forms 4684	**13**	
14 Combine the amounts from line 4 of all Forms 4684	**14**	
15 • If line 14 is **more than** line 13, enter the difference here and on Schedule D. Do not complete the rest of this section (see instructions). • If line 14 is **less than** line 13, enter -0- here and continue with the form. • If line 14 is **equal to** line 13, enter -0- here. Do not complete the rest of this section.	**15**	
16 If line 14 is **less than** line 13, enter the difference	**16**	
17 Enter 10% of your adjusted gross income (Form 1040, line 32). Estates and trusts, see instructions	**17**	
18 Subtract line 17 from line 16. If zero or less, enter -0-. Also enter result on Schedule A (Form 1040), line 19. Estates and trusts, enter on the ™Other deductions⌡ line of your tax return	**18**	

For Paperwork Reduction Act Notice, see page 1 of separate instructions.

Cat. No. 12997O

Form **4684** (1995)

A-6a Form 4684 — page 1.

Name(s) shown on tax return. Do not enter name and identifying number if shown on other side. | Identifying number

SECTION B–Business and Income-Producing Property (Use this section to report casualties and thefts of property used in a trade or business or for income-producing purposes.)

Part I **Casualty or Theft Gain or Loss** (Use a separate Part I for each casualty or theft.)

19   Description of properties (show type, location, and date acquired for each):

Property **A** ....

Property **B** ....

Property **C** ....

Property **D** ....

Properties (Use a separate column for each property lost or damaged from one casualty or theft.)

		A	B	C	D
20	Cost or adjusted basis of each property				
21	Insurance or other reimbursement (whether or not you filed a claim). See the instructions for line 3 . **Note:** *If line 20 is more than line 21, skip line 22.*				
22	Gain from casualty or theft. If line 21 is **more than** line 20, enter the difference here and on line 29 or line 34, column (**c**), except as provided in the instructions for line 33. Also, skip lines 23 through 27 for that column. See the instructions for line 4 if line 21 includes insurance or other reimbursement you did not claim, or you received payment for your loss in a later tax year				
23	Fair market value **before** casualty or theft . . .				
24	Fair market value **after** casualty or theft				
25	Subtract line 24 from line 23				
26	Enter the **smaller** of line 20 or line 25				
	Note: *If the property was totally destroyed by casualty or lost from theft, enter on line 26 the amount from line 20.*				
27	Subtract line 21 from line 26. If zero or less, enter -0-				

28   Casualty or theft loss. Add the amounts on line 27. Enter the total here and on line 29 **or** line 34 (see instructions). | 28 |

Part II **Summary of Gains and Losses** (from separate Parts I)

(a) Identify casualty or theft	(b) Losses from casualties or thefts		(c) Gains from casualties or thefts includible in income
	(i) Trade, business, rental or royalty property	(ii) Income-producing property	

Casualty or Theft of Property Held One Year or Less

29		( )	( )	
		( )	( )	
30	Totals. Add the amounts on line 29	30 ( )	( )	

31   Combine line 30, columns (b)(i) and (c). Enter the net gain or (loss) here and on Form 4797, line 15. If Form 4797 is not otherwise required, see instructions | 31 |

32   Enter the amount from line 30, column (b)(ii) here and on Schedule A (Form 1040), line 22. Partnerships, S corporations, estates and trusts, see instructions . . . . | 32 |

Casualty or Theft of Property Held More Than One Year

33	Casualty or theft gains from Form 4797, line 34	33		
34		( )	( )	
		( )	( )	
35	Total losses. Add amounts on line 34, columns (b)(i) and (b)(ii) . . .	35 ( )	( )	

36   Total gains. Add lines 33 and 34, column (c) . . . . | 36 |

37   Add amounts on line 35, columns (b)(i) and (b)(ii) . . . . | 37 |

38   If the loss on line 37 is **more than** the gain on line 36:

a   Combine line 35, column (b)(i) and line 36, and enter the net gain or (loss) here. Partnerships and S corporations see the note below. All others enter this amount on Form 4797, line 15. If Form 4797 is not otherwise required, see instructions . . . . | 38a |

b   Enter the amount from line 35, column (b)(ii) here. Partnerships and S corporations see the note below, enter this amount on Schedule A (Form 1040), line 22. Estates and trusts, enter on the ™Other deductions‖ line of your tax return | 38b |

39   If the loss on line 37 is **equal to** or **less than** the gain on line 36, combine these lines and enter here. Partnerships, see the note below. All others, enter this amount on Form 4797, line 3 . . . . | 39 |

Note: *Partnerships, enter the amount from line 38a, 38b, or line 39 on Form 1065, Schedule K, line 7. S corporations, enter the amount from line 38a or 38b on Form 1120S, Schedule K, line 6.*

A-6b Form 4684 — page 2.

<table>
<tr><td>Form 4797</td><td rowspan="2">Sales of Business Property
(Also Involuntary Conversions and Recapture Amounts
Under Sections 179 and 280F(b)(2))
▶ Attach to your tax return.　▶ See separate instructions.</td><td>OMB No. 1545-0184</td></tr>
<tr><td>Department of the Treasury
Internal Revenue Service　(99)</td><td>1995
Attachment
Sequence No. 27</td></tr>
</table>

Name(s) shown on return	Identifying number

1 Enter here the gross proceeds from the sale or exchange of real estate reported to you for 1995 on Form(s) 1099-S (or a substitute statement) that you will be including on line 2, 11, or 22 . . . . **1**

Part I **Sales or Exchanges of Property Used in a Trade or Business and Involuntary Conversions From Other Than Casualty or Theft–Property Held More Than 1 Year**

(a) Description of property	(b) Date acquired (mo., day, yr.)	(c) Date sold (mo., day, yr.)	(d) Gross sales price	(e) Depreciation allowed or allowable since acquisition	(f) Cost or other basis, plus improvements and expense of sale	(g) LOSS ((f) minus the sum of (d) and (e))	(h) GAIN ((d) plus (e) minus (f))
2							

3 Gain, if any, from Form 4684, line 39 . . . . **3**

4 Section 1231 gain from installment sales from Form 6252, line 26 or 37 . . . . **4**

5 Section 1231 gain or (loss) from like-kind exchanges from Form 8824 . . . . **5**

6 Gain, if any, from line 34, from other than casualty or theft **6**

7 Add lines 2 through 6 in columns (g) and (h) . . . . **7** ( )

8 Combine columns (g) and (h) of line 7. Enter gain or (loss) here, and on the appropriate line as follows: **8**

　Partnerships– Enter the gain or (loss) on Form 1065, Schedule K, line 6. Skip lines 9, 10, 12, and 13 below.

　S corporations– Report the gain or (loss) following the instructions for Form 1120S, Schedule K, lines 5 and 6. Skip lines 9, 10, 12, and 13 below, unless line 8 is a gain and the S corporation is subject to the capital gains tax.

　All others– If line 8 is zero or a loss, enter the amount on line 12 below and skip lines 9 and 10. If line 8 is a gain and you did not have any prior year section 1231 losses, or they were recaptured in an earlier year, enter the gain as a long-term capital gain on Schedule D and skip lines 9, 10, and 13 below.

9 Nonrecaptured net section 1231 losses from prior years (see instructions) **9**

10 Subtract line 9 from line 8. If zero or less, enter -0-. Also enter on the appropriate line as follows (see instructions): **10**

　S corporations– Enter this amount on Schedule D (Form 1120S), line 13, and skip lines 12 and 13 below.

　All others– If line 10 is zero, enter the amount from line 8 on line 13 below. If line 10 is more than zero, enter the amount from line 9 on line 13 below, and enter the amount from line 10 as a long-term capital gain on Schedule D.

Part II **Ordinary Gains and Losses**

11 Ordinary gains and losses not included on lines 12 through 18 (include property held 1 year or less):

12 Loss, if any, from line 8 . . . . **12**

13 Gain, if any, from line 8, or amount from line 9 if applicable **13**

14 Gain, if any, from line 33 . . . . **14**

15 Net gain or (loss) from Form 4684, lines 31 and 38a **15**

16 Ordinary gain from installment sales from Form 6252, line 25 or 36 . . . . **16**

17 Ordinary gain or (loss) from like-kind exchanges from Form 8824 . . . . **17**

18 Recapture of section 179 expense deduction for partners and S corporation shareholders from property dispositions by partnerships and S corporations (see instructions) **18**

19 Add lines 11 through 18 in columns (g) and (h) . . . . **19** ( )

20 Combine columns (g) and (h) of line 19. Enter gain or (loss) here, and on the appropriate line as follows: . . . **20**

　a For all except individual returns: Enter the gain or (loss) from line 20 on the return being filed.

　b For individual returns:

　　(1) If the loss on line 12 includes a loss from Form 4684, line 35, column (b)(ii), enter that part of the loss here and on line 22 of Schedule A (Form 1040). Identify as from ™Form 4797, line 20b(1).⌠ See instructions . . . . **20b(1)**

　　(2) Redetermine the gain or (loss) on line 20, excluding the loss, if any, on line 20b(1). Enter here and on Form 1040, line 14 . . **20b(2)**

For Paperwork Reduction Act Notice, see page 1 of separate instructions.　　Cat. No. 13086I　　Form **4797** (1995)

A-7a Form 4797—page 1.

Part III   Gain From Disposition of Property Under Sections 1245, 1250, 1252, 1254, and 1255

21	(a) Description of section 1245, 1250, 1252, 1254, or 1255 property:	(b) Date acquired (mo., day, yr.)	(c) Date sold (mo., day, yr.)
A			
B			
C			
D			

	Relate lines 21A through 21D to these columns ▶		Property A	Property B	Property C	Property D
22	Gross sales price (**Note:** *See line 1 before completing.*)	22				
23	Cost or other basis plus expense of sale	23				
24	Depreciation (or depletion) allowed or allowable	24				
25	Adjusted basis. Subtract line 24 from line 23	25				
26	Total gain. Subtract line 25 from line 22	26				
27	**If section 1245 property:**					
a	Depreciation allowed or allowable from line 24	27a				
b	Enter the **smaller** of line 26 or 27a	27b				
28	**If section 1250 property:** If straight line depreciation was used, enter -0- on line 28g, except for a corporation subject to section 291.					
a	Additional depreciation after 1975 (see instructions)	28a				
b	Applicable percentage multiplied by the **smaller** of line 26 or line 28a (see instructions)	28b				
c	Subtract line 28a from line 26. If residential rental property or line 26 is not more than line 28a, skip lines 28d and 28e	28c				
d	Additional depreciation after 1969 and before 1976	28d				
e	Enter the **smaller** of line 28c or 28d	28e				
f	Section 291 amount (corporations only)	28f				
g	Add lines 28b, 28e, and 28f	28g				
29	**If section 1252 property:** Skip this section if you did not dispose of farmland or if this form is being completed for a partnership.					
a	Soil, water, and land clearing expenses	29a				
b	Line 29a multiplied by applicable percentage (see instructions)	29b				
c	Enter the **smaller** of line 26 or 29b	29c				
30	**If section 1254 property:**					
a	Intangible drilling and development costs, expenditures for development of mines and other natural deposits, and mining exploration costs (see instructions)	30a				
b	Enter the **smaller** of line 26 or 30a	30b				
31	**If section 1255 property:**					
a	Applicable percentage of payments excluded from income under section 126 (see instructions)	31a				
b	Enter the **smaller** of line 26 or 31a (see instructions)	31b				

Summary of Part III Gains. Complete property columns A through D, through line 31b before going to line 32.

32	Total gains for all properties. Add property columns A through D, line 26	32	
33	Add property columns A through D, lines 27b, 28g, 29c, 30b, and 31b. Enter here and on line 14	33	
34	Subtract line 33 from line 32. Enter the portion from casualty or theft on Form 4684, line 33. Enter the portion from other than casualty or theft on Form 4797, line 6	34	

Part IV   Recapture Amounts Under Sections 179 and 280F(b)(2) When Business Use Drops to 50% or Less
     See instructions.

			(a) Section 179	(b) Section 280F(b)(2)
35	Section 179 expense deduction or depreciation allowable in prior years	35		
36	Recomputed depreciation. See instructions	36		
37	Recapture amount. Subtract line 36 from line 35. See the instructions for where to report	37		

A-7b Form 4797—page 2.

SCHEDULE E (Form 1040)	Supplemental Income and Loss	OMB No. 1545-0074

SCHEDULE E
(Form 1040)

Department of the Treasury
Internal Revenue Service   (99)

Supplemental Income and Loss

(From rental real estate, royalties, partnerships,
S corporations, estates, trusts, REMICs, etc.)

► **Attach to Form 1040 or Form 1041.**   ► **See Instructions for Schedule E (Form 1040).**

OMB No. 1545-0074

1995

Attachment
Sequence No. **13**

Name(s) shown on return

Your social security number

Part I   **Income or Loss From Rental Real Estate and Royalties**   **Note:** *Report income and expenses from your business of renting personal property on **Schedule C** or **C-EZ** (see page E-1). Report farm rental income or loss from **Form 4835** on page 2, line 39.*

1 Show the kind and location of each **rental real estate property:**	2 For each rental real estate property listed on line 1, did you or your family use it for personal purposes for more than the greater of 14 days or 10% of the total days rented at fair rental value during the tax year? (See page E-1.)	Yes	No
A	A		
B	B		
C	C		

Income:		Properties A	Properties B	Properties C	Totals (Add columns A, B, and C.)
3 Rents received	3		3		
4 Royalties received	4				4
Expenses:					
5 Advertising	5				
6 Auto and travel (see page E-2) .	6				
7 Cleaning and maintenance. . .	7				
8 Commissions	8				
9 Insurance	9				
10 Legal and other professional fees	10				
11 Management fees.	11				
12 Mortgage interest paid to banks, etc. (see page E-2)	12				12
13 Other interest	13				
14 Repairs	14				
15 Supplies	15				
16 Taxes	16				
17 Utilities	17				
18 Other (list) ►	18				
19 Add lines 5 through 18	19				19
20 Depreciation expense or depletion (see page E-2)	20				20
21 Total expenses. Add lines 19 and 20	21				
22 Income or (loss) from rental real estate or royalty properties. Subtract line 21 from line 3 (rents) or line 4 (royalties). If the result is a (loss), see page E-2 to find out if you must file **Form 6198** . . .	22				
23 Deductible rental real estate loss. **Caution:** *Your rental real estate loss on line 22 may be limited. See page E-3 to find out if you must file **Form 8582**. Real estate professionals must complete line 42 on page 2*	23 (	)(	)( )		
24 **Income.** Add positive amounts shown on line 22. **Do not** include any losses					24
25 **Losses.** Add royalty losses from line 22 and rental real estate losses from line 23. Enter the total losses here .					25 ( )
26 Total rental real estate and royalty income or (loss). Combine lines 24 and 25. Enter the result here. If Parts II, III, IV, and line 39 on page 2 do not apply to you, also enter this amount on Form 1040, line 17. Otherwise, include this amount in the total on line 40 on page 2					26

For Paperwork Reduction Act Notice, see Form 1040 instructions.   Cat. No. 11344L   Schedule E (Form 1040) 1995

A-8a Schedule E — page 1.

Name(s) shown on return. Do not enter name and social security number if shown on other side.　　　　**Your social security number**

Note: *If you report amounts from farming or fishing on Schedule E, you must enter your gross income from those activities on line 41 below. Real estate professionals must complete line 42 below.*

Part II	**Income or Loss From Partnerships and S Corporations**

Note: *If you report a loss from an at-risk activity, you MUST check either column (e) or (f) of line 27 to describe your investment in the activity. See page E-4. If you check column (f), you must attach Form 6198.*

27	(a) Name	(b) Enter P for partnership; S for S corporation	(c) Check if foreign partnership	(d) Employer identification number	Investment At Risk?	
					(e) All is at risk	(f) Some is not at risk
A						
B						
C						
D						
E						

	Passive Income and Loss		Nonpassive Income and Loss		
	(g) Passive loss allowed (attach Form 8582 if required)	(h) Passive income from Schedule K±1	(i) Nonpassive loss from Schedule K±1	(j) Section 179 expense deduction from Form 4562	(k) Nonpassive income from Schedule K±1
A					
B					
C					
D					
E					
28a Totals					
b Totals					

29	Add columns (h) and (k) of line 28a	29	
30	Add columns (g), (i), and (j) of line 28b	30	( )
31	Total partnership and S corporation income or (loss). Combine lines 29 and 30. Enter the result here and include in the total on line 40 below	31	

Part III	**Income or Loss From Estates and Trusts**

32	(a) Name	(b) Employer identification number
A		
B		

	Passive Income and Loss		Nonpassive Income and Loss	
	(c) Passive deduction or loss allowed (attach Form 8582 if required)	(d) Passive income from Schedule K±1	(e) Deduction or loss from Schedule K±1	(f) Other income from Schedule K±1
A				
B				
33a Totals				
b Totals				

34	Add columns (d) and (f) of line 33a	34	
35	Add columns (c) and (e) of line 33b	35	( )
36	Total estate and trust income or (loss). Combine lines 34 and 35. Enter the result here and include in the total on line 40 below	36	

Part IV	**Income or Loss From Real Estate Mortgage Investment Conduits (REMICs)—Residual Holder**

37	(a) Name	(b) Employer identification number	(c) Excess inclusion from Schedules Q, line 2c (see page E-4)	(d) Taxable income (net loss) from Schedules Q, line 1b	(e) Income from Schedules Q, line 3b

38	Combine columns (d) and (e) only. Enter the result here and include in the total on line 40 below	38	

Part V	**Summary**

39	Net farm rental income or (loss) from **Form 4835**. Also, complete line 41 below	39	
40	TOTAL income or (loss). Combine lines 26, 31, 36, 38, and 39. Enter the result here and on Form 1040, line 17 ▶	40	

41 **Reconciliation of Farming and Fishing Income.** Enter your **gross** farming and fishing income reported on Form 4835, line 7; Schedule K-1 (Form 1065), line 15b; Schedule K-1 (Form 1120S), line 23; and Schedule K-1 (Form 1041), line 13 (see page E-4) . . . . . . | **41** |

42 **Reconciliation for Real Estate Professionals.** If you were a real estate professional (see page E-3), enter the net income or (loss) you reported anywhere on Form 1040 from all rental real estate activities in which you materially participated under the passive activity loss rules . . . | **42** |

A-8b Schedule E—page 2.

SCHEDULES A&B
(Form 1040)

Department of the Treasury
Internal Revenue Service (99)

Schedule A—Itemized Deductions

(Schedule B is on back)

▶ **Attach to Form 1040.** ▶ **See Instructions for Schedules A and B (Form 1040).**

19 95

Attachment
Sequence No. **07**

Name(s) shown on Form 1040 | Your social security number

Medical and Dental Expenses		**Caution:** *Do not include expenses reimbursed or paid by others.*	
	1	Medical and dental expenses (see page A-1)	1
	2	Enter amount from Form 1040, line 32 . ⌊ 2 ⌋	
	3	Multiply line 2 above by 7.5% (.075)	3
	4	Subtract line 3 from line 1. If line 3 is more than line 1, enter -0-	4
Taxes You Paid (See page A-1.)	5	State and local income taxes	5
	6	Real estate taxes (see page A-2)	6
	7	Personal property taxes	7
	8	Other taxes. List type and amount ▶	8
	9	Add lines 5 through 8	9
Interest You Paid (See page A-2.)	10	Home mortgage interest and points reported to you on Form 1098	10
	11	Home mortgage interest not reported to you on Form 1098. If paid to the person from whom you bought the home, see page A-3 and show that person's name, identifying no., and address ▶	
Note: Personal interest is not deductible.		. . . .	11
	12	Points not reported to you on Form 1098. See page A-3 for special rules	12
	13	Investment interest. If required, attach Form 4952. (See page A-3.)	13
	14	Add lines 10 through 13	14
Gifts to Charity If you made a gift and got a benefit for it, see page A-3.	15	Gifts by cash or check. If you made any gift of $250 or more, see page A-3	15
	16	Other than by cash or check. If any gift of $250 or more, see page A-3. If over $500, you **MUST** attach Form 8283	16
	17	Carryover from prior year	17
	18	Add lines 15 through 17	18
Casualty and Theft Losses	19	Casualty or theft loss(es). Attach Form 4684. (See page A-4.)	19
Job Expenses and Most Other Miscellaneous Deductions (See page A-5 for expenses to deduct here.)	20	Unreimbursed employee expenses–job travel, union dues, job education, etc. If required, you **MUST** attach Form 2106 or 2106-EZ. (See page A-5.) ▶	20
	21	Tax preparation fees	21
	22	Other expenses–investment, safe deposit box, etc. List type and amount ▶	22
	23	Add lines 20 through 22	23
	24	Enter amount from Form 1040, line 32 . ⌊ 24 ⌋	
	25	Multiply line 24 above by 2% (.02)	25
	26	Subtract line 25 from line 23. If line 25 is more than line 23, enter -0-	26
Other Miscellaneous Deductions	27	Other–from list on page A-5. List type and amount ▶	27
Total Itemized Deductions	28	Is Form 1040, line 32, over $114,700 (over $57,350 if married filing separately)? **NO.** Your deduction is not limited. Add the amounts in the far right column for lines 4 through 27. Also, enter on Form 1040, line 34, the **larger** of this amount or your standard deduction. **YES.** Your deduction may be limited. See page A-5 for the amount to enter. ⎫⎬⎭ ▶	28

For Paperwork Reduction Act Notice, see Form 1040 instructions.   Cat. No. 11330X   **Schedule A (Form 1040) 1995**

A-9a Schedules A and B—page 1.

Name(s) shown on Form 1040. Do not enter name and social security number if shown on other side.

Your social security number

Schedule B–Interest and Dividend Income

Attachment
Sequence No. **08**

Part I
Interest
Income

(See
pages 15
and B-1.)

Note: If you
received a Form
1099-INT, Form
1099-OID, or
substitute
statement from
a brokerage firm,
list the firm's
name as the
payer and enter
the total interest
shown on that
form.

Note: *If you had over $400 in taxable interest income, you must also complete Part III.*

1   List name of payer. If any interest is from a seller-financed mortgage and the buyer used the property as a personal residence, see page B-1 and list this interest first. Also, show that buyer's social security number and address ▶

Amount

2   Add the amounts on line 1 . . . .   **2**

3   Excludable interest on series EE U.S. savings bonds issued after 1989 from Form 8815, line 14. You MUST attach Form 8815 to Form 1040 . . . .   **3**

4   Subtract line 3 from line 2. Enter the result here and on Form 1040, line 8a ▶   **4**

Part II
Dividend
Income

(See
pages 15
and B-1.)

Note: If you
received a Form
1099-DIV or
substitute
statement from
a brokerage
firm, list the
firm's name as
the payer and
enter the total
dividends
shown on that
form.

Note: *If you had over $400 in gross dividends and/or other distributions on stock, you must also complete Part III.*

5   List name of payer. Include gross dividends and/or other distributions on stock here. Any capital gain distributions and nontaxable distributions will be deducted on lines 7 and 8 ▶

Amount

6   Add the amounts on line 5 . . . .   **6**

7   Capital gain distributions. Enter here and on Schedule D* .   **7**

8   Nontaxable distributions. (See the inst. for Form 1040, line 9.)   **8**

9   Add lines 7 and 8 . . . .   **9**

10   Subtract line 9 from line 6. Enter the result here and on Form 1040, line 9 . ▶   **10**

If you do not need Schedule D to report any other gains or losses, see the instructions for Form 1040, line 13, on page 16.

Part III
Foreign
Accounts
and
Trusts

(See
page B-2.)

If you had over $400 of interest or dividends **or** had a foreign account or were a grantor of, or a transferor to, a foreign trust, you must complete this part.

Yes   **No**

11a   At any time during 1995, did you have an interest in or a signature or other authority over a financial account in a foreign country, such as a bank account, securities account, or other financial account? See page B-2 for exceptions and filing requirements for Form TD F 90-22.1 . . . .

b   If ™Yes,͵ enter the name of the foreign country▶

12   Were you the grantor of, or transferor to, a foreign trust that existed during 1995, whether or not you have any beneficial interest in it? If ™Yes,͵ you may have to file Form 3520, 3520-A, or 926.

For Paperwork Reduction Act Notice, see Form 1040 instructions.   Schedule B (Form 1040) 1995

A-9b Schedules A and B — page 2.

Department of the Treasury
Internal Revenue Service

Moving Expenses

▶ Attach to Form 1040.

▶ See instructions on back.

OMB No. 1545-0062

1995

Attachment
Sequence No. **62**

Name(s) shown on Form 1040 | Your social security number

Caution: *If you are a member of the armed forces, see the instructions before completing this form.*

1	Enter the number of miles from your **old home** to your **new workplace** . .	**1**	miles
2	Enter the number of miles from your **old home** to your **old workplace**. . .	**2**	miles
3	Subtract line 2 from line 1. Enter the result but not less than zero	**3**	miles

Is line 3 at least 50 miles?

Yes ▶ Go to line 4. Also, see **Time Test** in the instructions.

No ▶ You **cannot** deduct your moving expenses. Do not complete the rest of this form.

4	Transportation and storage of household goods and personal effects	**4**
5	Travel and lodging expenses of moving from your old home to your new home. **Do not** include meals	**5**
6	Add lines 4 and 5	**6**
7	Enter the total amount your employer paid for your move (including the value of services furnished in kind) that is **not** included in the wages box (box 1) of your W-2 form. This amount should be identified with code **P** in box 13 of your W-2 form.	**7**

Is line 6 more than line 7?

Yes ▶ Go to line 8.

No ▶ You **cannot** deduct your moving expenses. If line 6 is less than line 7, subtract line 6 from line 7 and include the result in income on Form 1040, line 7.

8	Subtract line 7 from line 6. Enter the result here and on Form 1040, line 24. This is your **moving expense deduction**	**8**

For Paperwork Reduction Act Notice, see back of form.

Cat. No. 12490K

Form **3903** (1995)

A-10a Form 3903—page 1.

Paperwork Reduction Act Notice

We ask for the information on this form to carry out the Internal Revenue laws of the United States. You are required to give us the information. We need it to ensure that you are complying with these laws and to allow us to figure and collect the right amount of tax.

The time needed to complete and file this form will vary depending on individual circumstances. The estimated average time is:

Recordkeeping. . . . . 33 min.
Learning about the
law or the form. . . . . . . .3 min.
Preparing the form . . . . . 13 min.
Copying, assembling, and
sending the form to the IRS. . . 20 min.

If you have comments concerning the accuracy of these time estimates or suggestions for making this form simpler, we would be happy to hear from you. You can write or call the IRS. See the Instructions for Form 1040.

General Instructions

Purpose of Form

Use Form 3903 to figure your moving expense deduction if you moved to a new principal place of work (workplace) within the United States or its possessions. If you qualify to deduct expenses for more than one move, use a separate Form 3903 for each move.

Note: *Use* **Form 3903-F,** *Foreign Moving Expenses, instead of this form if you are a U.S. citizen or resident alien who moved to a new principal workplace outside the United States or its possessions.*

Additional Information

For more details, get **Pub. 521,** Moving Expenses.

Other Forms You May Have To File

If you sold your main home in 1995, you must file **Form 2119,** Sale of Your Home, to report the sale.

Who May Deduct Moving Expenses

If you moved to a different home because of a change in job location, you may be able to deduct your moving expenses. You may be able to take the deduction whether you are self-employed or an employee. But you must meet certain tests explained next.

Distance Test.– Your new principal workplace must be at least 50 miles farther from your old home than your old workplace was. For example, if your old workplace was 3 miles from your old home, your new workplace must be at least 53 miles from that home. If you did not have an old workplace, your new workplace must be at least 50 miles from your old home. The distance between the two points is the shortest of the more commonly traveled routes between them.

Time Test.– If you are an employee, you must work full time in the general area of your new workplace for at least 39 weeks during

the 12 months right after you move. If you are self-employed, you must work full time in the general area of your new workplace for at least 39 weeks during the first 12 months and a total of at least 78 weeks during the 24 months right after you move.

You may deduct your moving expenses even if you have not met the time test before your return is due. You may do this if you expect to meet the 39-week test by the end of 1996 or the 78-week test by the end of 1997. If you deduct your moving expenses on your 1995 return but do not meet the time test, you will have to either:

● Amend your 1995 tax return by filing **Form 1040X,** Amended U.S. Individual Income Tax Return, or

● Report the amount of your 1995 moving expense deduction that reduced your 1995 income tax as income in the year you cannot meet the test. For more details, see **Time Test** in Pub. 521.

If you do not deduct your moving expenses on your 1995 return and you later meet the time test, you may take the deduction by filing an amended return for 1995. To do this, use Form 1040X.

Exceptions to the Time Test.– The time test does not have to be met if any of the following apply:

● Your job ends because of a disability.

● You are transferred for your employer's benefit.

● You are laid off or discharged for a reason other than willful misconduct.

● You meet the requirements (explained later) for retirees or survivors living outside the United States.

● You are filing this form for a decedent.

Members of the Armed Forces

If you are in the armed forces, you do not have to meet the **distance and time tests** if the move is due to a permanent change of station. A permanent change of station includes a move in connection with and within 1 year of retirement or other termination of active duty.

How To Complete the Form.– First, complete lines 4 through 6 using your actual expenses. **Do** not reduce your expenses by any reimbursements or allowances you received from the government in connection with the move. Also, do not include any expenses for moving services that were provided by the government. If you and your spouse and dependents are moved to or from different locations, treat the moves as a single move.

Next, enter on line 7 the total reimbursements and allowances you received from the government in connection with the expenses you claimed on lines 4 and 5. **Do not** include the value of moving services provided by the government. Then, complete line 8 if applicable.

Retirees or Survivors Living Outside the United States

If you are a retiree or survivor who moved to a home in the United States or its possessions and you meet the following requirements, you are treated as if you moved

to a new workplace located in the United States. You are subject to the distance test. Use this form instead of Form 3903-F to figure your moving expense deduction.

Retirees.– You may deduct moving expenses for a move to a new home in the United States when you actually retire if both your old principal workplace and your old home were outside the United States.

Survivors.– You may deduct moving expenses for a move to a home in the United States if you are the spouse or dependent of a person whose principal workplace at the time of death was outside the United States. In addition, the expenses must be for a move (1) that begins within 6 months after the decedent's death, and (2) from a former home outside the United States that you lived in with the decedent at the time of death.

Reimbursements

If your employer paid for any part of your move, your employer must give you a statement showing a detailed breakdown of reimbursements or payments for moving expenses. Your employer may use **Form 4782,** Employee Moving Expense Information, or his or her own form.

You may choose to deduct moving expenses in the year you are reimbursed by your employer, even though you paid the expenses in a different year. However, special rules apply. See **How To Report** in Pub. 521.

Moving Expenses Incurred Before 1994.– If you were reimbursed for moving expenses incurred before 1994 and you did not deduct those expenses on a prior year's return, you may be able to deduct them on **Schedule A,** Itemized Deductions. But you must use the **1994** Form 3903 to do so. You can get the 1994 Form 3903 by calling 1-800-TAX-FORM (1-800-829-3676).

Specific Instructions

You may deduct the following expenses you incurred in moving your family and dependent household members. Do not deduct expenses for employees such as a maid, nanny, or nurse.

Line 4.– Enter the actual cost to pack, crate, and move your household goods and personal effects. You may also include the cost to store and insure household goods and personal effects within any period of 30 days in a row after the items were moved from your old home and before they were delivered to your new home.

Line 5.– Enter the costs of travel from your old home to your new home. These include transportation and lodging on the way. Include costs for the day you arrive. Although not all the members of your household have to travel together or at the same time, you may only include expenses for one trip per person.

If you use your own car(s), you may figure the expenses by using either:

● Actual out-of-pocket expenses for gas and oil, or

● Mileage at the rate of 9 cents a mile.

You may add parking fees and tolls to the amount claimed under either method. Keep records to verify your expenses.

A-10b Form 3903—page 2.

Form **8829**

Department of the Treasury
Internal Revenue Service (99)

Expenses for Business Use of Your Home

▶ File only with Schedule C (Form 1040). Use a separate Form 8829 for each home you used for business during the year.

▶ See separate instructions.

OMB No. 1545-1266

1995

Attachment
Sequence No. **66**

Name(s) of proprietor(s)

Your social security number

Part I Part of Your Home Used for Business

1	Area used regularly and exclusively for business, regularly for day care, or for inventory storage. See instructions	1	
2	Total area of home	2	
3	Divide line 1 by line 2. Enter the result as a percentage	3	%

• For day-care facilities not used exclusively for business, also complete lines 4–6.

• All others, skip lines 4–6 and enter the amount from line 3 on line 7.

4	Multiply days used for day care during year by hours used per day	4	hr.	
5	Total hours available for use during the year (365 days × 24 hours). See instructions	5	8,760 hr.	
6	Divide line 4 by line 5. Enter the result as a decimal amount	6	.	
7	Business percentage. For day-care facilities not used exclusively for business, multiply line 6 by line 3 (enter the result as a percentage). All others, enter the amount from line 3 ▶	7		%

Part II Figure Your Allowable Deduction

		(a) Direct expenses	(b) Indirect expenses	
8	Enter the amount from Schedule C, line 29, **plus** any net gain or (loss) derived from the business use of your home and shown on Schedule D or Form 4797. If more than one place of business, see instructions			8

See instructions for columns (a) and (b) before completing lines 9–20.

			(a) Direct expenses	(b) Indirect expenses	
9	Casualty losses. See instructions	9			
10	Deductible mortgage interest. See instructions	10			
11	Real estate taxes. See instructions	11			
12	Add lines 9, 10, and 11	12			
13	Multiply line 12, column (b) by line 7	13			
14	Add line 12, column (a) and line 13				14
15	Subtract line 14 from line 8. If zero or less, enter -0-				15
16	Excess mortgage interest. See instructions	16			
17	Insurance	17			
18	Repairs and maintenance	18			
19	Utilities	19			
20	Other expenses. See instructions	20			
21	Add lines 16 through 20	21			
22	Multiply line 21, column (b) by line 7	22			
23	Carryover of operating expenses from 1994 Form 8829, line 41	23			
24	Add line 21 in column (a), line 22, and line 23				24
25	Allowable operating expenses. Enter the **smaller** of line 15 or line 24				25
26	Limit on excess casualty losses and depreciation. Subtract line 25 from line 15				26
27	Excess casualty losses. See instructions	27			
28	Depreciation of your home from Part III below	28			
29	Carryover of excess casualty losses and depreciation from 1994 Form 8829, line 42	29			
30	Add lines 27 through 29				30
31	Allowable excess casualty losses and depreciation. Enter the **smaller** of line 26 or line 30				31
32	Add lines 14, 25, and 31				32
33	Casualty loss portion, if any, from lines 14 and 31. Carry amount to **Form 4684,** Section B				33
34	Allowable expenses for business use of your home. Subtract line 33 from line 32. Enter here and on Schedule C, line 30. If your home was used for more than one business, see instructions ▶				34

Part III Depreciation of Your Home

35	Enter the **smaller** of your home's adjusted basis or its fair market value. See instructions	35	
36	Value of land included on line 35	36	
37	Basis of building. Subtract line 36 from line 35	37	
38	Business basis of building. Multiply line 37 by line 7	38	
39	Depreciation percentage. See instructions	39	%
40	Depreciation allowable. Multiply line 38 by line 39. Enter here and on line 28 above. See instructions	40	

Part IV Carryover of Unallowed Expenses to 1996

41	Operating expenses. Subtract line 25 from line 24. If less than zero, enter -0-	41	
42	Excess casualty losses and depreciation. Subtract line 31 from line 30. If less than zero, enter -0-	42	

For Paperwork Reduction Act Notice, see page 1 of separate instructions.

Cat. No. 13232M   Form **8829** (1995)

A-11 Form 8829.

Form 2688

Department of the Treasury
Internal Revenue Service

Application for Additional Extension of Time To File U.S. Individual Income Tax Return

▶ See instructions on back.
▶ You MUST complete all items that apply to you.

OMB No. 1545-0066

1995

Attachment
Sequence No. **59**

Please type or print.	Your first name and initial	Last name	Your social security number
File the original and one copy by the due date for filing your return.	If a joint return, spouse's first name and initial	Last name	Spouse's social security number
	Home address (number, street, and apt. no. or rural route). If you have a P.O. box, see the instructions.		
	City, town or post office, state, and ZIP code		

1 I request an extension of time until ...., 19..., to file Form 1040EZ, Form 1040A, Form 1040 or Form 1040-T for the calendar year 1995, or other tax year ending ...., 19........ .

2 Explain why you need an extension. All individuals filing this form must give an adequate explanation ▶ ....
....
....
....
....

3 Have you filed Form 4868 to request an extension of time to file for this tax year? . . . . ☐ Yes ☐ No
If you checked ™No, we will grant your extension only for undue hardship. Fully explain the hardship in item 2. Attach any information you have that helps explain the hardship.

If you expect to owe gift or generation-skipping transfer (GST) tax, complete line 4.

4 If you or your spouse plan to file a gift or GST tax return (Form 709 or 709-A) for 1995, generally due by April 15, 1996, see the instructions and check here . . . . } **Yourself** . . ▶ ☐ **Spouse** . . ▶ ☐

Signature and Verification

Under penalties of perjury, I declare that I have examined this form, including accompanying schedules and statements, and to the best of my knowledge and belief, it is true, correct, and complete; and, if prepared by someone other than the taxpayer, that I am authorized to prepare this form.

Signature of taxpayer ▶ _____ Date ▶ _____

Signature of spouse ▶ _____ Date ▶ _____
(If filing jointly, BOTH must sign even if only one had income)

Signature of preparer
other than taxpayer ▶ _____ Date ▶ _____

File original and one copy. The IRS will show below whether or not your application is approved and will return the copy.

Notice to Applicant–To Be Completed by the IRS

☐ We **HAVE** approved your application. Please attach this form to your return.

☐ We **HAVE NOT** approved your application. However, we have granted a 10-day grace period from the later of the date shown below or the due date of your return. This grace period is considered to be a valid extension of time for elections otherwise required to be made on a timely return. Please attach this form to your return.

☐ We **HAVE NOT** approved your application. After considering your reasons stated in item 2 above, we cannot grant your request for an extension of time to file. We are not granting a 10-day grace period.

☐ We cannot consider your application because it was filed after the due date of your return.

☐ We **HAVE NOT** approved your application. The maximum extension of time allowed by law is 6 months.

☐ Other ....

Director

By _____

_____ Date

Please type or print	Name
	Number and street (include suite, room, or apt. no.) or P.O. box number if mail is not delivered to street address
	City, town or post office, state, and ZIP code

If you want the copy of this form returned to you at an address other than that shown above or to an agent acting for you, enter the name of the agent and/or the address to which the copy should be sent.

For Paperwork Reduction Act Notice, see back of form.   Cat. No. 11958F   Form **2688** (1995)

A-12a Form 2688—page 1.

Paperwork Reduction Act Notice

We ask for the information on this form to carry out the Internal Revenue laws of the United States. You are required to give us the information. We need it to ensure that you are complying with these laws and to allow us to figure and collect the right amount of tax.

The time needed to complete and file this form will vary depending on individual circumstances. The estimated average time is: **Learning about the law or the form,** 7 min.; **Preparing the form,** 10 min.; and **Copying, assembling, and sending the form to the IRS,** 20 min.

If you have comments concerning the accuracy of these time estimates or suggestions for making this form simpler, we would be happy to hear from you. You can write to the Tax Forms Committee, Western Area Distribution Center, Rancho Cordova, CA 95743-0001. **DO NOT** send the form to this address. Instead, see **Where To File** below.

General Instructions

Note: *Form 1040-T references are to a new form sent to certain individuals on a test basis.*

Purpose of Form

Use Form 2688 to ask for more time to file **Form 1040EZ, Form 1040A, Form 1040,** or **Form 1040-T.** Generally, use it only if you already asked for more time on **Form 4868** (the ™automatic⌡ extension form) and that time was not enough. We will make an exception **only** for undue hardship.

To get the extra time you **MUST:**

● Complete and file Form 2688 on time, **AND**

● Have a good reason why the first 4 months were not enough. Explain the reason in item 2.

Generally, we will not give you more time to file just for the convenience of your tax return preparer. But if the reasons for being late are beyond his or her control or, despite a good effort, you cannot get professional help in time to file, we will usually give you the extra time.

Caution: *If we give you more time to file and later find that the statements made on this form are false or misleading, the extension is null and void. You will owe the late filing penalty explained on this page.*

You cannot have the IRS figure your tax if you file after the regular due date of your return.

Note: *An extension of time to file your 1995 calendar year income tax return also extends the time to file a gift or generation-skipping transfer (GST) tax return (Form 709 or 709-A) for 1995.*

If You Live Abroad.– U.S. citizens or resident aliens living abroad may qualify for special tax treatment if they meet the required foreign residence or presence tests. If you do not expect to meet either of those tests by the due date of your return, request an extension to a date after you expect to qualify. Ask for it on **Form 2350,** Application for Extension of Time To

File U.S. Income Tax Return. Get **Pub. 54,** Tax Guide for U.S. Citizens and Resident Aliens Abroad.

Total Time Allowed

Generally, we cannot extend the due date of your return for more than 6 months. This includes the 4 extra months allowed by Form 4868. There may be an exception if you live abroad. See the previous discussion.

When To File

If you filed Form 4868, file Form 2688 by the extended due date of your return. For most people, this is August 15, 1996. If you didn't file Form 4868 first because of undue hardship, file Form 2688 by the due date of your return. The due date is April 15, 1996, for a calendar year return. Be sure to fully explain in item 2 why you are filing Form 2688 first. Also, file Form 2688 early so that if your request is not approved, you can still file your return on time.

Out of the Country.– You may have been allowed 2 extra months to file if you were a U.S. citizen or resident out of the country. ™Out of the country⌡ means either**(a)** you live outside the United States and Puerto Rico **and** your main place of work is outside the United States and Puerto Rico, **or (b)** you are in military or naval service outside the United States and Puerto Rico.

Where To File

Mail the original **AND** one copy of Form **2688** to the Internal Revenue Service Center where you file your return.

Filing Your Tax Return

You may file your tax return any time before the extension expires. But remember, Form 2688 does not extend the time to pay taxes. If you do not pay the amount due by the regular due date, you will owe interest. You may also be charged penalties.

Interest.– You will owe interest on any tax not paid by the regular due date of your return. The interest runs until you pay the tax. Even if you had a good reason for not paying on time, you will still owe interest.

Late Payment Penalty.– The penalty is usually ¹§ of 1% of any tax (other than estimated tax) not paid by the regular due date. It is charged for each month or part of a month the tax is unpaid. The maximum penalty is 25%. You might not owe this penalty if you have a good reason for not paying on time. Attach a statement to your return, not Form 2688, explaining the reason.

Late Filing Penalty.– A penalty is usually charged if your return is filed after the due date (including extensions). It is usually 5% of the tax not paid by the regular due date for each month or part of a month your return is late. Generally, the maximum penalty is 25%. If your return is more than 60 days late, the minimum penalty is $100 or the balance of tax due on your return, whichever is smaller. You might not owe the penalty if you have a good reason for

filing late. Attach a statement to your return, not Form 2688, explaining the reason.

How To Claim Credit for Payment Made With This Form.– Include any payment you sent with Form 2688 on the appropriate line of your tax return. If you file Form 1040EZ, the instructions for line 9 of that form will tell you how to report the payment. If you file Form 1040A, see the instructions for line 29d. If you file Form 1040, enter the payment on line 58. If you file Form 1040-T, enter the payment on line 35.

If you and your spouse each filed a separate Form 2688 but later file a joint return for 1995, enter the total paid with both Forms 2688 on the appropriate line of your joint return.

If you and your spouse jointly filed Form 2688 but later file separate returns for 1995, you may enter the total amount paid with Form 2688 on either of your separate returns. Or you and your spouse may divide the payment in any agreed amounts. Be sure each separate return has the social security numbers of both spouses.

Specific Instructions

Name, Address, and Social Security Number (SSN).– Enter your name, address, and SSN. If you plan to file a joint return, also enter your spouse's name and SSN. If the post office does not deliver mail to your street address and you have a P.O. box, enter the box number instead.

Note: *If you changed your mailing address after you filed your last return, you should use Form 8822, Change of Address, to notify the IRS of the change. Showing a new address on Form 2688 will not update your record. You can get Form 8822 by calling 1-800-829-3676.*

Item 2.– Clearly describe the reasons that will delay your return. We cannot accept incomplete reasons, such as ™illness⌡ or ™practitioner too busy,⌡ without adequate explanations. If it is clear that you have no important reason but only want more time, we will deny your request. The 10-day grace period will also be denied.

Line 4.– If you or your spouse plan to file Form 709 or 709-A for 1995, check whichever box applies. But if your spouse files a separate Form 2688, do not check the box for your spouse.

Your Signature.– This form must be signed. If you plan to file a joint return, both of you should sign. If there is a good reason why one of you cannot, the other spouse may sign for both. Attach a statement explaining why the other spouse cannot sign.

Others Who Can Sign for You.– Anyone with a power of attorney can sign. But the following can sign for you without a power of attorney:

● Attorneys, CPAs, and enrolled agents.

● A person in close personal or business relationship to you who is signing because you cannot. There must be a good reason why you cannot sign, such as illness or absence. Attach an explanation.

A-12b Form 2688 — page 2.

Form **8822**

(Rev. May 1995)

Department of the Treasury
Internal Revenue Service

Change of Address

▶ Please type or print.

▶ See instructions on back.　　▶ Do not attach this form to your return.

OMB No. 1545-1163

Part I | **Complete This Part To Change Your Home Mailing Address**

Check **ALL** boxes this change affects:

1 ☐ Individual income tax returns (Forms 1040, 1040A, 1040EZ, 1040NR, etc.)

　　▶ If your last return was a joint return and you are now establishing a residence separate
　　　from the spouse with whom you filed that return, check here . . . . ▶ ☐

2 ☐ Employment tax returns for household employers (Forms 942, 940, 940-EZ, etc.)

　　▶ Enter your employer identification number here . . . . ▶ _____

3 ☐ Gift, estate, or generation-skipping transfer tax returns (Forms 706, 709, etc.)

　　▶ For Forms 706 and 706-NA, enter the decedent's name and social security number below.

　　▶ Decedent's name　　　　　　　　　　　　　　　▶ Social security number

4a Your name (first name, initial, and last name)	4b Your social security number
5a Spouse's name (first name, initial, and last name)	5b Spouse's social security number

6　Prior name(s). See instructions.

7a Old address (no., street, city or town, state, and ZIP code). If a P.O. box or foreign address, see instructions.	Apt. no.
7b Spouse's old address, if different from line 7a (no., street, city or town, state, and ZIP code). If a P.O. box or foreign address, see instructions.	Apt. no.
8 New address (no., street, city or town, state, and ZIP code). If a P.O. box or foreign address, see instructions.	Apt. no.

Part II | **Complete This Part To Change Your Business Mailing Address or Business Location**

Check **ALL** boxes this change affects:

9 ☐ Employment, excise, and other business returns (Forms 720, 941, 990, 1041, 1065, 1120, etc.)

10 ☐ Employee plan returns (Forms 5500, 5500-C/R, and 5500-EZ). See instructions.

11 ☐ Business location

12a Business name	12b Employer identification number
13 Old address (no., street, city or town, state, and ZIP code). If a P.O. box or foreign address, see instructions.	Room or suite no.
14 New address (no., street, city or town, state, and ZIP code). If a P.O. box or foreign address, see instructions.	Room or suite no.
15 New business location (no., street, city or town, state, and ZIP code). If a foreign address, see instructions.	Room or suite no.

Part III | **Signature**

Daytime telephone number of person to contact (optional) ▶ ( )

**Please
Sign
Here**

Your signature	Date	If Part II completed, signature of owner, officer, or representative	Date
If joint return, spouse's signature	Date	Title	

For Privacy Act and Paperwork Reduction Act Notice, see back of form.　　　Cat. No. 12081V　　　Form **8822** (Rev. 5-95)

A-13a Form 8822—page 1.

Privacy Act and Paperwork Reduction Act Notice

We ask for this information to carry out the Internal Revenue laws of the United States. We may give the information to the Department of Justice and to other Federal agencies, as provided by law. We may also give it to cities, states, the District of Columbia, and U.S. commonwealths or possessions to carry out their tax laws. And we may give it to foreign governments because of tax treaties they have with the United States.

If you fail to provide the Internal Revenue Service with your current mailing address, you may not receive a notice of deficiency or a notice and demand for tax. Despite the failure to receive such notices, penalties and interest will continue to accrue on the tax deficiencies.

The time needed to complete and file this form will vary depending on individual circumstances. The estimated average time is 16 minutes.

If you have comments concerning the accuracy of this time estimate or suggestions for making this form simpler, we would be happy to hear from you. You can write to the **Internal Revenue Service,** Attention: Tax Forms Committee, PC:FP, Washington, DC 20224. **DO NOT** send the form to this address. Instead, see **Where To File** on this page.

Purpose of Form

You may use Form 8822 to notify the Internal Revenue Service if you changed your home or business mailing address or your business location. Generally, complete only one Form 8822 to change your home and business addresses. If this change also affects the mailing address for your children who filed income tax returns, complete and file a separate Form 8822 for each child. If you are a representative signing for the taxpayer, attach to Form 8822 a copy of your power of attorney.

Note: *If you moved after you filed your return and you are expecting a refund, also notify the post office serving your old address. This will help forward your check to your new address.*

Prior Name(s)

If you or your spouse changed your name because of marriage, divorce, etc., complete line 6. Also, be sure to notify the **Social Security Administration** of your new name so that it has the same name in its records that you have on your tax return. This prevents delays in processing your return and issuing refunds. It also safeguards your future social security benefits.

P.O. Box

If your post office does not deliver mail to your street address and you have a P.O. box, show your box number instead of your street address.

Foreign Address

If your address is outside the United States or its possessions or territories, enter the information in the following order: number, street, city, province or state, postal code, and country. **Do not** abbreviate the country name. Be sure to include any apartment, room, or suite number in the space provided.

Employee Plan Returns

A change in the mailing address for employee plan returns must be shown on a separate Form 8822 unless the **Exception** below applies.

Exception. If the employee plan returns were filed with the same service center as your other returns (individual, business, employment, gift, estate, etc.), you do not have to use a separate Form 8822. See **Where To File** below.

Where To File

Send this form to the **Internal Revenue Service Center** shown below for your old address. But if you checked the box on line 10 (employee plan returns), send it to the address shown in the far right column.

If your old address was in:	Use this address:
Florida, Georgia, South Carolina	Atlanta, GA 39901
New Jersey, New York (New York City and counties of Nassau, Rockland, Suffolk, and Westchester)	Holtsville, NY 00501
New York (all other counties), Connecticut, Maine, Massachusetts, New Hampshire, Rhode Island, Vermont	Andover, MA 05501
Alaska, Arizona, California (counties of Alpine, Amador, Butte, Calaveras, Colusa, Contra Costa, Del Norte, El Dorado, Glenn, Humboldt, Lake, Lassen, Marin, Mendocino, Modoc, Napa, Nevada, Placer, Plumas, Sacramento, San Joaquin, Shasta, Sierra, Siskiyou, Solano, Sonoma, Sutter, Tehama, Trinity, Yolo, and Yuba), Colorado, Idaho, Montana, Nebraska, Nevada, North Dakota, Oregon, South Dakota, Utah, Washington, Wyoming	Ogden, UT 84201
California (all other counties), Hawaii	Fresno, CA 93888

Indiana, Kentucky, Michigan, Ohio, West Virginia	Cincinnati, OH 45999
Kansas, New Mexico, Oklahoma, Texas	Austin, TX 73301
Delaware, District of Columbia, Maryland, Pennsylvania, Virginia	Philadelphia, PA 19255
Alabama, Arkansas, Louisiana, Mississippi, North Carolina, Tennessee	Memphis, TN 37501
Illinois, Iowa, Minnesota, Missouri, Wisconsin	Kansas City, MO 64999
American Samoa	Philadelphia, PA 19255
Guam: Permanent residents	Department of Revenue and Taxation Government of Guam 378 Chalan San Antonio Tamuning, GU 96911
Guam: Nonpermanent residents Puerto Rico (or if excluding income under section 933) Virgin Islands: Nonpermanent residents	Philadelphia, PA 19255
Virgin Islands: Permanent residents	V. I. Bureau of Internal Revenue Lockhart Gardens No. 1-A Charlotte Amalie, St. Thomas, VI 00802
Foreign country: U.S. citizens and those filing Form 2555, Form 2555-EZ, or Form 4563	Philadelphia, PA 19255
All APO and FPO addresses	

Employee Plan Returns ONLY (Form 5500 series)

If the principal office of the plan sponsor or the plan administrator was in:	Use this address:
Connecticut, Delaware, District of Columbia, Maine, Maryland, Massachusetts, New Hampshire, New Jersey, New York, Pennsylvania, Puerto Rico, Rhode Island, Vermont, Virginia	Holtsville, NY 00501
Alabama, Alaska, Arkansas, California, Florida, Georgia, Hawaii, Idaho, Louisiana, Mississippi, Nevada, North Carolina, Oregon, South Carolina, Tennessee, Washington	Atlanta, GA 39901
Arizona, Colorado, Illinois, Indiana, Iowa, Kansas, Kentucky, Michigan, Minnesota, Missouri, Montana, Nebraska, New Mexico, North Dakota, Ohio, Oklahoma, South Dakota, Texas, Utah, West Virginia, Wisconsin, Wyoming	Memphis, TN 37501
Foreign country	Holtsville, NY 00501
All Form 5500-EZ filers	Andover, MA 05501

A-13b Form 8822 — page 2.

Department of the Treasury—Internal Revenue Service

**Interest and Dividend Income
for Form 1040A Filers**   (99)   **1995**

OMB No. 1545-0085

Name(s) shown on Form 1040A	Your social security number
	: :

Part I

Interest income

(See pages 28 and 71.)

Note: *If you received a Form 1099–INT, Form 1099–OID, or substitute statement from a brokerage firm, enter the firm's name and the total interest shown on that form.*

1 List name of payer. If any interest is from a seller-financed mortgage and the buyer used the property as a personal residence, see page 71 and list this interest first. Also, show that buyer's social security number and address.

	Amount
1	

2 Add the amounts on line 1.   **2** _____

3 Excludable interest on series EE U.S. savings bonds issued after 1989 from Form 8815, line 14. You **must** attach Form 8815 to Form 1040A.   **3** _____

4 Subtract line 3 from line 2. Enter the result here and on Form 1040A, line 8a.   **4** _____

Part II

Dividend income

(See pages 28 and 72.)

Note: *If you received a Form 1099–DIV or substitute statement from a brokerage firm, enter the firm's name and the total dividends shown on that form.*

5 List name of payer

	Amount
5	

5/15AAA

6 Add the amounts on line 5. Enter the total here and on Form 1040A, line 9.   **6** _____

For Paperwork Reduction Act Notice, see Form 1040A instructions.   Cat. No. 12075R   **1995 Schedule 1 (Form 1040A) page 1**

Printed on recycled paper

A-14 Schedule 1.

Form **2106**

Department of the Treasury
Internal Revenue Service (99)

Employee Business Expenses

▶ See separate instructions.

▶ Attach to Form 1040 or Form 1040-T.

OMB No. 1545-0139

1995

Attachment
Sequence No. **54**

Your name	Social security number	Occupation in which expenses were incurred

Part I   Employee Business Expenses and Reimbursements

STEP 1   Enter Your Expenses

			Column A		Column B	
			Other Than Meals and Entertainment		Meals and Entertainment	
1	Vehicle expense from line 22 or line 29	**1**				
2	Parking fees, tolls, and transportation, including train, bus, etc., that **did not** involve overnight travel	**2**				
3	Travel expense while away from home overnight, including lodging, airplane, car rental, etc. **Do not** include meals and entertainment	**3**				
4	Business expenses not included on lines 1 through 3. **Do not** include meals and entertainment	**4**				
5	Meals and entertainment expenses (see instructions)	**5**				
6	**Total expenses.** In Column A, add lines 1 through 4 and enter the result. In Column B, enter the amount from line 5	**6**				

Note: *If you were not reimbursed for any expenses in Step 1, skip line 7 and enter the amount from line 6 on line 8.*

STEP 2   Enter Amounts Your Employer Gave You for Expenses Listed in STEP 1

7	Enter amounts your employer gave you that were **not** reported to you in box 1 of Form W-2. Include any amount reported under code ™LĴ in box 13 of your Form W-2 (see instructions) . . .	**7**				

STEP 3   Figure Expenses To Deduct on Schedule A (Form 1040) or Form 1040-T, Section B

8	Subtract line 7 from line 6	**8**				
	Note: *If **both columns** of line 8 are zero, **stop here.** If Column A is less than zero, report the amount as income on Form 1040, line 7, or Form 1040-T, line 1.*					
9	In Column A, enter the amount from line 8 (if zero or less, enter -0-). In Column B, multiply the amount on line 8 by 50% (.50) .	**9**				
10	Add the amounts on line 9 of both columns and enter the total here. **Also, enter the total on Schedule A (Form 1040), line 20, or Form 1040-T, Section B, line n.** (Qualified performing artists and individuals with disabilities, see the instructions for special rules on where to enter the total.) . . . . ▶	**10**				

For Paperwork Reduction Act Notice, see instructions.   Cat. No. 11700N   Form **2106** (1995)

A-15a Form 2106 — page 1.

Part II   Vehicle Expenses (See instructions to find out which sections to complete.)

Section A.–General Information

			(a) Vehicle 1	(b) Vehicle 2
11	Enter the date vehicle was placed in service	11	/ /	/ /
12	Total miles vehicle was driven during 1995	12	miles	miles
13	Business miles included on line 12	13	miles	miles
14	Percent of business use. Divide line 13 by line 12	14	%	%
15	Average daily round trip commuting distance	15	miles	miles
16	Commuting miles included on line 12	16	miles	miles
17	Other personal miles. Add lines 13 and 16 and subtract the total from line 12	17	miles	miles

18 Do you (or your spouse) have another vehicle available for personal purposes? . . . . ☐ Yes ☐ No

19 If your employer provided you with a vehicle, is personal use during off-duty hours permitted? ☐ Yes ☐ No ☐ Not applicable

20 Do you have evidence to support your deduction? . . . . ☐ Yes ☐ No

21 If ™Yes,ʃ is the evidence written? . . . . ☐ Yes ☐ No

Section B.–Standard Mileage Rate (Use this section only if you own the vehicle.)

22	Multiply line 13 by 30¢ (.30). Enter the result here and on line 1. (Rural mail carriers, see instructions.)	22	

Section C.–Actual Expenses

			(a) Vehicle 1		(b) Vehicle 2	
23	Gasoline, oil, repairs, vehicle insurance, etc.	23				
24a	Vehicle rentals	24a				
b	Inclusion amount (see instructions)	24b				
c	Subtract line 24b from line 24a	24c				
25	Value of employer-provided vehicle (applies only if 100% of annual lease value was included on Form W-2–see instructions)	25				
26	Add lines 23, 24c, and 25 . .	26				
27	Multiply line 26 by the percentage on line 14 . . .	27				
28	Depreciation. Enter amount from line 38 below	28				
29	Add lines 27 and 28. Enter total here and on line 1	29				

Section D.–Depreciation of Vehicles (Use this section only if you own the vehicle.)

			(a) Vehicle 1		(b) Vehicle 2	
30	Enter cost or other basis (see instructions)	30				
31	Enter amount of section 179 deduction (see instructions) .	31				
32	Multiply line 30 by line 14 (see instructions if you elected the section 179 deduction) . . .	32				
33	Enter depreciation method and percentage (see instructions) .	33				
34	Multiply line 32 by the percentage on line 33 (see instructions) . .	34				
35	Add lines 31 and 34	35				
36	Enter the limitation amount from the table in the line 36 instructions	36				
37	Multiply line 36 by the percentage on line 14 . . .	37				
38	Enter the **smaller** of line 35 or line 37. Also, enter this amount on line 28 above	38				

A-15b Form 2106—page 2.

Glossary

accrued interest Accrued interest is interest that is earned but not paid until the maturity of the loan. For example, a balloon loan is structured to have the interest on the note accrue until maturity. During the term of the balloon loan, the borrower does not make interest payments. The loan amount builds up interest, at the rate described in the loan agreement, to be paid when the loan is paid off.

acquisition cost The cost of obtaining a property.

addendum A document that becomes a part of another document, such as a lease. Addendums may be used to alter the terms and conditions of a lease. When properly prepared, addendums become an integral part of an existing agreement.

amenities Benefits or objects provided with a property. Examples of amenities include swimming pools, tennis courts, fitness rooms, etc.

amortization The method of designing a payment schedule over an agreed term to repay a debt.

amortization schedule A table that defines periodic payment amounts for principal and interest payments required to repay a debt.

annual debt service The amount of principal and interest payments required on an annual basis to repay a loan.

annual percentage rate (APR) The effective rate of interest charged over a period of one year for a loan amount. The APR is normally higher than the note rate of a loan. For example, if your monthly payments are calculated with an interest rate of 10%, the APR might be higher. If discount points are paid, they increase the note rate of a loan to balance a higher annual percentage rate.

appraisal An estimated value of a property.

appraiser A person who estimates the value of a property.

appurtenance Items outside of a property but considered a part of the real estate. Examples of appurtenances include garages and private storage areas.

arm's-length transaction A transaction between parties seeking to fulfill their personal best interests. A transaction between husband and wife, parent and child, or corporate divisions is not considered an arm's-length transaction. There must be no combined interest in benefits derived from the transaction to either party when creating an arm's-length transaction.

as is A term used to acknowledge a sale or rental that is made based on existing conditions, without a warranty of the sale. The term is meant to mean a property is accepted in its present condition, with no warranty or guarantee.

asking price The price advertised or presented in the initial offering of the sale or rental of a property.

assessed value Usually used in conjunction with property taxes, when an assessor assigns a tax value to a property. Assessed values are normally lower than appraised values.

assessment The amount of tax levied by a municipality or local authority for property tax.

assessment ratio A formula used to determine a property's assessed value, based on the property's market value. For example, if the assessment ratio is 85%, a property with a market value of $100,000 would have an assessed value of $85,000.

assessor An individual with the responsibility of determining the assessed value of real property.

assignee A person or entity to whom a contract is assigned. For example, if you sold your rental property and assigned the existing leases to the purchaser of the property, that purchaser would be the assignee.

assignment The act of transferring rights or interests in a contract to another party.

assignor A person or entity who assigns rights or a contractual interest to another party.

attachment The result of a legal act of seizing property to secure or force payment of a debt.

attorney-in-fact A person or entity authorized to act for another. This arrangement is commonly referred to as a power of attorney. The authorization might be limited to certain aspects or dealings. It might also be general in its scope with all aspects and dealings included.

balance sheet A financial report that shows assets, equity, and liabilities. These reports are laid out in double columns. At the end of the report, the two columns should have identical balances.

balloon loan A loan with a balloon payment due at a specified time. The balloon payment is usually large and must be paid in one lump sum.

balloon payment A lump sum payment paid to satisfy a balloon loan.

bankruptcy A court action exercised to protect debtors. In many cases, people filing bankruptcy are insolvent and forfeit any assets they have to the court for dispersement to creditors.

bilateral contract A contractual agreement requiring both parties of the contract to promise performance.

blind pool A term used to describe a group of investors placing funds in a program to buy unknown properties.

broker A person properly licensed to act on the behalf of others for a fee. Real estate brokers and mortgage brokers are two examples of brokers.

brokerage A business involving the use of brokers.

building codes Rules and regulations adopted by a local jurisdiction to maintain an established minimum level of consistency in building practices.

building permit A license issued by an authorized agency to allow a person to build or alter a building.

cash flow The money received during the life of an investment.

certificate of insurance Physical evidence from an insurer proving the type and amount of coverage held by the insured. All subcontractors should present a certificate of insurance to the person they will be working for before work is started.

certificate of occupancy A certificate issued by the code-enforcement office allowing a property to be occupied by humans. Without a certificate of occupancy, a property is considered uninhabitable.

certificate of title An opinion of title. This opinion of title is generally provided by an attorney to address the status of a property's title based on recorded public records.

chain of title A report that discloses the history of all acts affecting the title of a property.

chattel Personal property. For example, a range and refrigerator are chattel. If you are buying or selling a rental property, all chattel that will convey with the transaction must be detailed in the purchase and sale agreement. When left silent in a contract, chattel does not convey with the sale of real property.

chattel mortgage A mortgage loan secured by personal property. For example, if you own a furnished apartment building, you might pledge the furniture as security for a loan under a chattel mortgage. The chattel mortgage does not encumber the real estate, but it can use all personal property as security.

clear title A title free of clouds or liens. To be marketable, a title should be clear.

cloud of title A dispute, encumbrance, or pending lawsuit that, if valid or perfected, affects the value and marketability of the title.

collateral The property or goods pledged to secure a loan.

common area An area of a property used by all people involved with the property, such as tenants. Frequent common areas include halls, parking areas, laundry rooms, and the grounds of the property.

consideration An object of value given when entering into a contract. Examples of consideration include an earnest money deposit, love and affection, or a promise for a promise.

contractor A person or entity contracting to provide goods or services for an agreed-upon fee to another person or entity.

convey To transfer an object to another.

conveyance The act of conveying rights or objects to another.

covenants Promises or rules written into deeds and leases or placed on public record to require or prohibit certain items or acts. For example, a covenant in a lease might prevent tenants from housing pets in rental units.

default The breach of agreed-upon terms.

defect of title A recorded encumbrance prohibiting the transfer of a free and clear title.

deferred payment A payment made at a later date.

deficiency judgment A court action requiring a debtor to repay the difference between a defaulted debt and the value of the security pledged to the debt. For example, if you had a tenant cause $2000 in damages to your property, but you only had a $600 damage deposit, you could sue the tenant for a deficiency judgment for the remaining $1400 to repair the damages.

demographic study The research done to establish characteristics of the population of an area, such as sex, age, size of families, and occupations.

discrimination The act of showing special treatment, good or bad, to an individual based on the person's race, religion, or sex.

due-on-sale clause A clause found in modern loans forbidding the owner from financing the sale of the property until the existing loan is paid in full. These clauses can be triggered by some lease-purchase agreements. The clause gives a lender the right to demand the existing mortgage be paid in full on demand.

duplex A residential property with the provisions to house two residential dwellings.

dwelling A place of residency in a residential property.

equitable title The interest held by a purchaser of a property that has been placed under contract but not yet closed on.

equity The amount of value between the market value of a property and the outstanding liens against it.

escrow The placement of certain money or documents in the hands of a neutral third party for safekeeping until the transaction can be completed. Security and damage deposits are frequently placed in escrow accounts.

escrow agent A person or entity receiving escrows for deposit and disbursement.

eviction The legal method for a property owner to regain possession of his real property from tenants in default of lease agreements.

fair market rent The amount of money a rental property may command in the present economy.

fair market value The amount of money a property may be sold for in the present economy.

feasibility study A study done to determine if a venture is viable.

first mortgage A mortgage with priority over all other mortgages.

hypothecate To pledge an item as security for a loan without relinquishing possession of the item.

income property Real estate used to produce rental income for the owner.

insurable title A title to property that is capable of being insured by a title insurance company.

landlord A person who leases property to another.

leasehold The interest a tenant holds in rental property he or she is paying rent to occupy.

lessee A person renting property from a landlord.

lessor A person renting property to a tenant.

letter of credit A document from a lender acknowledging that lender's promise to provide credit for a customer.

leverage The use of equity and borrowed money to increase buying power.

lien A notice filed against property to secure a debt or other financial obligation.

life estate An interest in real property that terminates upon the death of the holder or other designated individual.

life tenant An individual allowed to use a property until the death of a designated individual.

line of credit An agreement from a lender to loan a specified sum of money, upon demand, to a borrower, without further loan application or approval.

MAI An appraisal designation meaning "Member, Appraisal Institute." Appraisers with this designation are well-regarded professionals.

marketable title A title to real property that is free from defects and enforceable by a court decision.

mortgage banker Someone who originates, sells, and services mortgage loans.

mortgage broker Someone who arranges financing for others for a fee.

mortgagee A person or entity holding a mortgage against real property. For example, if a bank loans a landlord money to buy a building, the bank would be the mortgagee and the landlord would be the mortgagor.

mortgagor A person or entity pledging property as security for a loan.

net income The amount of money remaining from gross income after all expenses, including taxes, are paid.

net worth The amount of equity remaining when all liabilities are subtracted from all assets.

net yield The return on an investment after all fees and expenses are subtracted.

novation An agreement in which one individual is released from an obligation through the substitution of another party.

passive investor An investor who provides money but does not provide personal services in a business endeavor.

pro-forma statement A a report projecting the outcome of an investment.

secondary mortgage market The market where mortgages are bought and sold by investors. A majority of the loans originated in banks are sold in the secondary mortgage market.

zoning The legal regulations by approved authorities for the use of private land.

Index

A

Accounting (*see* Financial management)
Accounts payable form, 154
Accounts receivable form, 159
Advertising, 3, 10, 23, 29, 103, 281–286
  (*see also* Sales and marketing)
  creative methods, 283
  credit accounts for, 56, 60
  direct-mail, 103–104, 283
  direct sale activity with, 285
  expenses, 256
  flyers, 282
  handouts, 282
  name recognition through, 284–285
  newspaper ads, 282
  pamphlets, 282
  phone directory, 27, 281–282
  promotional activities, 285–286
  radio, 282
  rate of return on, 283–284
  television commercials, 282–283
  using for multiple purposes, 284
Annuities, 317
Answering machines, 79–80, 272
Answering services, 28, 79–80
Appointments, scheduling, 133–134
Appraisals, 4, 239
Automobiles, 245
  company, 25
  credit accounts for, 57
  disabled, 300
  expenses incurred with, 255
  leasing, 184–185
  purchasing, 185

B

Balloon promissory note, 122
Bankruptcy, 263
Banks and banking (*see* Financial
  management)
Bidding (*see* Job bids)
Blueprints, 4 (*see also* House plans)

Bonding, 226
Bridge loans, 67–68
Broker commission arrangement form, 110
Brokers, real estate (*see* Real estate brokers)
Building codes
  code violation notification form, 221
  enforcement rejections, 300
  inspections, 220
  officers, 220
Building lots, 81–88
  access, 85
  acquiring, 96
  big deals, 98–99
  controlling
    through purchase options, 90–93
    with take-down schedules, 89–90
    within subdivisions, 89–94
  developing your own, 95–99
  engineering studies, 96
  filing fees, 97
  hard costs associated with, 97
  housing development dues and
    assessments, 85
  land characteristics, 84–85
    flood zones, 84–85
    rock, 84
    soggy ground, 84
    trees, 85
  locating, 86–88
    small publications, 87
    through brokers, 87–88
    unadvertised specials, 86–87
  mid-sized deals, 98
  restrictions, 86
  road maintenance agreements, 85
  small deals, 97–98
  survey studies, 96
  utility hook-ups, 82–84
    electricity, 83
    septic systems, 84
    sewer systems, 83
    water systems, 82–83
    wells, 84

Business
  adjusting to changing economic
    conditions, 31–33
  allowing employees to manage, 314
  cash reserves, 18–19
  closing, 315
  diversity, 32–33
  documenting activities, 164–175 (see also
    Contracts)
  expenses (see Expenses)
  home building (see Home builders)
  passing it onto your children, 314
  preparing for growth of, 32
  selling, 314–315
Business goals, 13–19
  customer base, 16–17
  diversity, 16
  focal point, 17–18
  future prospects in five years, 15
  size and growth, 15–16
  what are your expectations, 14–15
Business image, 271–286
  building demand for services, 278–279
  changing, 276–277
  company colors, 277–278
  company logos, 274–275
  company names, 272–274
  fee schedule and, 276
  importance of a good, 272
  joining clubs and organizations, 279
  setting yourself apart, 277–278
  slogans, 278
  visual perception, 275
Business office (see Office)
Business structures, 35–40
  choosing one, 36–38
  corporations, 35
    advantages, 38
    disadvantages, 38–39
  liability concerns, 37–38
  partnerships, 36, 38
    advantages, 39
    disadvantages, 39–40
  sole proprietorships, 36
    advantages, 40
    disadvantages, 40
  subchapter S corporations, 36
    advantages, 39
    disadvantages, 39

C

CAD, 145
Call-back procedures, 300–301
Cash receipts, 179

Cash receipts form, 157, 165
CD-ROM, 148
Cellular phones, 135
Certificate of deposit (CD), 60
Change orders, 48, 167, 176
  written, 47
Closing costs form, 120
Clubs, 279
Code officers, 220
Code violation notification form, 221
Colors, choosing, 105
Commercial office, 76
Company image (see Business image)
Competition, pricing and, 241–242, 246
Completion dates, 227
Computer systems, 137–151
  advantages, 137–138
  applications, 138–142
    bank balances, 140
    budget tracking, 139
    check writing, 140
    customer service, 140
    database, 140
    estimating, 140
    inventory control, 140–141
    job costing, 139
    marketing, 138–139
    payroll preparation, 139
    tax liability projections, 140
    word-processing, 140
  building credibility with customer, 142–143
  drives
    CD-ROM, 148
    floppy, 148
    hard, 147–148
  memory, 146–147
  modems, 149
  monitors, 148–149
    CGA, 148
    EGA, 148
    SVGA, 148–149
    VGA, 148–149
  mouse, 149–150
  package deals, 150
  printers, 149
    dot-matrix, 149
    ink-jet, 149
    laser, 149
    thermal, 149
  software
    choosing, 150–151
    computer-aided design (CAD), 145
    database, 144
    integrated programs, 144–145
    spreadsheet, 143–144

Computer systems, software, (*Cont.*)
  word-processing, 144
  speed, 147
  types of, 145–146
    desktop computers, 145–146
    laptop computers, 146
    notebook computers, 146
Computer-aided design (CAD), 145
Construction loans, 67, 158, 160
Contingency release form, 118
Contracts, 165–172, 175–176
  addendums, 167, 176
  extension form, 124
  for sale of real estate, 114–117
  getting it in writing, 43–44
  liability waivers, 172
  negotiating with subcontractors, 202
  oral, 165
  proposal form, 166–167
  remodeling, 168–170
  subcontract agreement, 203–205
  subcontractor liability clause, 206
  writing good, 181
  written, 165
Corporations, 35
  advantages, 38
  disadvantages, 38–39
  subchapter S, 36
Cost projections form, 238–239
Counteroffer form, 123
Covenants, 45
Credit (*see* Financial management)
Credit applications, 175, 179
Credit cards, secured, 61
Customer base, 9
Customer relations, 46–47, 261–270 (*see also*
  Public relations)
  attracting new, 224–225
  calming disgruntled customers, 266
  communications, 264–265
  defusing tense situations, 265–266
  determining who should handle, 265
  ensuring satisfaction from, 224
  finding credit for, 63, 66–68
  getting referrals from, 223–224, 267
  give-and-take principles, 263–264
  guidelines for, 263
  liens rights and waivers, 267–270
  listening to, 71
  meeting on their level, 261
  qualifying, 262–263
    bankruptcy, 263
    deadbeats, 262
    death, 262–263
    dissatisfied customers, 262

Customer relations, qualifying, (*Cont.*)
    loan denial, 262
  satisfying your customers, 263
  selecting a target group, 16–17
  skills, 264
  solidifying plans/specifications, 270
  training employees to deal with, 298

D

Dental insurance, 28, 307–308
Deposits
  eliminating subcontractor, 156
  obtaining contract, 156
Directory advertising, 27
Disability insurance, 28, 308
Dot-matrix printers, 149
Dress code, 245, 261
Drywall bid sheet, 217

E

Earnest money deposit receipt, 119
Economy, 30
Education, continuing, 32
Electrical bid sheet, 218
Electricity, 83
Employees (*see* Personnel)
Employer ID number, 294
Engineering studies, 96
Equipment
  expenses, 255 (*see also* Expenses)
  financial justification for purchasing, 186
  leasing vs. purchasing, 183–185
  office (*see* Office equipment)
  renting tools, 184
  weighing need vs. desire, 185–186
Estimates, 46
  forms for estimating
    appliances, 234
    exterior work, 233, 235
    heating, 233
    interior work, 232, 235
    options, 236, 237
    plumbing, 233
    prepurchase phase, 230
    site work, 231
  using effective techniques for, 246
  written, 172, 178
Exercise of option form, 93
Expenses, 21–29
  advertising, 23, 27, 29, 256
  answering services, 28, 79–80
  bid-sheet subscriptions, 29
  cash receipts, 26

Expenses, (*Cont.*)
  company vehicles, 25
  credit bureaus, 29
  cutting, 21, 25–26
  cutting the wrong, 27–29
  determining future, 29–31
  employee, 255
  equipment, 255
  field, 255
  field supervisors, 25
  growth, 256
  insurance, 23, 255 (*see also* Insurance)
    dental, 28
    disability, 28
    health, 28
  inventory assets, 28
  leftover materials, 25–26
  legal fees, 24
  loans, 256
  office, 254
  office equipment, 24–25
  office furniture, 24
  office help, 24
  office space, 78–79
  office supplies, 24
  overhead, 21–25, 42
  professional fees, 23–24
  rent, 22–23
  retirement, 28–29, 256–257
  salary, 254
  sales force, 29
  taxes, 256
  telephone bills, 23
  testing justifiability of, 26–29
  travel, 26
  utilities, 23
  vehicle, 255

F

Federal unemployment tax (FUTA), 294
Fee schedule (*see* Pricing)
Field supervisors, 25
Financial management, 153–181
  adjusting to changing economic
    conditions, 31–33
  banking, 55–68
    establishing good credit, 55–57
    choosing a lending institution, 57–58
  bankruptcy, 263
  budgets
    maintaining, 257
    personal statement, 65
    projecting your business, 254–257
    projecting your job, 253–254

Financial management, (*Cont.*)
  cash flow, 153, 155, 226–227
  cash reserves, 18–19, 179
  certificate of deposit (CD), 60
  choosing an accountant, 163–164
  computerized (*see* Computer systems)
  credit
    applications for customers, 175, 179
    establishing, 58–60
    extending to customers, 158
    finding for customers, 63, 66–68
    successful, 62–63
    keeping accounts under control, 153
  credit accounts for
    advertising, 56, 60
    equipment, 57
    fuel, 57
    office supplies, 57
    suppliers, 56–57, 59–60
    vehicles, 57
  credit cards, secured, 61
  credit reports
    erroneous, 61
    explanation letters for poor, 62
    getting a copy of, 62
    overcoming poor, 60–62
  deposits
    eliminating subcontractor, 156
    obtaining contract, 156
  expenses (*see* Expenses)
  financial statement, 64
  forms
    accounts payable, 154
    accounts receivable, 159
    cash receipts, 157, 165
    cost projections, 238–239
    petty-cash record, 160
  future planning, 305–318
  future projections, 155–156
  getting paid, 227
  learning to stretch your money, 94, 158
  line of credit, 43
  loans (*see* Loans)
  money needed to get started, 42
  operating capital, 56
  past-due accounts, collecting, 158, 160
  payroll, 289–290
  personal financial protection, 37–38
  purchase options, 90–93
    basic, 92–93
    form for, 91
    short-term, 92
    exercise of, 93
  retirement plans, 315–318
  start-up money, 55–56

Financial management, (*Cont.*)
  take-down schedules, 89–90
  taxes (*see* Taxes)
Financing, 10 (*see also* Financial
    management; Loans)
Flood zones, 84–85
Flooring bid sheet, 214
Floppy disk drives, 148
Forms
  accounts payable, 154
  accounts receivable, 159
  balloon promissory note, 122
  bid request, 211
  broker commission arrangement, 110
  cash receipts, 157, 165
  certificate of subcontractor completion
    acceptance, 251
  change order, 48
  code violation notification, 221
  commencement and completion
    schedule, 206
  contingency release, 118
  contract extension form, 124
  contract for sale of real estate, 114–117
  contractor rating sheet, 201
  cost projections, 238–239
  counteroffer, 123
  drywall bid sheet, 217
  earnest money deposit receipt, 119
  electrical bid sheet, 218
  estimate of purchaser's closing costs, 120
  estimating forms for
    appliances, 234
    exterior work, 233, 235
    heating, 233
    interior work, 232, 235
    options, 236, 237
    plumbing, 233
    prepurchase phase, 230
    site work, 231
  exercise of option, 93
  financial statement, 64
  flooring bid sheet, 214
  framing bid sheet, 214
  heating bid sheet, 217
  insulation bid sheet, 213
  inventory control for trucks, 188
  inventory log, 180
  job cost log, 240
  letter of engagement, 44
  long-form lien waiver, 269
  material order log, 196
  material quote solicitation, 194
  personal budget, 65
  petty-cash record, 161

Forms, (*Cont.*)
  plumbing bid sheet, 216
  promissory note, 121
  proposal, 166–167
  purchase option, 91
  quote, 174
  remodeling contract, 168–170
  request for substitutions, 177
  roofing bid sheet, 219
  seller's disclosure form, 125–128
  service call ticket, 172
  short-form lien waiver, 268
  siding bid sheet, 215
  site work bid sheet, 218
  subcontractor agreement, 203–205
  subcontractor liability clause, 206
  subcontractor questionnaire, 199–200
  subcontractor schedule, 248
  take-off, 229
  telephone log, 195
  time log, 132
  tree clearing bid sheet, 216
  trim bid sheet, 219
  truck inventory, 189
  well bid sheet, 215
  work estimate, 173
Framing bid sheet, 214
Fuel, credit accounts for, 57

H

Hard disk drives, 147–148
Health insurance, 28, 306–307
  copayments, 307
  deductible payments, 306
  dependent coverage, 307
  group advantages, 307
  limits of standard coverage, 306
  pre-existing conditions, 306
  rate increases, 307
  waiting period, 307
Heating bid sheet, 217
Home builders
  building on speculation (*see* Speculation
    builders)
  employer ID number, 294
  future planning, 305–318
  getting started, 1–5, 7–11
    advertising, 3
    avoiding dangers, 4
    choosing your house plans, 8
    creating your edge, 9–10
    determining how much money you
      make, 4–5
    determining number of houses to build, 4

Home builders, getting started, (*Cont.*)
  determining what type of home to build, 8
  educational resources, 7
  gaining knowledge, 3
  getting references, 2–3
  identifying your basic needs, 2
  money for, 55–56 (*see also* Financial
    management)
  mistakes made by, 41–48
  purchase options for property, 90–93
    (*see also* Building lots)
  staying busy in slow times, 286
Home office, 22, 76
Homes, model (*see* Model homes)
House plans (*see also* Blueprints)
  choosing, 8, 101–102

I

Ink-jet printers, 149
Inspections, 46
  avoiding rejected code-enforcement, 220
Insulation bid sheet, 213
Insurance, 23, 46, 255
  choosing a company for, 311
  dental, 28, 307–308
  disability, 28, 308
  health, 28, 306–307
    copayments, 307
    deductible payments, 306
    dependent coverage, 307
    group advantages, 307
    limits of standard coverage, 306
    pre-existing conditions, 306
    rate increases, 307
    waiting period, 307
  liability, 312
  life, 308–311
    additional options, 311
    amount needed, 309–310
    key-man, 310–311
    term life, 310
    universal/variable policies, 310
    whole-life, 310
  worker's compensation, 312–313
Inventory, 28, 186–189
  control form for trucks, 188
  for planned jobs, 186–187
  saving time with, 187
  shop stock, 187
  stocking trucks, 189
  theft of, 187
  truck inventory form, 189
Inventory logs, 179–180

J

Job bids (*see also* Estimates)
  bid sheets
    beating competition with, 228
    definition/description, 225
    drywall bid sheet, 217
    electrical bid sheet, 218
    flooring bid sheet, 214
    framing bid sheet, 214
    government, 225
    heating bid sheet, 217
    insulation bid sheet, 213
    obtaining, 225
    plumbing bid sheet, 216
    roofing bid sheet, 219
    siding bid sheet, 215
    site work bid sheet, 218
    tree clearing bid sheet, 216
    trim bid sheet, 219
    types of jobs on, 225
    well bid sheet, 215
  bidder agencies, 225
  forms, job cost log, 240
  large projects, 226–227
  making in-person, 228
  participating in the bid process, 228
  presentations, 244–245
    being confident, 245
    by mail, 244
    by telephone, 244–245
    dress code for, 245
    vehicle you drive, 245
  pricing (*see* Pricing)
  referrals, word-of-mouth, 223–224
  request form, 211
  take-offs, 228–229, 237
Job costing, 257–259
  long-term success at, 259–260
  percentage method, 258–259
  simple method, 258
  tracking profitability from, 259
Job cost log, 240
Job pricing (*see* Pricing)
Job specifications, 175, 178–179

K

Keogh plans, 316

L

Labor laws, 295
Land (*see* Building lots)
Laser printers, 149

Leasing
  vehicles, 184–185
  vs. purchasing equipment, 183–185
Legal issues, 163–164
  choosing an attorney, 164
  fees for, 24
  labor laws, 295
  protection from lawsuits, 37
Letter of engagement, 44
Liability insurance, 312
Liability waivers, 172, 178
Lien rights, 267–270
Lien waivers, 267–270
  long form for, 269
  short form for, 268
Life insurance, 308–311
  additional options, 311
  amount needed, 309–310
  key-man, 310–311
  term life, 310
  universal/variable policies, 310
  whole-life, 310
Loans (*see also* Financial management)
  application needs, 66, 256
  bridge, 67–68
  CD loans, 61
  collateral for, 58
  construction, 67, 158, 160
  customer denial of, 262
  from major lenders, 60
  home equity, 58
  lien rights and waivers, 267–270
  mortgage, 66
  secured, 58
Logos, 274–275
Lots (*see* Building lots)

M

Maintenance agreements, roadways, 85
Management
  financial (*see* Financial management)
  time, 130–135, 187 (*see also* Schedules
    and scheduling)
    areas of waste, 131–132
    controlling employee conversations,
      132–133
    eliminating lost, 260
    inventory and, 187
    log for, 132
    reducing lost time in field, 134–135
    reducing lost time in office, 134
    setting appointments, 133–134
    using a mobile phone, 135

Management, time, (*Cont.*)
  using a tape recorder, 134–135
Marketing (*see* Sales and marketing)
Materials (*see also* Products; Suppliers)
  avoiding delivery delays, 193–197
  cost projections form, 238–239
  credit accounts for, 56–57, 59–60
  estimating forms (*see* Estimates, forms for
    estimating)
  finding suppliers, 192
  leftover, 25–26
  order log, 196
  pricing, 239–242
    marking up, 241
  purchasing, 192
  quote solicitation form, 194
  recordkeeping, 237–238
  take-offs, 228–229, 237
    building in a margin of error, 237
    forms for, 229
    keeping track of materials, 237
    recordkeeping, 237–238
Memory, computer, 146
Model homes, 49–53
  advantages, 49–51
    credibility, 50
    ease of selling, 49–50
    extra attention to builder, 50–51
  disadvantages, 51–52
    money to build, 51–52
    reality vs. dreams, 51
  feasability, 52
  in subdivisions, 52
  personal opinion of, 52–53
Modems, 149
Monitors, computer, 148–149
Mortgage loans, 66
Mouse system, 149–150

O

Occupational Safety and Health
  Administration (OSHA), 295
Office, 75–80
  affordability, 78–79
  commercial, 76
  computerized, 137–151
  home, 22, 76
  location, 75–76, 77–78
  needs assessment, 76–77
    space, 76–77
    visibility, 77
    warehouse space, 77
  renting and leasing, 78
  storefront requirements, 75

Office equipment, 24
Office furniture, 24
Office supplies, 24
 credit accounts for, 57
OSHA, 295

P

Partnerships, 36, 38
 advantages, 39
 disadvantages, 39–40
Payroll, 289–290
 computerized preparation, 139
Pension plans, 316–317
Personnel, 287–303
 awards and certificates, 302–303
 benefits, 297, 311–312
 bonus incentives, 301
 call-back procedures, 300–301
 establishing cost of, 299
 expenses with having, 255
 forms
  employee-file checklist, 296
  weekly work history, 296
 hourly, 289
 independent contractors, 289
 insurance costs, 289
 keeping good, 295–297
 labor laws, 295
 locating, 291–292
  hand-picking, 292
  in classified ads, 291
  in unemployment office listing, 292
  using employment agencies, 291
  word-of-mouth referrals, 291
 managing field, 302
 managing office, 301–302
 motivating, 302–303
 paid vacations, 290
 paperwork, 293–295
  1099 forms, 293
  applications, 293
  I-9 forms, 293
  taxes, 294–295
  W-2 forms, 293
  W-4 forms, 293
 payroll, 139, 289–290
 performance ratings, 303
 production downtime and, 299–300
 quality control (QC) and, 297
 reasons for hiring, 287–289
 sick leave, 290
 taxes, 290
 terminating, 295
 theft and, 297

Personnel, (*Cont.*)
 titles for, 303
 training, 298
 transportation costs, 289
 using employees vs. subcontractors, 290
Petty-cash record, 160
Plumbing bid sheet, 216
Pricing, 4, 9, 44–45
 competitors and, 241–242, 246
 fee schedule, 276
 guidelines, 242–243
 job cost log, 240
 job costing and, 257–259
 materials, 239–242
  marking up, 241
 researching, 243
 your services, 239–242
Printers, 149
 dot-matrix, 149
 ink-jet, 149
 laser, 149
 thermal, 149
Production schedules (*see* Schedules
 and scheduling)
Products, choosing, 105, 192–193 (*see also*
 Materials; Suppliers)
Professional fees, 23–24
Profit-sharing plans, 317
Promissory note, 121
 balloon, 122
Proposal form, 166–167
Public relations (*see also* Customer relations)
 business image and, 271–272
 promotional activities, 285–286
 skills, 264
Purchase option form, 91

Q

QC, 297
Quality control (QC), 297
Quote form, 174
Quotes, 46

R

Radio advertising, 282
Real estate brokers, 87–88, 107–128
 asking questions, 111–113
 commissions, 112
 forms
  balloon promissory note, 122
  broker commission arrangement, 110
  contingency release, 118
  contract extension, 124

Real estate brokers, forms, (*Cont.*)
   contract for sale of real estate, 114–117
   counteroffer, 123
   earnest money deposit receipt, 119
   estimate of purchaser's closing costs, 120
   promissory note, 121
   seller's disclosure form, 125–128
  letting them handle deals, 113
  listing with big names, 111
  selling through buyer's, 109
  selling through seller's, 109–113
  using multiple listing services, 112
References
  customer, 2–3, 267
  subcontractor, 200
Regulations
  covenants, 45
  restrictions, 45
  zoning, 45
Remodeling contract, 168–170
Rent, 22–23
Rental properties, 315–316
Request for substitutions form, 177
Restrictions, 45, 86
Retirement plans, 256–257, 315–318
  annuities, 317
  funding for, 28–29
  future investments, 317–318
  Keogh plans, 316
  pension plans, 316–317
  profit-sharing plans, 317
  rental properties, 315–316
  social security, 317
Rocky ground, 84
Roofing bid sheet, 219

S

Sales and marketing, 69–73 (*see also*
  Advertising; Customer relations)
  commissioned sales, 280–281
  computerized, 138–139
  finding opportunities for making, 279–280
  getting down to business, 71–73
  in-house sales staff, 108–109
  key elements, 73
  learning the basics, 69–70
  listening to the customer, 71
  portfolio and props, 72
  sales force, 29, 280
  selecting a target group of customers, 16–17
  selling houses yourself, 108
  selling through buyer's brokers, 109
  selling through seller's brokers, 109–113
  targeting your market, 102–103

Sales and marketing, (*Cont.*)
  tips on, 70
  using real estate brokers (*see* Real estate
   brokers)
  word-of-mouth referrals, 223–224
Schedules and scheduling
  adjusting, 251–253
  appointments, 133–134
  bad weather and, 299–300
  commencement dates, 206
  completion dates, 206, 227
  eliminating lost time, 260
  example of work schedule, 249
  forms
   certificate of subcontractor completion
    acceptance, 251
   subcontractor schedule, 248
  keeping subcontractors in line, 250
  keeping suppliers in line, 250
  making confirmation calls, 250
  production, 247–253
  projecting your job budget, 253–254
  staying on track, 250–252
  time management (*see* Time management)
  tracking production, 250
Self-employment tax, 294–295
Seller's disclosure form, 125–128
Septic systems, 84
Service call ticket, 172
Service orders, 172, 176, 178
Sewer systems, 83
Siding bid sheet, 215
Site work bid sheet, 218
Slogans, 278
Software (*see also* Computer systems)
  choosing, 150–151
  computer-aided design (CAD), 145
  database, 144
  integrated programs, 144–145
  spreadsheet, 143–144
  word-processing, 144
Sole proprietorships, 36
  advantages, 40
  disadvantages, 40
Speculation builders, 101–105
  advertising, 103
  choosing colors and products, 105
  choosing house plans, 101–102
  creating a safety net, 104–105
  picking lots, 101–102
  selling your houses, 105
  targeting your market, 102–103
State tax, 295
Subchapter S corporations, 36
  advantages, 39

Subchapter S corporations, (*Cont.*)
disadvantages, 39
Subcontractors
analyzing, 200–202
application forms, 198
avoiding problems with, 213
bid sheets (*see* Job bids)
business procedures, 207
choosing, 197–207
contract negotiations, 202
controlling, 209
credit checks, 200
eliminating deposits to, 156
employee checks, 209
forms
commencement and completion
schedule, 206
liability clause, 206
questionnaire, 199–200
rating sheet, 201
scheduling, 248
subcontractor agreement, 203–205
hiring, 191–192
initial contact with, 197–198
insurance coverage verification, 208
interviewing, 198
license verification, 208
maintaining relationships with, 202
rating, 207–209
references, 200
schedules with, 250
selecting, 43
setting guidelines for, 202
specialties, 208
term negotiations, 202
tools and equipment, 208
work history evaluations, 207
Subdivisions, model homes in, 52
Supervisors, field, 25
Suppliers, 209–213 (*see also* Materials;
Products)
avoiding problems with, 213
expediting materials from, 212–213
finding, 192
negotiating with, 210
schedules with, 250
Surveying, 96

T

Take-offs, 228–229, 237
building in a margin of error, 237
forms for, 229

Take-offs, (*Cont.*)
keeping track of materials, 237
recordkeeping, 237–238
Tape recorders, 134–135
Taxes, 161–162, 256
cash receipts, 179
employee withholdings, 294
Federal unemployment (FUTA), 294
IRS audits, 162–163
payroll, 290
payroll-tax deposits, 294
planning for deductions, 161–162
self-employment, 294–295
state, 295
Telephone directory, advertising, 281–282
Telephones
answering machines, 79–80, 272
answering services, 28, 79–80
bills for, 23
cellular, 135
confirmation calls, 250
log form, 195
making job bids over, 245
Television commercials, 282–283
Thermal printers, 149
Time and material (T&M) prices, 44–45
Time management, 130–135
areas of waste, 131–132
controlling employee conversations,
132–133
eliminating lost, 260
inventory and, 187
log for, 132
reducing lost time in field, 134–135
reducing lost time in office, 134
setting appointments, 133–134
using a mobile phone, 135
using a tape recorder, 134–135
Tools
buying specialty, 184
renting, 184
Training, personnel, 298
Travel expenses, 26
Tree clearing bid sheet, 216
Trees, 85
Trim bid sheet, 219

U

Utilities, 23, 82–84
electricity, 83
septic systems, 84
sewer systems, 83

Utilities, (*Cont.*)
  water, 82–83
  wells, 84

V

Vehicles, 245
  company, 25
  credit accounts for, 57
  disabled, 300
  expenses incurred with, 255
  leasing, 184–185
  purchasing, 185

W

Warranty work, 300–301
Water systems, 82–83
  well bid sheet, 215
  wells, 84
Wells, 84
  bid sheet, 215
Work estimate form, 173
Worker's compensation insurance, 312–313

Z

Zoning regulations, 45

About the Author

R. Dodge Woodson is a licensed building contractor and master plumber. He has worked in the building trades for more than 20 years, and as a home builder has built as many as 60 homes a year. He owns a plumbing company and the Mid-Coast Training and Development Center, as well as a real estate brokerage. Mr. Woodson has also been an instructor at Central Maine Technical College, where he has taught plumbing code and apprenticeship courses. He is the author of 48 books, including *Builder's Guide to Change-of-Use Properties*, *National Plumbing Code Handbook*, and *The Plumber's Apprentice Handbook*.